Free Energy Relationships in Organic and Bio-organic Chemistry

This book is dedicated to William P. Jencks
and the late Myron L. Bender

Free Energy Relationships in Organic and Bio-organic Chemistry

Andrew Williams
University of Kent, Canterbury, UK

advancing the chemical sciences

Published by The Royal Society of Chemistry,
Thomas Graham House, Science Park, Milton Road,
Cambridge CB4 0WF, UK

Registered Charity Number 207890

For further information see our web site at www.rsc.org

Typeset by RefineCatch Limited, Bungay, Suffolk
Printed by TJ International Ltd, Padstow, Cornwall

Preface

Proper interpretation of free energy relationships can provide detailed information involving charge distribution in the transition structure of a reaction mechanism. Free energy relationships are thus of enormous significance in studies of mechanism and this text is offered as a course-book for senior undergraduates and postdoctoral workers engaged in such investigations. Since publication in 1973 of previous monographs at this level there have been substantial developments in the theory and application of free energy relationships and their application to reaction mechanism. This is especially true of the concepts of *effective charge* and *similarity* and the text demonstrates how these concepts can be used to elucidate physico-chemical attributes of transition structures. These concepts are therefore likely to be of significance as vehicles for the treatment of free energy relationships in undergraduate courses at the senior level and in postgraduate courses.

Results from free energy relationships are generally easy to achieve but, since they are probably the most difficult to interpret, the technique has been the subject of fierce critical appraisal. The most extreme criticism has used the existence of apparent anomalies to deny any use-fulness of the relationships. These criticisms have led to an undeservedly poor appreciation of free energy relationships; this, together with fashion, has ensured that free energy relationships scarcely figure in contemporary chemistry courses at any level and has denied students access to a major technique for studying mechanism. On the contrary the apparent anomalies of free energy relationships provide extremely important insights into mechanism, in particular regarding the relative timing of the fundamental processes of a reaction under investigation. Free energy relationships continue to provide one of the major routes *via* which knowledge of reaction mechanism can be derived.

The text is a source for the most commonly used free energy relation-ships; it is meant to be self contained and while most of the terms employed in it are standard or are defined in the text reference can be usefully made to the IUPAC glossary of terms used in physical organic chemistry.[a] Sufficient data are provided in the form of empirical

---

[a] P. Müller, *Pure Appl. Chem.*, 1994, **66**, 1077; http://www.chem.qmul.ac.uk/iupac/gtpoc

equations and tables of parameters for the reader to calculate physico-chemical constants. The book can be used without attempting the problems, in which case the answers in the appendix provide further illustrative material.

For clarity, the examples discussed in the text and used as problems do not as a rule include contributions from other techniques. One of the most serious problems associated with studies of reaction mechanism is over-confidence in the power of a single technique or theory. By its very nature a monograph focusses on a particular subject, and in this case the text should not be construed as an argument that the technique of free energy relationships is the *only* tool worthy of use in mechanistic studies.

I am delighted to acknowledge the help of the following colleagues who have read parts of the text in its early form or who have made useful suggestions: Stefano Biagini (Kent), Bill Bentley (Swansea), Howard Maskill (Newcastle), Herbert Mayr (Munich), Mike Page (Huddersfield), François Terrier (Versailles) and Sergio Thea (Genoa). I am also grateful to Katrina Turner and the staff of the Royal Society of Chemistry at Cambridge for their help in the preparation of the book and to the School of Physical Sciences at the University of Kent for providing me with facilities.

Andrew Williams
Canterbury, July 2003

Contents

Chapter 1	**Free Energy Relationships**	**1**

	1.1	Mechanism and Structure	3
		1.1.1 Interconversion of States – Reaction and Encounter Complexes	4
	1.2	Universal Measure of Polarity	5
	1.3	Classes of Free Energy Relationship	6
	1.4	Origin of Free Energy Relationships	7
	1.5	Similarity and the Leffler α-Parameter	10
	1.6	Bonding in Transition Structures	12
	1.7	Further Reading	13
	1.8	References	15

Chapter 2	**The Equations**	**17**

	2.1	The Hammett Equation (Class II)	17
	2.2	The Taft Equation (Class I)	19
		2.2.1 Steric Demand at Reaction Centres	23
	2.3	Inductive Polar Constants σ_I	24
		2.3.1 Choice of σ Value	26
	2.4	The Brønsted Equation (Class I)	27
	2.5	The Extended Brønsted Equation (Class II)	30
	2.6	Equations for Nucleophilic Substitution	32
		2.6.1 The Swain–Scott Equation (Class II)	32
		2.6.2 The Ritchie Equation (Class II)	34
	2.7	Solvent Equations	35
		2.7.1 The Grunwald–Winstein Equation (Class II)	36
		2.7.2 The Kosower Z and Reichardt E_T Scales (Class II)	39
		2.7.3 The Hansch Equation (Class II)	40
	2.8	Other Standard Processes	41
		2.8.1 Spectroscopic Transitions	41
		2.8.2 Acidity and Basicity Functions	42
		2.8.3 Molar Refractivity	43
	2.9	Further Reading	44
	2.10	Problems	45
	2.11	References	51

Chapter 3   Effective Charge   55

3.1   Equilibria   55
      3.1.1 Measuring Effective Charge in Equilibria   58
      3.1.2 A Group Transfer of Biological Interest   63
      3.1.3 Additivity of Effective Charge in
             Reactants and Products   65
3.2   Rates   65
      3.2.1 Additivity of Effective Charge in Tran-
             sition Structures   67
      3.2.2 Effective Charge Maps for More Than
             Two Bond Changes   67
3.3   Further Reading   69
3.4   Problems   69
3.5   References   73

Chapter 4   Multiple Pathways to the Reaction Centre   75

4.1   Strength of the Interaction   75
4.2   Additivity of Inductive Substituent Effects   78
4.3   Multiple Interactions   79
      4.3.1 The Jaffé Relationship   80
      4.3.2 Resonance Effects   83
      4.3.3 The Yukawa–Tsuno Equation   86
      4.3.4 Other Two-parameter Equations   88
             4.3.4.1   The Dewar–Grisdale Equation   88
             4.3.4.2   The Swain–Lupton Equation   89
             4.3.4.3   The Edwards Equation for
                       Nucleophilic Aliphatic
                       Substitution   92
             4.3.4.4   The Mayr Equation   93
4.4   Multi-parameter Solvent Equations   94
      4.4.1 Extension of the Grunwald–Winstein
             Equation   94
      4.4.2 The Swain Equation   94
      4.4.3 Extended Hansch Equations   96
      4.4.4 Solvent Equations Based on Macro-
             scopic Quantities   96
4.5   Further Reading   97
4.6   Problems   97
4.7   References   103

Chapter 5   **Coupling Between Bonds in Transition Structures of Displacement Reactions**   **107**

5.1   Cross and Self Interaction Between Two or More Substituent Effects   107

    5.1.1   Hammond and Cordes–Thornton Coefficients   109

5.2   Shapes of Energy Surfaces from Coupling Coefficients   110

    5.2.1   General Acid-catalysed Addition of Thiol Anions to Acetaldehyde   115

    5.2.2   Reaction of Amines with Phosphate Monoesters   116

    5.2.3   Elimination Reactions of 2-Arylethylquinuclidinium Ions   117

5.3   Magnitude and Sign of Cordes–Thornton Coefficients   118

5.4   Identity Reactions   119

5.5   Further Reading   123

5.6   Problems   124

5.7   References   126

Chapter 6   **Anomalies, Special Cases and Non-linearity**   **129**

6.1   Free Energy Relationships are not Always Linear   129

    6.1.1   Marcus Curvature   131

6.2   The Reactivity–Selectivity Postulate   135

6.3   Bordwell's Anomaly   138

6.4   Microscopic Medium Effects and Deviant Points   140

6.5   Statistical Treatment of Brønsted Plots   144

6.6   Are Free Energy Relationships Statistical Artifacts?   145

6.7   The *Ortho* Effect   146

6.8   Temperature Effects   147

6.9   Further Reading   148

6.10   Problems   149

6.11   References   154

Chapter 7   Applications   158

7.1   Diagnosis of Mechanism   158
        7.1.1 Mechanism by Comparison   158
        7.1.2 Effective Charge Distribution of the
                  Transition Structure   160
                  7.1.2.1 Effective Charge Maps and
                              Conservation of Effective Charge 160
                  7.1.2.2 Charge and Bond Order Balance   161
7.2   Demonstration of Intermediates   163
        7.2.1 Curvature in pH-profiles   166
7.3   Demonstration of Parallel Reactions   166
        7.3.1 Concave Curvature in pH-profiles   168
7.4   Demonstration of Concerted Mechanisms   169
        7.4.1 Technique of Quasi-symmetrical
                  Reactions   169
        7.4.2 Criteria for Observing the Break   173
                  7.4.2.1 Effective Charge Map for a
                              Putative Stepwise Process   173
7.5   Calculation of Physico-Chemical Constants   174
        7.5.1 Dissociation and Equilibrium Constants   174
        7.5.2 Rate Constants   179
        7.5.3 Partition Coefficients and Other
                  Constitutive Molecular Properties   180
7.6   Resolution of Kinetically Equivalent
        Mechanisms   181
        7.6.1 General Base versus Nucleophilic
                  Reaction   181
        7.6.2 General Acid Catalysis Versus
                  Specific Acid–General Base
                  Catalysis   182
7.7   Some Biological Applications   184
7.8   Further Reading   187
7.9   Problems   189
7.10 References   196

Appendix 1   Equations and the More O'Ferrall–Jencks Diagram   201

A1.1 Summary of Equations   201
        A1.1.1 The Morse Equation   201
        A1.1.2 Derivation of Equation 7 (Chapter 1)   202

A1.1.3 The Effect of ΔE_o on the Leffler α
Parameter 202
A1.1.4 Fitting Data to Theoretical Equations 202
A1.1.4.1 Linear Equation 203
A1.1.4.2 Bilinear Equation 203
A1.1.4.3 Global Analysis of Two
Linear Equations 204
A1.1.4.4 Binomial Equation 205
A1.1.4.5 Cross-correlation 206
A1.1.5 Effective Charge from Hammett Slopes 206
A1.2 The More O'Ferrall–Jencks Diagram 207

Appendix 2 **Answers to Problems** **210**

Appendix 3 **Tables of Structure Parameters** **258**

Appendix 4 **Some Linear Free Energy Equations** **282**

Subject Index **290**

CHAPTER 1

Free Energy Relationships

There are many techniques of varying degrees of generality for the study of mechanisms and that of free energy relationships is the most readily applicable and general. Free energy relationships comprise the simplest and easiest of techniques to use but the results are probably the trickiest to interpret of all the mechanistic tools.

The quantitative study of free energy relationships was first introduced by Brønsted[1] and Hammett[2] in the early part of the twentieth century. As well as playing a major rôle in understanding and deducing mechanism, free energy relationships have been incorporated into software to provide computational methods for calculating physico-chemical parameters of molecules. Such techniques are of substantial technical importance in drug design[3] where parameters such as solubility, pK_a, *etc.* can be useful indicators of the biological activity of chemical structures of potentially therapeutic compounds. Free energy relationships may be employed in the design of synthetic routes.[4]

The most recent undergraduate texts devoted to free energy relationships were published in 1973.[5,6] Since that time the application of free energy relationships to the study of organic and bio-organic mechanisms has undergone substantial transformation and fierce critical appraisal. The slopes of linear free energy correlations of polar substituent effects are related directly to changes in electronic charge (defined as *effective charge*)[7] at a reaction centre. Since charge difference is a function of bonding change, polar substituent effects are related to bonding in the transition structure[a] relative to that in the reactant;[8] it has to be

---

[a] The term *transition state* refers to a hypothetical quasi-thermodynamic state. *Transition structure* (formerly called the *activated complex*) refers to the molecular structure at the potential energy maximum on the reaction path which, unlike reactants and products, has no stability (see for example H. Maskill, *Mechanisms of Organic Reactions*, Oxford Science Publications, Oxford, 1996).

recognised that this relationship is often not a simple one. Effective charge is measured by comparing the polar substituent effect with that of a standard equilibrium, such as the (heterolytic) dissociation of an acid, where the charge difference is arbitrarily defined. Comparison of the polar substituent effect with that of a standard dissociation is a special case of the *similarity* concept where the change in free energy in a reaction or process is compared with that of a *similar standard process*. This way of utilising the polar substituent effect provides a direct relationship between an empirical parameter (the slope of the free energy relationship) and an easily understood physical quantity, namely charge, developing in the reference process.

Polar substituent effects are observed when there is a change in the polarity of the substituent in the system. The observer's tool (the substituent change) is part of the system being observed and will thus affect the results. The observational effect is inherent in all measurements (which must be made by an observer) but is usually negligible, for example in chemical experiments involving relatively non-invasive techniques, such as electromagnetic radiation, as the means of observation. In free energy relationships the effect could be serious unless it is recognised. Indeed, as we shall see in Chapters 5 and 6 the curvature sometimes observed in free energy correlations is a direct consequence of the connection between observational tool and observed reaction centre. The additional complication that changing the substituent could have a *gross* effect on the mechanism is a very useful tool as explained in Chapter 7.

We shall be discussing the following main attributes of the technique of free energy relationships throughout the text:

- *Linear* free energy relationships are empirical observations which can be derived when the shapes of the potential energy surfaces of a reaction are not substantially altered by varying the substituent.
- The slopes of *linear* free energy relationships for rate constants are related to transition structure.
- Transition structure is obtained from knowledge of the bond orders of the major bonding changes as well as of solvation.
- Changes in the shape of the energy surface of a reaction give rise to predictable changes in the slopes of free energy relationships.
- Polar substituent effects arise from changes in electronic charge in a reaction brought about by bond order and solvation changes.

1.1 MECHANISM AND STRUCTURE

The mechanism of a reaction is described as the structure and energy of a system of molecules in progress from reactants to products. This representation gives rise to an expectation that experimental methods should indicate the structures as shown in static diagrams; it obscures the fact that the experimental techniques refer to assemblies of structures each of which has rotational, vibrational and translational activity.

An assembly of reactant molecules is converted into product molecules through an assembly of transient structures which are *not* discrete molecules. Even though the structures making up the transition state assembly are not interconvertible within the period of their life-times (10^{-13} to 10^{-14} seconds) the assembly is postulated to have a normal thermodynamic distribution of energies.[9] A definition of mechanism which can be fulfilled experimentally, at least in principle, is a description of any intermediates and the average transition structures intervening between these intermediates, reactants and products. This definition of mechanism is currently employed in mechanistic studies. More funda-mental knowledge, such as that of the structure of the system between reactant and transition structure, is at present largely in the realm of theoretical chemistry.[b]

Structure requires a description of the relative positions of nuclei and the electron density distribution in a system. The existence of zero-point vibrations indicates that even at absolute zero the exact positions of atoms in a molecule are uncertain. As the temperature is increased, higher quantum states are occupied for each degree of freedom so that fluctuations around the mean positions of atoms increase. In order to discuss the mechanism of any reaction, it is necessary to bear in mind precisely what the term *structure* means. A pure compound is usually visualised as an assembly of molecules each comprising atoms which have identical topology relative to their neighbours. The exact relative position of each atom is time-dependent, and a bond length, measured on a collection of molecules, is an *average* quantity. The two-dimensional static structures possessing bonds represented by lines (Lewis bonds) are hypothetical *models*. These structures, commonly designated Kekulé structures, are very convenient for visualising organic molecules as they are easily comprehensible and are easily represented graphically in two dimensions. Although they are often taken for reality (particularly in their three-dimensional form) they are simply *representations* of hypotheses which fit experimental knowledge of the compounds.

[b] See however A.H. Zewail, *Science*, 1988, **242**, 1645 and A.H. Zewail (ed.), *The Chemical Bond, Structure and Dynamics*, Academic Press, New York, 1992.

The *reaction coordinate*[10] is a measure of the average structure taken by a single molecular system on passage from reactants to products. Structure usually equates with potential energy. The transition structure is not a molecular species and corresponds to the maximum of the Gibbs free energy in the system on progression from reactant to product state. The free energy is distributed amongst the available quantum states of the various degrees of freedom of the collection of pseudo-molecules of the transition state.

The structure of the molecular species in the transition state is thus an average but it should be stressed that this is no different in principle from that of a discrete molecule which is also an average.[c]

1.1.1 Interconversion of States – Reaction and Encounter Complexes

In a bimolecular reaction in solution reactants diffuse through the assembly of solute and solvent molecules (Scheme 1) and collide to form an *encounter complex* within a solvent cage. Reaction is not possible until any necessary changes occur in the ionic atmosphere to form an *active complex* and in solvation (such as desolvation of lone pairs) to form a *reaction complex* in which bonding changes take place. The encounter complex remains essentially intact for the time period of several collisions because of the protecting effect of the solvent surrounding the molecules once they have collided. The products of the subsequent reaction could either return to reactants or diffuse into the bulk solvent.

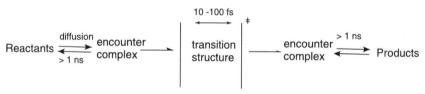

Scheme 1   *Bimolecular mechanism in solution (ns = 10^{-9} sec; fs = 10^{-15} sec)*

A similar description applies to a unimolecular reaction except that formation of the transition structure is initiated by energy accumulation in the solvated reactant by collision.

Scheme 2 gives typical half-lives for reactant molecules *destined* to react.[11] Many encounters do not lead to reaction and only a small fraction of the complexes will have the appropriate transition state solvation in place to form a *reaction complex* and for reaction to proceed.

---

[c] Even the most precise X-ray crystallographic measurements of structure associate the positions of atoms in a molecule with error volumes.

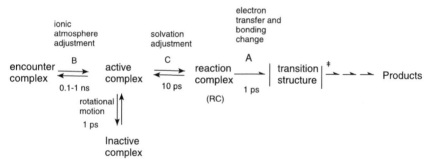

Scheme 2 *Range of lifetimes of processes in a bimolecular mechanism in solution* $(ps = 10^{-12}\,sec)$

Extracting conclusions from experimental results derived from reactions of collections of molecules requires the interpretation of measured
global energy changes in these collections as a function of "average"
structures. Free energy relationships provide an important technique
for carrying out this interpretation. In general polar substituents will
have their greatest effect on the process between reaction complexes
(A in Scheme 2). Solvent change is largely effective on processes B and C
but there is some effect on A due to solvent stabilisation of charge and
dipole caused by bonding change.

1.2 UNIVERSAL MEASURE OF POLARITY

It is commonplace that some groups withdraw electrons from a reaction
centre whereas others donate electrons. For example, a halo group
is expected to attract electrons relative to hydrogen, and alkyl groups
to donate electrons. Free energy relationships provide a means of
quantifying polar effects. The concept of relative polarity is not simply
that of an *inductive* effect (I) through σ-orbitals, because transmission
of charge can occur through space as a *field* effect (F) or through a
conjugated system of π-orbitals as a *resonance* effect (R). The extent
to which these three transmission vectors contribute to the overall
polarity of a substituent group is dependent on the system to which it is
attached.

The original measure of substituent polarity was chosen to be the
dissociation constant of a substituted benzoic acid in water (Equation
1).[12,13] It is readily confirmed that the pK_a of the benzoic acid is raised by
an electron donating substituent and lowered by an electron withdrawing
substituent.

$$X\!\!-\!\!\langle \bigcirc \rangle\!\!-\!\!CO_2H \;+\; H_2O \rightleftharpoons X\!\!-\!\!\langle \bigcirc \rangle\!\!-\!\!CO_2^- \;+\; H_3O^+ \qquad (1)$$

A polar substituent changes the energy of the transition state of a reaction by modifying the change in charge brought about by the bond changes (Hine).[14,15] Thus substituents that withdraw electrons and *disperse* negative charge will tend to stabilise a structure if there is an increase in negative charge at the reaction centre (and *vice versa*). The polar substituent must be suitably placed in the molecule for the effect to be transmitted efficiently to the reaction centre through the bond framework, space or solvent.

Most experimental techniques measure differences; a change in the rate constant of a reaction brought about by a change of substituent represents the *difference* in the effect on reactants compared with that on the transition state. An increase in rate constant could result from the reactant becoming less stable, a more stable transition state or a combination of these effects. The observed effect is due to the difference between the effects on the reactant (∂G_r) and transition $(\partial G^{\ddagger})$ states as illustrated in Scheme 3.

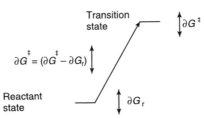

Scheme 3 *The source of the substituent effect $(\partial\Delta G^{\ddagger})$ on a rate process*

1.3 CLASSES OF FREE ENERGY RELATIONSHIP

A free energy relationship is defined by Equation (2) where the parameter *a* is called the *similarity* coefficient.

$$\Delta G = a\Delta G_s + b \qquad (2)$$

The term ΔG is the free energy of a process such as a rate or equilibrium and ΔG_s is the free energy of a standard process, often an equilibrium, which could be the process under investigation or some other standard reaction.

Equation (2) holds for many processes over relatively large changes in free energy and is usually written in terms more directly related to those which have been experimentally determined such as the logarithms (to base 10) of rate (k) or equilibrium (K) constants (Equations 3–5).[d]

$$\log K = a_1 \log K_s + b_1 \tag{3}$$

$$\log k = a_2 \log K_s + b_2 \tag{4}$$

$$\log k = a_3 \log k_s + b_3 \tag{5}$$

Class I free energy relationships compare a rate constant with the equilibrium constant of the same process. The Brønsted and Leffler equations (see Chapter 2) are examples of this class of free energy correlation.

The rate or equilibrium constant in Class II is related to the rate or equilibrium constant of an unconnected but (often) similar process. Class II free energy relationships are in general more common than those of Class I because equilibrium constants are more difficult to measure than rate constants (except in certain cases such as dissociation constants). The Hammett equation is the best-known Class II free energy relationship.

1.4 ORIGIN OF FREE ENERGY RELATIONSHIPS

The existence of Class I free energy relationships can be deduced from the energy profile of a reaction.[e] Figure 1 illustrates what happens to the profile as a substituent is changed. We shall assume that the reaction (Equation 6) consists of a single bond fission (A–B $\rightarrow$ B + A)

$$\text{A–B} + \text{C} \rightarrow \text{A–C} + \text{B} \tag{6}$$

and a single bond formation (C + A $\rightarrow$ A–C) related to inter-atom distance by Morse curves (see Appendix 1, Section A1.1). The intersecting point (Figure 1) represents the transition structure where cross-over from one system to the other results in a smooth junction between the two processes (as shown). For small changes in overall energy the variation in ΔE_a is linearly related to the change in ΔE_o by Equation (7) (see Appendix 1, Section A1.2, for derivation).

---

[d] This is because free energy is related to equilibrium and rate constants by logarithmic equations $\Delta G = -2.303 RT \log K$ and $\Delta G^{\ddagger} = -2.303 RT \log k + 2.303\,RT \log \kappa k_B T/h$ where κ is the transmission coefficient and k_B and h are Boltzmann's and Planck's constants.

[e] The Class I relationship has been derived by the application of molecular orbital theory and for details the reader is referred to M.J.S. Dewar, *The Molecular Orbital Theory of Organic Chemistry*, McGraw-Hill, New York, 1969; R.F. Hudson, *Angew. Chem. In. Ed.*, 1973, **85**, 63; M.J.S. Dewar and R.C. Dougherty, *The PMO Theory of Organic Chemistry*, Plenum Press, New York, 1975; A. Warshel, *Computer Modelling of Chemical Reactions in Enzymes and Solutions*, Wiley-Interscience, New York, 1991, Section 3.7.

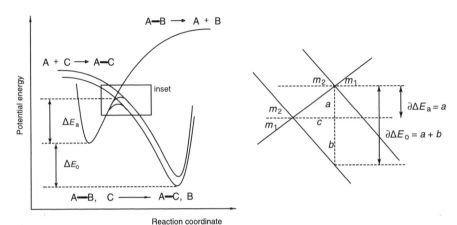

Figure 1   *Two-dimensional potential energy diagram for Equation (6) formed from combined Morse curves for the elementary processes $A + C \rightarrow A{-}C$ and $A{-}B \rightarrow A + B$*

$$\partial\Delta E_{a} = \frac{m_{1}}{m_{1} + m_{2}}\,\partial\Delta E_{o} \qquad (7)$$

Over the small range of ΔE_{o} the entropic component of the free energy can be assumed not to change so that $\partial\Delta G$ and $\partial\Delta G^{\ddagger}$ are proportional to $\partial\Delta E$ and $\partial\Delta E^{\ddagger}$ respectively and Equation (8) follows.

$$\partial\Delta G^{\ddagger} = \frac{m_{1}}{m_{1} + m_{2}}\,\partial\Delta G_{o} \qquad (8)$$

Integration and rearrangement of this equation gives rise to the Class I free energy relationship (Equation 9).[16]

$$\log k = \frac{m_{1}}{m_{1} + m_{2}}\,\log K + b = a\log K + b \qquad (9)$$

This equation fails if the slopes m_{1} and m_{2} change owing to large changes in ΔG_{o}. The Class I relationship can alternatively be derived from the Marcus-type function (Chapter 6).

In the meantime it is necessary to define what is meant by the term *reaction coordinate*. The reaction coordinate for a dissociation process of a diatomic species is simply the distance between the atoms as the reaction proceeds. The majority of reactions are not of diatomic molecules and there is usually more than one bond undergoing a major bonding change, so that the reaction coordinate can no longer be regarded as a simple distance between atoms in a two-dimensional representation. For the purposes of Figure 1, the reaction coordinate has the dimension of interatomic distance for each component process, even though the identity of the major bonding change alters as the system progresses from reactants to products.

The origins of the Class II free energy relationships are not so easy to visualise as those of Class I in terms of a molecular model, because there is no direct atomic interaction between the systems giving rise to the two free energy changes (as in Figure 1). In this case it is necessary to invoke chemical intuition to indicate that substituent effects on chemically similar processes will be related to each other. No such assumption is necessary in the Class I case.

Leffler and Grunwald devised a treatment which derives both Class I and Class II free energy relationships and is more general than that using crossed energy surfaces.[17] The equation, $\partial \Delta G_R = a.\partial G_{Rs}$, relating the free energy of reaction of the reactant XR with that of the standard reaction XR_s can be derived using the assumption that free energies of substances are additive functions. Let us consider a transformation (Equation 10) of a system (XR) comprising a single interaction between variable, non-reacting, substituent (X) and the reacting group (R) and product group (P).

$$XR \rightleftharpoons XP \tag{10}$$

The free energies of XR and XP are assumed to be additive functions:

$$G^\circ_{XR} = G^\circ_X + G^\circ_R + I_{X,R} \qquad \text{and} \qquad G^\circ_{XP} = G^\circ_X + G^\circ_P + I_{X,P}$$

where $I_{X,R}$, $I_{X,P}$ are interaction terms between substituent X and the molecule. Therefore

$$\Delta G^\circ_R = G^\circ_{XP} - G^\circ_{XR} = (G^\circ_P - G^\circ_R) + (I_{X,P} - I_{X,R})$$

The *standard* or *reference* reaction (Equation 11)

$$XR_s \rightleftharpoons XP_s \tag{11}$$

has a free energy of reaction of:

$$\Delta G^\circ_{Rs} = G^\circ_{XPs} - G^\circ_{XRs} = (G^\circ_{Ps} - G^\circ_{Rs}) + (I_{X,Ps} - I_{X,Rs})$$

An assumption, the postulate of *separability*, is necessary so that the interaction terms ($I_{X,R}$) may be factored:

$$I_{X,R} = I_X I_R$$

and the energies can therefore be written:

$$\Delta G^\circ_R = (G^\circ_P - G^\circ_R) + I_X(I_P - I_R)$$

and

$$\Delta G^\circ_{Rs} = (G^\circ_{Ps} - G^\circ_{Rs}) + I_X(I_{Ps} - I_{Rs})$$

Partial differentiation of these equations against the variable X leads to

$$\partial \Delta G^\circ_R = \partial I_X(I_P - I_R)$$

and

$$\partial \Delta G^\circ_{Rs} = \partial I_X(I_{Ps} - I_{Rs})$$

and Equation (12) results which is analogous to Equation (8).

$$\partial \Delta G^\circ_R = \left\{ \frac{I_P - I_R}{I_{Ps} - I_{Rs}} \right\} \partial \Delta G^\circ_{Rs} = a \partial \Delta G^\circ_{Rs} \tag{12}$$

The assumption of a single interaction pathway and the *separability* postulate are only likely to hold if the changes made in X are small. Equation (12) could fail when there are large changes in X leading to large changes in the standard or reference reaction (Chapter 6).

1.5 SIMILARITY AND THE LEFFLER α-PARAMETER

The similarity coefficient "*a*" in the free energy equation (Equation 2) compares the change in the process under investigation (the unknown system) with that for a standard or reference (the known) system. The coefficient measures the *similarity* between the two systems. In the case of Class I relationships the similarity coefficient measures the extent to which the transition state resembles the products compared with the reactants. The similarity coefficient for Class II systems measures the extent to which conversion to the transition state resembles the conversion of reactants into products in the standard or reference system.

Free energy relationships are not confined to chemical reactivity or equilibria and can arise from comparisons of rate constants with solvent parameters such as partition coefficients, concentrations or spectral frequencies, all of which relate to energy changes. NMR chemical shifts have been studied extensively as a function of substituent change[18] as have polarographic half-wave potentials and oxidation–reduction potentials.[19] Steric similarity coefficients also enable steric requirements of reactions to be quantified in terms of the steric requirements of a standard reaction.

The reference reaction in the case of Class I systems is the equilibrium reaction of the system under investigation and therefore offers the best possible comparison for determining a transition structure. For this reason the Leffler equation (Class I) is commonly used to indicate the structure of a transition state.[20]

Introduction of a substituent perturbs the energy of a reactant or product. Change in the free energy of the transition state is assumed to be a linear combination of the corresponding changes in reactants and products (Equation 13)[21] which can be expressed as Equation (14) with $0 < \alpha < 1$ and used to derive Equation (16) as below.

$$\partial G^{\ddagger} = a\partial G^{\circ}_{XP} + b\partial G^{\circ}_{XR} \tag{13}$$

$$\partial G^{\ddagger} = \alpha\partial G^{\circ}_{XP} + (1-\alpha)\partial G^{\circ}_{XR} \tag{14}$$

$$\partial G^{\ddagger} - \partial G^{\circ}_{XR} = \alpha(\partial G^{\circ}_{XP} - \delta G^{\circ}_{XR}) \tag{15}$$

Let

$$\Delta G^{\ddagger} = G^{\ddagger} - G^{\circ}_{XR} \qquad \text{and} \qquad \Delta G^{\circ}_{R} = G^{\circ}_{XP} - G^{\circ}_{XR}$$

Therefore

$$\partial\Delta G^{\ddagger} = \partial G^{\ddagger} - \partial G^{\circ}_{XR} \qquad \text{and} \qquad \partial\Delta G^{\circ}_{R} = \partial G^{\circ}_{XP} - \partial G^{\circ}_{XR}$$

Substituting for $\Delta G^{\ddagger}$ and ΔG°_{R} into Equation (15) gives:

$$\partial\Delta G^{\ddagger} = \alpha\partial\Delta G^{\circ}_{R} \tag{16}$$

The congruence between this equation and Equations (8) and (12) gives some credence to the assumption of a linear combination (Equation 14), and Equation (16) may be re-written as the Leffler Equation (17)

$$\partial\Delta G^{\ddagger}/\delta\Delta G^{ST} = \alpha\partial\Delta G^{\circ}_{R}/\partial\Delta G^{ST} \tag{17}$$

where ΔG^{ST} refers to any standard equilibrium. The Leffler parameter α may then be written in terms of Hammett (ρ), Brønsted (β) or general similarity coefficients (a) as in Equation (18) where ρ or β refer to substituent effects on the rate constant and ρ_{eq} or β_{eq} refer to substituent effects on the equilibrium constant of the reaction.

$$\alpha = \rho/\rho_{eq} = \beta/\beta_{eq} = a/a_{eq} \qquad (18)$$

Since the transition structure lies between that of reactant (XR) and product (XP) it is likely that *changes* in its free energy will be intermediate between the corresponding changes for reactant and product (although the standard free energy of the transition state is larger than those of reactant and product). This can be expressed mathematically in Equation (16) using a Leffler α-parameter which has a value between 0 and 1. These relationships can be derived graphically as described in Appendix 1, Section A1.3.

1.6 BONDING IN TRANSITION STRUCTURES

The strength of a bond undergoing a major change in a reaction may be defined by its interatomic distance relative to that of reactant or product in the reaction complex (RC, Scheme 2).[22] The distance between the atoms in a bond undergoing major change is only experimentally accessible for reactants, products and intermediates but not for any of the other structures, including the transition structures, on the path from reactant to product. Although interatomic distance is employed extensively in theoretical studies other measures of bonding must be sought for use in experimental studies, and one of these is the polar substituent effect.[f]

The polar substituent effect relates to changes in charge or dipole in the system that derive from the difference in electronic structure between ground and transition state and for this reason is a measure of bonding. Great care must be taken in interpreting polar substituent effects in terms of the extent of bonding. The shorthand notation used in illustrating reaction mechanisms tacitly implies that bonding change is the only component to the energy in a reaction. Simply changing the solvent can have enormous energy effects: for example, the gas-phase identity displacement at methyl chloride by chloride ion has been shown to possess no energy of activation whereas in solution the solvent interactions give rise to a substantial barrier.[23]

Considerable effort has been expended in discussion of polar

---

[f] For other approaches to the study of bond strength in reaction mechanisms see R.L. Schowen and R.D. Gandour (eds.), *Transition States of Biochemical Processes*, Plenum Press, New York, 1978.

substituent effects as direct measures of bonding in the transition state whereas, although these effects result from differences in electronic structure between ground and transition state, they do not directly indicate the strengths of individual bonds.

One of the difficulties in the past application of free energy relationships has been the attempt to define a *global* description of the transition state in reactions which involve two or more major bonding changes. A substituent effect, from which the free energy relationship derives, usually arises from interaction with a single bond change. Thus, the description derived from a free energy relationship would only refer to this bond in the transition structure and not to the whole system. It is therefore wrong in this case to describe the transition structure as being close to reactants or products. It is only possible to describe an individual bond undergoing change in the transition structure as being close to that in reactants or products. The dissection of the substituent effects in a reaction into their component bonding changes is one of the major themes in this book.

In the application of free energy relationships to the study of bond order in a reaction the following considerations need to be addressed:[24]

- The free energy relationship for a rate process involves a reference equilibrium process.
- The polar effect (slope of a linear free energy relationship) for the rate of a reaction refers to a single rate-limiting transition structure.
- The relationship between charge development and bond order is not necessarily linear.
- Solvation changes, including hydrogen-bonding, may not be in step with bonding changes.
- A reaction does not necessarily involve a single bonding change – indeed there are often two or more bonds undergoing change in any reaction.
- Bonding changes involved in a reaction may not be synchronised.

In spite of the theoretical problems connecting similarity coefficients with bonding in the transition structure, they are used extensively in More O'Ferrall–Jencks diagrams (see Chapter 6) connecting energy and bonding for reactions with two major bonding changes.

1.7 FURTHER READING

W.J. Albery, Transition-state Theory Revisited, *Adv. Phys. Org. Chem.*, 1993, **28**, 139.

E.V. Anslyn and D.A. Dougherty, *Modern Physical Organic Chemistry*, University Science Books, Mill Valley, CA, 2004.

R.P. Bell, *The Proton in Chemistry*, Methuen, London, 1959.

E.F. Caldin and V. Gold (eds.), *Proton Transfer Reactions*, Chapman & Hall, London, 1975.

E.F. Caldin, *The Mechanisms of Fast Reactions in Solution*, IOS Press, Amsterdam, 2001.

N.B. Chapman and J. Shorter (eds.), *Advances in Linear Free Energy Relationships*, Plenum Press, New York, 1972.

N.B. Chapman & J. Shorter (eds.), *Correlation Analysis in Chemistry*, Plenum Press, New York, 1978.

O. Exner, *Correlation Analysis of Chemical Data*, Plenum Press, New York, 1988.

S.J. Formosinho, Reactivity and Selectivity. An Intersecting State View, *J. Chem. Soc. Perkin Trans. 2*, 1988, 839.

C.A. Grob, Polar Effects in Organic Reactions, *Angew. Chem.*, 1976, **15**, 569.

E. Grunwald, Linear Free Energy Relationships: A Historical Perspective, *Chemtech*, 1984, 698.

L.P. Hammett, Some Relations Between Reaction Rates and Equilibrium Constants, *Chem. Rev.*, 1935, **17**, 125.

L.P. Hammett, *Physical Organic Chemistry*, McGraw-Hill, New York, 1st edition, 1940.

L.P. Hammett, Physical Organic Chemistry in Retrospect, *J. Chem. Ed.*, 1966, **43**, 464.

J. Hine, *Structural Effects on Equilibria in Organic Chemistry*, John Wiley, New York, 1975.

W.P. Jencks, Are Structure–Reactivity Correlations Useful? *Bull. Soc. Chim. France*, 1988, 218.

C.D. Johnson and K. Schofield, A Criticism of the use of the Hammett Equation in Structure–Reactivity Correlations, *J. Am. Chem. Soc.*, 1973, **95**, 270.

M. Kamlet and R.W. Taft, Linear Solvation Free Energy Relationships. Local Empirical Rules or Fundamental Laws of Chemistry? *Acta Chem. Scand.*, 1985, **B39**, 611.

C.E. Klots, The Reaction Coordinate and its Limitations: an Experimental Perspective, *Accs. Chem. Res.*, 1988, **21**, 16.

J.E. Leffler and E. Grunwald, *Rates and Equilibria of Organic Reactions*, John Wiley, New York, 1963.

E.S. Lewis, Linear Free Energy Relationships, in *Investigation of Rate and Mechanisms of Reactions Part 1*, C.F. Bernasconi (ed.) Wiley-Interscience, New York, 1986, 4th edition, p. 871.

T.H. Lowry and K.S. Richardson, *Mechanism and Theory in Organic Chemistry*, Harper and Row, New York, 1987, 3rd edition, p. 596.

H. Maskill, *The Physical Basis of Organic Chemistry*, Oxford University Press, Oxford, 1985.

H. Maskill, *Mechanisms of Organic Reactions*, Oxford Science Publications, Oxford, 1996.

J. Shorter, *Prog. Phys. Org. Chem.*, 1990, **17**, 1.

M. Sjostrom and S. Wold, Linear Free Energy Relationships. Local Empirical Rules or Fundamental Laws of Chemistry? *Acta Chem. Scand.*, 1981, **B35**, 537.

P.R. Wells, *Linear Free Energy Relationships*, Academic Press, New York, 1968.

S. Wold and M. Sjostrom, Linear Free Energy Relationships. Local Empirical Rules or Fundamental Laws of Chemistry. A Reply to Kamlet and Taft, *Acta Chem. Scand.*, 1986, **B40**, 270.

1.8 REFERENCES

1. J.N. Brønsted, Acid and Base Catalysis, *Chem. Rev.*, 1928, **5**, 23.
2. L.P. Hammett, The Effect of Structure Upon the Reactivity of Organic Compounds, *J. Am. Chem. Soc.*, 1937, **59**, 96.
3. D. Livingstone, *Data Analysis for Chemists*, Oxford University Press, Oxford, 1995.
4. H. Mayr *et al.*, Linear Free Energy Relationships: A Powerful Tool for the Design of Organic and Organo-metallic Synthesis, *J. Phys. Org. Chem.*, 1998, **11**, 642.
5. C.D. Johnson, *The Hammett Equation*, Cambridge University Press, Cambridge, 1973.
6. J. Shorter, *Correlation Analysis in Organic Chemistry*, Clarendon Press, Oxford, 1973.
7. A. Williams, Effective Charge and the Transition State Structure in Solution, *Adv. Phys. Org. Chem.*, 1991, **27**, 1.
8. P. Müller, Glossary of Terms used in Physical Organic Chemistry, *Pure Appl. Chem.*, 1994, **66**, 1077.
9. D.G. Truhlar *et al.*, Current Status of Transition State Theory, *J. Phys. Chem.*, 1996, **100**, 1977.
10. E. Grunwald, Reaction Coordinate and Structure–Energy Relationships, *Prog. Phys. Org. Chem.*, 1990, **17**, 55.
11. A.H. Zewail, Laser Femtosecond Chemistry, *Science*, 1988, **242**, 1645.
12. L.P. Hammett, *Physical Organic Chemistry*, McGraw-Hill, New York, 2nd edition, 1970.

13. H.H. Jaffé, A Re-examination of the Hammett Equation, *Chem. Rev.*, 1953, **53**, 191.
14. J. Hine, Polar Effects on Rates and Equilibria, *J. Am. Chem. Soc.*, 1959, **81**, 1126.
15. J. Hine, Polar Effects on Rates and Equilibria (III), *J. Am. Chem. Soc.*, 1960, **82**, 4877.
16. R.P. Bell, *The Proton in Chemistry*, Chapman & Hall, London, 2nd edition, 1973.
17. J.E. Leffler and E. Grunwald, *Rates and Equilibria of Organic Reactions*, John Wiley, New York, 1963.
18. D.F. Ewing, Correlation of NMR Chemical Shifts with Hammett σ Values and Analogous Parameters, in *Correlation Analysis in Chemistry. Recent Advances*, N.B. Chapman & J. Shorter (eds.), Plenum Press, New York, 1978.
19. P. Zuman, *Substituent Effects in Organic Polarography*, Plenum Press, New York, 1967.
20. A. Williams, Effective Charge and Leffler's Index as Mechanistic Tools for Reactions in Solution, *Acc. Chem. Res.*, 1984, **17**, 425.
21. J.E. Leffler, Parameters for the Description of Transition States, *Science*, 1953, **117**, 340.
22. W.J. Albery and M.M. Kreevoy, Methyl Transfer Reactions, *Adv. Phys. Org. Chem.*, 1978, **16**, 87.
23. J. Chandrasekhar and W.L. Jorgensen, Energy Profile for a Non-concerted S_N2 Reaction in Solution, *J. Am. Chem. Soc.*, 1985, **107**, 2974.
24. A. Pross and S.S. Shaik, Brønsted Coefficients. Do they Measure Transition State Structure? *New J. Chem.*, 1989, **13**, 427.

CHAPTER 2

The Equations

Changes in energy of the transition state or of the product state relative to the reactant state, can be brought about by modifying the polar, solvent or steric parameters of the system, and give rise to free energy relationships. Many empirical equations have been advanced for polar substituent effects but none is intrinsically different from any other; each is suitable for a particular purpose. For example, the Brønsted equation is useful for studying proton transfer reactions or nucleophilicity in a series of structurally similar nucleophiles and the Hammett equation for reactions involving substituent change on an aromatic nucleus.

The following sections provide a conventional summary of commonly used free energy relationships and modifications of these equations are described in Chapter 4. A multiplicity of equations were developed between 1950 and 1970 to attempt to obtain the "best" or "preferred" sets of polarity constants, σ, (see Section 2.1). Each of these polarity constants has its own symbolism (σ_m, σ', σ_I, σ'', etc.) which can lead to substantial confusion if care is not taken to choose the appropriate parameter for the system under investigation.

2.1 THE HAMMETT EQUATION (CLASS II)

Linear correlations (Equation 1) between the logarithms of rate constants for reactions of *meta-* and *para*-substituted phenyl derivatives ($\log k_X$) and pK_a values of the correspondingly substituted benzoic acids were reported by Hammett.[1] The equation for the unsubstituted derivative (H, Equation 2) can be combined with Equation (1) for the X-substituted derivative to yield Equations (3) and (4) where $\sigma = pK_a^H - pK_a^X$.

$$\log k_X = \rho \log K_a^X + C \tag{1}$$

$$\log k_H = \rho \log K_a^H + C \tag{2}$$

$$\log k_X = \rho \left(\log K_a^X - \log K_a^H\right) + \log k_H \tag{3}$$

$$= \rho \, \sigma + \log k_H \tag{4}$$

The dissociation of benzoic acids is a reasonable model for the bimolecular reaction of benzoate esters with phenolate anion and a plot of $\log k_{OPh}$ versus σ obeys Equation (4) (Figure 1).

Figure 2 illustrates the Hammett dependences of dissociation equilibria for various substituted phenylcarboxylic acids.

The ρ-value for a reaction measures its susceptibility to change in substituent and a positive ρ-value registers that electron withdrawing substituents increase the rate or equilibrium constant consistent with an increase in negative charge at the reaction centre. The ρ-value is therefore a measure of the change in charge in the system *relative* to that in the dissociation of benzoic acids where ρ is defined as 1.0 and change in charge in the carboxyl group is -1.0. Values of σ for individual substituents are tabulated in the appendices together with ρ values for rates and equilibria for a selection of reactions.

The Hammett relationship only applies to systems where substituents are attached to the reaction centre *via* aromatic rings and are situated *meta* or *para*; *ortho* substituents are *not* included in the Hammett equation owing to steric and through-space interactions (Chapter 6).

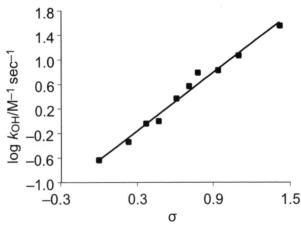

Figure 1   *Reaction of phenoxide ion with 4-nitrophenyl-substituted benzoate esters*[2]

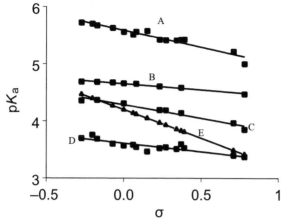

Figure 2 *Hammett dependence for the dissociation of substituted phenyl carboxylic acids. (A) arylmercaptoacetic (50% dioxan/water), (B) arylpropionic, (C) arylacetic, (D) arylmercaptoacetic (water) and (E) substituted benzoic acids (the standard or reference reaction)*

2.2 THE TAFT EQUATION (CLASS II)

The use of Hammett's σ often gives uninterpretable results in reactions where the substituent is located on aliphatic carbon and transmission of charge is through sigma bonds. Another polarity parameter, analogous to the Hammett σ, must therefore be sought for aliphatic systems.

Taft's polar substituent constant, σ^*, applicable to aliphatic systems is based on the notion that free energy relationships can exist between rate processes as well as between rate and equilibrium processes (Equations 3–5, Chapter 1). A rate process can thus be employed as a reference reaction. A simple example of such a Class II rate–rate free energy relationship is shown in Figure 3 where the reaction of phenoxide ion with 4-nitrophenyl-substituted benzoate esters is compared with the corresponding reaction of hydroxide ion, which therefore becomes the reference free energy change.

In Taft's treatment the acid- and base-catalysed hydrolysis of substituted esters X–COOR′ are employed as combined reference reactions for defining polar and steric effects for the substituent X.[3] Equation (5) governs the combined inductive, steric (S) and resonance (R) effects on the rate of a reaction.[4,5] Equations (6) and (7) can be written for base- and acid-catalysed hydrolysis using methyl (Me) as the standard substituent.

$$\log(k^X/k^{Me}) = \rho^*\sigma^* + S + R \tag{5}$$

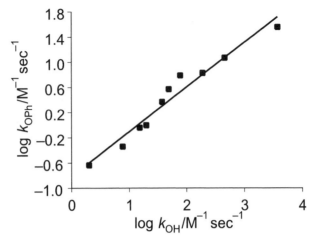

Figure 3 *Reaction of phenoxide ion and hydroxide ion with 4-nitrophenyl-substituted benzoate esters*[2]

$$\log(k_{OH}{}^X/k_{OH}{}^{Me}) = \rho_{OH}{}^*\sigma^* + S_{OH} + R_{OH} \qquad (6)$$

$$\log(k_H{}^X/k_H{}^{Me}) = \rho_H{}^*\sigma^* + S_H + R_H \qquad (7)$$

If S and R are the same for acid and base reactions, Equations (6) and (7) can be combined to give Equation (8):

$$\log(k_{OH}{}^X/k_{OH}{}^{Me}) - \log(k_H{}^X/k_H{}^{Me}) = (\rho_{OH}{}^* - \rho_H{}^*)\sigma^* \qquad (8)$$

The σ^* values may be defined by Equation (9) where $\rho^*_{OH} - \rho_H{}^*$ has been set to 2.48 in order that σ^* values approximate to the corresponding Hammett σ constants. The value 2.48 is the difference between the Hammett ρ values for base- and acid-catalysed hydrolysis of ethyl benzoates. Thus:

$$\sigma^* = [\log(k_{OH}{}^X/k_{OH}{}^{Me}) - \log(k_H{}^X/k_H{}^{Me})]/2.48 \qquad (9)$$

Reasonable assumptions necessary for the Taft analysis are:

- That the additivity expressed by Equation (5) is valid. This assumption is not self-evident but is attested by the consistency of the results.
- The assumption ($S = S_H = S_{OH}$) is reasonable since the size of the proton is very small and would lead to closely similar steric requirements for the transition structures (Equations 10 and 11). Perusal

of the structures indicates that any resonance interactions, R, from the substituent in X will be similar for both acid and base reactions.

- Polar effects of substituents are markedly greater in the basic than in the acidic reaction. This assumption is attested by the values of the Hammett ρ values for the corresponding reactions of ethyl benzoates. The ratios $k_{OH}{}^X/k_{OH}{}^{Me}$ and $k_H{}^X/k_H{}^{Me}$ are independent of R' as long as it is the same for the given pair of rate constants.
- The inductive component in Equation (5) ($\rho^*\sigma^*$) is the sum of the through-space (field effect, F) and through-bond inductive effect (I). The significance of the field effect (F) is discussed later.

$$\text{(10)}$$

$$\text{(11)}$$

The Hammett relationship for the acid-catalysed hydrolysis of ethyl benzoates has negligible slope so that the inductive and resonance (R) effects for the acid-catalysed reaction are deduced to be negligible. Equation (7) thus reduces to Equation (12) from which the value of S may be determined.

$$\log(k_H{}^X/k_H{}^{Me}) = S \qquad (12)$$

Equation (12) is a special case of Equation (13) where δ is a similarity coefficient for the steric requirement of the reaction relative to the reference addition of water to the formate ester (which has δ set at unity).

$$\log(k_H{}^X/k_H{}^{Me}) = \delta E_S \qquad (13)$$

E_S is a measure of the bulk of the substituent compared with that of Me which has E_S set at zero.

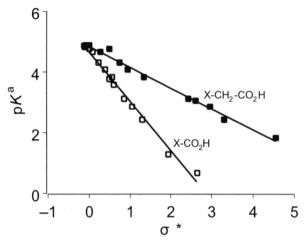

Figure 4   *Taft plot for the dissociation of carboxylic acids (X-CO_2H $\rho^* = -1.62$ and X-CH_2-CO_2H $\rho^* = -0.67$) (E_S is negligible)*

Values of E_S and σ^* are tabulated in Appendix 3, Table 1, and Figure 4 illustrates Taft plots for the dissociation of substituted carboxylic acids (X-CO_2H and X-CH_2-CO_2H).

The Taft analysis has the working disadvantage that it requires two measurements to define a σ^* value, and it also suffers from the problem that the parameters of some substituents cannot be obtained either because the ester decomposes too quickly for measurements to be made or because it would not decompose by ester hydrolysis. Such restrictions apply to the halogen substituents, nitrile or the nitro group, which would require study of such compounds as Hal-CO-OR, NC-CO-OR and O_2N-CO-OR. This problem can be solved by use of Taft analysis of esters of the type X-CH_2-CO-OR instead of X-CO-OR. In this analysis the similarity coefficient, ρ^*, for the sub-stituted acetic acid derivatives is attenuated by 0.41 from the set value of $\rho^* = 2.48$ for the formic acid derivatives (Equations 8 and 9). The σ^* constants based on formic acid derivatives are recorded in Table 1 in Appendix 3.[a]

The σ^* constants for substituents inaccessible from the formic acid standard (such as Cl-, NC-, O_2N- *etc.*) may be defined by a more

---

[a]   The reader should be aware that a set of σ^* constants has been defined from Taft analysis of acetic acid derivatives with ρ^* set at 2.48. There is an obvious problem with comparing the definitions as the formate standard refers to σ_{Me}^* as zero ($\sigma_H^* = 0.49$) and the acetate standard refers to σ_H^* as zero ($\sigma_{Me}^* = -0.11$). Indeed the simplest way of deciding which σ^* is tabulated is to check its magnitude for hydrogen as a substituent.

convenient secondary standard. Taft noted that the pK_a of acetic acids is linearly related to σ^* (Figure 4) and this enables the estimation of inaccessible σ^* values simply by determination of the pK_a values of acetic acids. The Taft equations for the dissociation of substituted acetic and formic acids shown in Figure 4 have ρ^* values of -0.67 and -1.62 respectively (σ^* from Table 1 in Appendix 3). Thus σ^* based on acetic acid derivatives will be 0.41 of that based on formic acid derivatives (Equation 14).

$$\sigma^*(CH_2X) = 0.41\sigma^*(X) \tag{14}$$

2.2.1 Steric Demand at Reaction Centres

Assuming that the resonance contribution is negligible, Equation (5) can be written as the Pavelich–Taft equation (Equation 15).[6]

$$\log(k_X/k_{Me}) = \rho^*\sigma^* + \delta E_S \tag{15}$$

Values of $\delta > 1$ for a reaction indicate that the reaction centre is more sterically crowded than that for hydrolysis of ethyl formates whereas $\delta < 1$ indicates a smaller steric requirement.

Until recently, bilinear equations such as the Pavelich–Taft or Jaffé (see Chapter 4) equations were fitted by rearranging them to a linear form. This procedure has the disadvantage that some data points are given more weight than others. Fitting to theoretical equations without mathematical rearrangement can be carried out by use of statistical procedures (see Appendix 1, Section A1.1.4). The fit of the data to the equation is demonstrated by plotting the observed data against values calculated from the theoretical, fitted, equation. Figure 5 illustrates the fit to Equation (15) for the hydrolysis of menthyl esters of carboxylic acids.[6]

An alternative measure of steric requirements was suggested by Charton[7] who defined a steric scale v_x which is the difference between the van der Waals radius for a symmetrical substituent and hydrogen ($v_x = r_x - 1.20$). The steric parameter can be incorporated into the free energy relationship by the additive function (Equation 16) where ψ is the similarity coefficient, analogous to δ, for the steric effect.

$$\log(k_x/k_H) = \rho_I.\sigma_I + \psi.v_x \tag{16}$$

Values of the Charton v_x parameter are tabulated in Appendix 3.

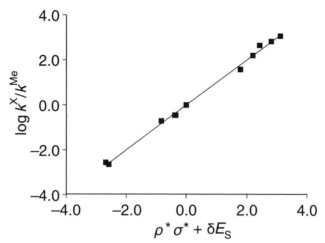

Figure 5   *Reaction of methoxide ion with menthyl esters, X–COO–menthyl.*
$p^* = 2.58$, $\delta = 1.24$

2.3 INDUCTIVE POLAR CONSTANTS σ_I

Assuming that resonance, steric and inductive effects are additive, an inductive polar substituent constant (σ_I)[8] may be defined by employing standard reactions reasonably assumed to possess no resonance transmission or steric effect.

Roberts and Moreland measured the dissociation constants for a series of 4-substituted bicyclo[2.2.2]octane-1-carboxylic acids (Equation 17) and defined an inductive polar substituent constant σ' by Equation (18).[9]

$$X\!\!-\!\!\underset{}{\bigcirc}\!\!-\!\!CO_2H \;\rightleftharpoons\; X\!\!-\!\!\underset{}{\bigcirc}\!\!-\!\!CO_2^- + H^+ \qquad (17)$$

$$\sigma' = (pK_a^H - pK_a^X)/1.65 \qquad (18)$$

The factor, 1.65, is employed to scale the σ' values so that they approximate the Hammett σ constants for the substituents in question.

An inductive polar substituent constant (σ_I^Q) based on the dissociation of 4-substituted quinuclidines (Equations 19 and 20) was introduced by Grob;[10] no scaling factor is used in this analysis to relate the values to Hammett's σ.

$$X\!\!-\!\!\underset{}{\bigcirc}\!\!N^+\!\!-\!\!H \;\rightleftharpoons\; X\!\!-\!\!\underset{}{\bigcirc}\!\!N \; + \; H^+ \qquad (19)$$

$$\sigma_I^Q = pK_a^H - pK_a^X \tag{20}$$

While σ' parameters from the dissociation of 4-substituted bicyclo-[222]octane-1-carboxylic acids are excellent measures of the inductive polar effect they suffer from the problem that the carboxylic acids cannot be synthesised readily for a wide range of X-substituents. A more substantial (but not universal) range of σ_I^Q constants is known than for σ' parameters.[10]

Taft defined a σ_I parameter for substituents by use of the alkaline and acid hydrolysis rate constants of acetate esters (X–CH$_2$CO–OEt, Equation 21) where σ_I is directly related to σ^*.

$$\sigma_I = [\log(k_{OH}^X/k_{OH}^{Me}) - \log(k_H^X/k_H^{Me})]6.25 \tag{21}$$

Charton[11] made use of the dissociation constants of the readily available substituted acetic acids (Equation 22) stable for a wide range of X-substituents. There are negligible resonance and steric components to the transmission of the polar effect to the reaction centre in the equilibrium of Equation (22) and the pK_a values fit Equation (23). The σ_I value may be determined according to Equation (23) obtained from the dissociation which is very accurately documented for a very large number and variety of X-substituents and $\rho_I = 3.95$. Other equilibria and rates (such as in Equations 17, 19 and 21) may be used as secondary standards to define σ_I on the basis of Equation (23).

$$X-CH_2-CO_2H \rightleftharpoons X-CH_2-CO_2^- + H^+ \tag{22}$$

$$(pK_a^H - pK_a^X) = \rho_I\sigma_I \tag{23}$$

Taft showed that the magnetic shielding of the ^{19}F-nucleus is linearly related to σ_I (Equation 24); this relationship may be employed to calculate a further set of σ_I values from the ^{19}F-NMR chemical shift relative to that of fluorobenzene for *meta*-substituted fluorobenzenes.[12]

$$(H_A - H_B) \times 10^{-6}/H_B = 7.30\,\sigma_I - 1.168 \tag{24}$$

where H_A = applied field at the resonance of the 3-substituted fluorobenzene and H_B is that of 1,4-difluorobenzene.

A relatively comprehensive table of σ_I constants based on Equation (23) is included in Appendix 3. It should be noted that some substituted acetic acids are unstable at the pH values required to measure dissociation constants, and σ_I constants for inaccessible substituents are measured using secondary standard dissociation equilibria and phenomena such as ^{19}F-magnetic resonance. The constant F of Swain and Lupton (Chapter 4) is also a good approximation to σ_I.

The transmission of charge between substituent and reaction centre *via* inductive and through-space pathways (field effect) is not readily separated and it is the convention that these effects are combined under the field effect mantra. More details of the separation of field and *through-bond* inductive effects are discussed in the review of Bowden and Grubbs.[13]

2.3.1 Choice of σ Value

Bearing in mind that free energy relationships are essentially comparisons of unknown systems with known systems it is advantageous to employ the σ or polarity value from the reference reaction or process which most closely resembles the reaction under investigation. Thus, nucleophilic reactions of the nitrogen of substituted pyridines with electrophiles are best modelled using values of pK_a of the conjugate acids of pyridines rather than Hammett's σ; Hammett's σ would be appropriate for electrophilic ring substitution at substituted pyridines. A comparison of the various σ values for standard substituents is shown in Table 1 in Appendix 3.

Inductive σ values for substituents may be obtained by fitting Hammett Equations to a standard series of reactions which are considered to possess no resonance interaction; the average values of σ from these correlations provides a set of coefficients (σ^n)[14] which correspond to purely inductive transmission of the polar effect. The coefficients derived from the ionisation of $XC_6H_4CH_2CO_2H$, $XC_6H_4CH_2CH_2CO_2H$ and the alkaline hydrolysis of $XC_6H_4CH_2CO_2C_2H_5$ and $XC_6H_4CH_2CH_2CO_2C_2H_5$ (σ^o) are regarded as purely inductive and field effect constants.[15]

Since all the types of σ value recorded here are meant to register the absolute polar effect of a substituent it would appear superfluous to have such a wide choice of parameter. The main objective of the technique is to seek information regarding the mechanism of a reaction rather than the polar effect *per se* of an individual substituent; the identity of the reference reaction for σ is of the utmost importance. Most types of σ covered in this text are employed in practice and the choice depends on the resemblance between the reaction under investigation and the standard reaction by which the σ value is defined. The regular Hammett σ is generally preferred for reactions where substituent change is on a benzenoid system whereas Taft σ* values are useful in studies of aliphatic systems. The substantial number of Charton σ_I values available from the pK_a values of substituted acetic acids (see Appendix 3) makes them very useful parameters for aliphatic systems.

Jaffé proposed a definition of the Hammett substituent constant σ which utilises the entire body of experimental data.[16] This definition gives an average value for a given σ but it suffers from the operational problem that σ must be revised periodically as time progresses and as new data accrues. Owing to the ready availability of fast computers with large memory this process is possible and has been achieved. Such a procedure has the advantage that it tends to smooth out irregularities in σ caused by the fact that results from pK_a measurements can vary quite considerably because conditions are not always standard.[17]

Primary standard σ values are obtained from experimental determination of the dissociation constants of benzoic acids given that ρ is defined as unity. *Secondary* standard σ values (Equation 25) can be obtained from any reaction series provided its ρ value is obtained on the basis of *primary* standard σ values.

$$\sigma = (\log k_x - \log k_H)/\rho \tag{25}$$

2.4 THE BRØNSTED EQUATION (CLASS I)

The proton transfer reaction to a substrate (S) from an acid acting as a variable proton donor (k_{HA}, Equation 26) exhibits a linear free energy relationship (Equation 27) and a similar equation often holds for the case where a base acts as a variable proton acceptor (k_B, Equations 28 and 29).[18] These free energy relationships are conventional *Brønsted equations* and the slopes are classically given the symbols α and β, for general acid and base catalysis respectively.[b] The slopes compare the change in energy of the transition state relative to reactants with change in the acid dissociation reaction. The iodination of acetone (Figure 6) provides a good example of a Brønsted equation for base catalysis.

$$S \xrightarrow[k_{HA}]{H-A} P \tag{26}$$

$$\log k_{HA} = -\alpha p K_a^{HA} + C \tag{27}$$

$$S-H \xrightarrow[k_B]{B} P \tag{28}$$

$$\log k_B = \beta p K_a^{HB} + C \tag{29}$$

---

[b] The reader should be aware of the potential confusion between the identity of the Brønsted α and the Leffler α.

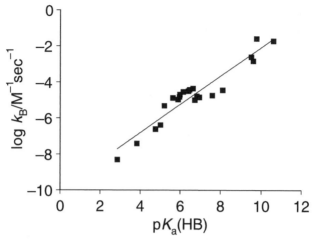

Figure 6 *Iodination of acetone catalysed by general bases*[19]

The Brønsted equation is a Class I free energy relationship and this may be shown by considering as an example the acid-catalysed dehydration of acetaldehyde hydrate (Equation 30). This reaction also provides a good example of an acid-catalysed reaction following a Brønsted equation (Figure 7).

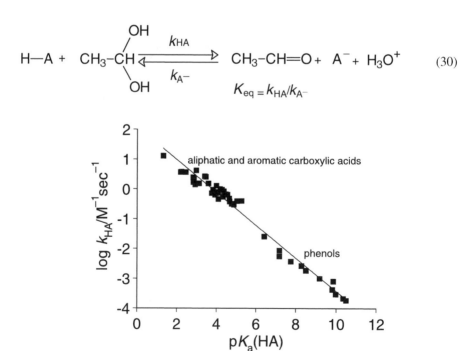

Figure 7 *Dehydration of acetaldehyde hydrate catalysed by acids*[20]

The slope of the plot of $\log k_{HA}$ versus pK_a^{HA} is the same as that against $-\log K_{eq}$, because pK_{HS} is constant and independent of the acid (HA). This relationship is very important as it is often very difficult to measure an equilibrium constant K_{eq} explicitly and changes would be even more difficult to determine accurately. It is usually a simple matter to determine dissociation constants accurately and moreover there are large databases of measured pK_a values in existence. The equilibria (Equations 31–33) sum to give the equilibrium of the initially formed intermediate (MeCHOH$^+$) from HA and acetaldehyde hydrate.

$$HA + H_2O \xrightleftharpoons{K_a^{HA}} A^- + H_3O^+ \tag{31}$$

$$CH_3CH(OH)_2 + H_3O^+ \xrightleftharpoons{K_1} CH_3CH(OH)(OH_2)^+ + H_2O \tag{32}$$

$$CH_3CH(OH)(OH_2)^+ \xrightleftharpoons{K_2} CH_3CHOH^+ + H_2O \tag{33}$$

$$CH_3CHOH^+ + H_2O \xrightleftharpoons{K_a^{HS}} CH_3CHO + H_3O^+ \tag{34}$$

Summing Equations (31) to (34) yields Equation (30) and $K_{eq} = K_1.K_2.K_a^{HA}K_a^{HS}$ whence Equation (35):

$$\log K_{eq} = \log K_1 + \log K_2 - pK_a^{HS} - pK_a^{HA} \tag{35}$$

Differentiation leads to

$$\partial \log K_{eq} = -\partial pK_a^{HA}$$

and

$$-\alpha = \partial \log k_{HA}/\partial pK_a^{HA}$$

$$= - \partial \log k_{HA}/\partial \log K_{eq}.$$

The Brønsted α for a proton transfer reaction is thus the same as the Leffler α (Chapter 1) although with opposite sign. Similar arguments show that the Brønsted β for a proton transfer reaction is the same as the Leffler α for that reaction.

The principle of microscopic reversibility[21] dictates that the reverse step of a general acid-catalysed reaction[c] is general base catalysed. There is a simple relationship between values of the Brønsted coefficients α and β for the forward and reverse reactions respectively, and this may be derived using the example of enolisation of acetone (Equation 36). The equilibrium constant for the formation of enolate ion is given by the equation $K_{eq} = k_A/k_{HA}$ $(= K_a^{HS}/K_a^{HA})$ so that the Brønsted relationships (Equations 37 and 38) may be combined.

$$A^- + CH_3COCH_3 \underset{K_{HA}}{\overset{K_A}{\rightleftharpoons}} CH_3COCH_2^- + HA \tag{36}$$

$$\log k_A = \beta p K_a^{HA} + C \tag{37}$$

$$\log k_{HA} = -\alpha p K_a^{HA} + D \tag{38}$$

$$\log K_{eq} = \log k_A - \log k_{HA} = \beta p K_a^{HA} + C + \alpha p K_a^{HA} - D$$

$$= (\beta + \alpha) p K_a^{HA} + C - D$$

$$\partial \log K_{eq} = (\beta + \alpha)\partial p K_a^{HA} \tag{39}$$

Since $K_{eq} = K_{HS}/K_a^{HA}$, $\partial \log K_{eq} = \partial p K_a^{HA}$ and thus

$$\beta + \alpha = 1 \tag{40}$$

This equation can also be derived in a different way (see Problem 2, Chapter 5).

Brønsted and Hammett equations become identical when substituted benzoic acids or phenols act as proton donors or, in their conjugate base form, as proton acceptors.

2.5 THE EXTENDED BRØNSTED EQUATION (CLASS II)

The Brønsted relationship for proton transfer reactions is directly related to the dissociation process which formally involves the addition or removal of a unit charge. Bases may also act as nucleophiles by

---

[c] General acid catalysis occurs when the rate law includes a concentration term due to added acid (rate = k_{HA}[HA]). Specific acid catalysis involves a rate law with only the oxonium ion (rate = k_H[H$_3$O$^+$]). Similar definitions apply to general base and specific base catalysis involving base and hydroxide ion respectively.

donating an electron pair to carbon or other electrophilic centres. Formally, this process could involve the addition or removal of unit charge and it is therefore not surprising that there are often good Class II correlations between the logarithms of the rate constants and the pK_a values of the conjugate acids of nucleophiles.[22] This relationship is called an *extended Brønsted* equation; it often breaks down when the gross structure of the nucleophile is changed, such as from carboxylate ions to amines, azide ion or halide ions. The reverse process, the expulsion of a leaving group from a carbon or other centre, generates a base. Nucleophilic substitution of 2-(4-nitrophenoxy)-4,6-dimethoxy-1,3,5-triazine by phenolate ions (Equation 41) may be correlated with the dissociation of the phenol (Figure 8).

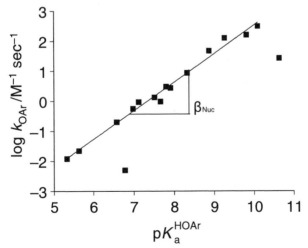

(41)

In view of the substantial use of the extended Brønsted correlation it is always the convention to name the Brønsted exponent as β with a subscript to indicate which reaction is being measured and to distinguish it from the parameters of the standard acid-base correlations.

Figure 8 *Reaction of substituted phenoxide ions with a 4-nitrophenyl triazine.[23] The significance of the deviant points is discussed in Chapters 6 and 7*

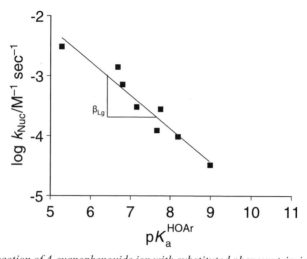

Figure 9   *Reaction of 4-cyanophenoxide ion with substituted phenoxy triazines*[23]

Thus β_f, β_1 and β_{Nuc} refer to forward reactions and β_r, β_{-1} and β_{Lg} refer to their microscopic reverse. Figure 9 illustrates the extended Brønsted relationship for the reaction of Equation (41) when the leaving group basicity is varied. Unlike the case of proton transfer the values β_{Nuc} or β_{Lg} are not the same as the Leffler α. Moreover the value of the β_{Lg} is quoted with its sign unlike that of the Brønsted or Leffler α.

2.6 EQUATIONS FOR NUCLEOPHILIC SUBSTITUTION

A number of equations have been developed to model the nucleophilic substitution reaction at tetrahedral carbon. These equations use reference reactions ranging from other S_N2 displacement reactions to physical processes such as electron transfer.

2.6.1 The Swain–Scott Equation (Class II)

The Swain–Scott equation (Equation 42)[24] compares the reactivity of nucleophiles in an S_N2 mechanism with the reaction of nucleophiles with methyl bromide (Equation 42) as a reference reaction. Water is taken as the standard nucleophile ($n = 0$) and the equation can be written where the nucleophilicity (n) of a given nucleophile is defined by setting the reaction constant (s) for the reference reaction at *unity* (Equation 43).

$$\log(k^{Nuc}/k^{water}) = s.n \tag{42}$$

$$n = \log k_{\text{MeBr}}{}^{\text{Nuc}} - \log k_{\text{MeBr}}{}^{\text{water}} \tag{43}$$

Since the reaction of nucleophiles with methyl iodide can be measured much more easily and accurately by spectroscopic analysis of the product iodide ion, a subsequent set of n values was devised using Equation (44). The n value of methanol with methyl iodide is set at zero and the reaction constant, s, at unity. Values of n derived from the methyl iodide reaction are tabulated in Appendix 3, Table 5.

$$n_{\text{MeI}} = \log k_{\text{MeI}}{}^{\text{Nuc}} - \log k_{\text{MeI}}{}^{\text{MeOH}} \tag{44}$$

Figure 10 illustrates the application of the Swain–Scott equation to the reaction of nucleophiles with *N*-methoxymethyl-*N*,*N*-dimethyl-3-nitroanilinium ion (Equation 45).[25]

$$
\begin{array}{ccc}
& \overset{\displaystyle \text{CH}_3}{\underset{\displaystyle \text{CH}_3}{\text{N}^+\text{—CH}_2\text{OCH}_3}} & \xrightarrow{\;k_{\text{Nuc}}\; \text{Nu}^-\;} \quad \overset{\displaystyle \text{CH}_3}{\underset{\displaystyle \text{CH}_3}{\text{N}}} \;+\; \text{NuCH}_2\text{OCH}_3
\end{array} \tag{45}
$$

The slope of the Swain–Scott correlation, $s = 0.26$, indicates that the reaction is about one quarter as sensitive to change in the nucleophile as is the standard reaction with methyl halides (where $s = 1.0$). This result is consistent with the small value of β_{Nuc} (0.4) for the extended Brønsted slope for reaction of the substrate with primary aliphatic amines and a

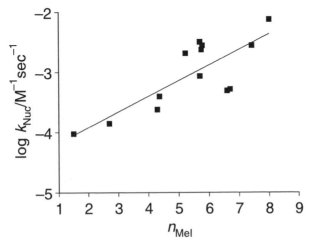

Figure 10   *Reaction of nucleophiles with N-methoxymethyl-N,N-dimethyl-3-nitro-anilinium ion;*[25] *s = 0.26*

small amount of bond formation in the transition state. The scatter of the data in Figure 10 is not unexpected because of the structural diversity of the nucleophiles in use (see Chapter 6, Section 4).

2.6.2 The Ritchie Equation (Class II)

The addition of nucleophiles to a carbenium ion (4-nitromalachite green, Equation 46) provides a useful reference reaction for nucleophilic attack. Parameters for nucleophiles (N_+) are defined by Equation (47) (the Ritchie equation)[26] for reactions of nucleophile with malachite green and 4-nitromalachite green (see structures in Equation 46). N_+ values are tabulated in Appendix 3, Table 4.

1 4-nitromalachite green

$$\log(k_{\mathrm{Nuc}}/k_{\mathrm{water}}) = N_+ \tag{47}$$

Equation (47) is of significance in mechanistic studies of nucleophiles with carbenium ions and it fits a wide range of anion–cation combination reactions over a wide range of reactivity contrary to the reactivity–selectivity principle (see Chapter 6). The N_+ values in Appendix 3, Table 4 are obtained from reaction of either malachite green or 4-nitromalachite green. Figure 11 illustrates plots of $\log(k_{\mathrm{Nuc}}/k_{\mathrm{water}})$ for a number of carbenium ions (**1–4**).

As the reactivity (N_+) of the nucleophile increases the rate constant for combination approaches a diffusion-controlled limit[d] and for the reaction of anions with 4-nitrobenzenediazonium ion (**4**) a break in the plot starts to occur (Figure 12).

---

[d] In a diffusion-controlled reaction the steps (Scheme 1, Chapter 1) subsequent to the formation of the encounter complex become faster than the diffusion rate constant ($\sim 1\ \mathrm{ns}^{-1}$).

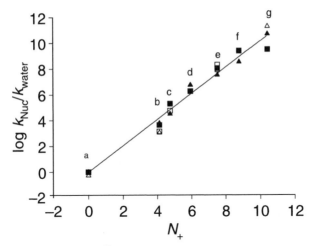

Figure 11   *The Ritchie equation;[27] △, (1); ▲, (2); □, (3); ■, (4). The slope of the line
is 1.05. Nucleophiles are: (a) H₂O; (b) HO⁻; (c) CN⁻; (d) MeO⁻; (e) N₃⁻;
(f) CNO⁻; (g) PhS⁻*

2 pyronin-Y

3

4

2.7 SOLVENT EQUATIONS

The previous sections are about free energy relationships where the
changes in energy are caused by altering the structure of the molecules
undergoing reaction. Solvents can also cause changes in free energy by
interacting with changing electrostatic charge or dipole. Reactivity is
decreased by increasing the polarity of the solvent for a transition
state where dispersal of charge is occurring; localisation of charge in a
transition state results in accelerated rates as a result of increased solvent
polarity.

It is unrealistic to expect that a single universal parameter would
successfully predict the variation of a rate or equilibrium for solvents
that have grossly different structures. The best that can be expected

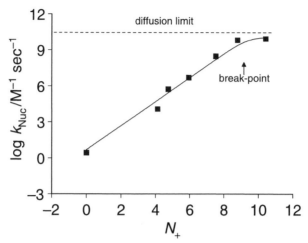

Figure 12 *Diffusion-controlled limit of combination of anions with 4-nitrobenzene diazonium ion. The data fit an equation* $k = k_{max}/(1 + 10^{-N}/10^{-N})$ *where* $N = 9.39$ *and* $\log k_{max} = 10.08$[27]

is a parameter which fits an effect caused by small changes in the composition of say a binary solvent mixture. This seems to be borne out in practice where linear free energy correlations are observed for changes in binary solvent composition whereas solvents of different structural types have dispersed lines.

2.7.1 The Grunwald–Winstein Equation (Class II)

The extent to which a solvent stabilises ions could be measured by comparing the equilibrium constant for a standard dissociation with the value for that in a standard solvent. The Grunwald–Winstein[28] equation (Equation 48) uses as its standard the solvolysis of *t*-butyl chloride (Equation 49) and assumes that the transition state has almost complete carbenium ion character and that the formation of the ion-pair is rate determining. The rate constant would therefore measure the energy of formation of the free carbenium ion. Equation (48) defines the solvent parameter, Y, where k_S is the solvolysis rate constant of *t*-butyl chloride in the solvent (S) and the standard solvent (ss) is 80% EtOH/H$_2$O. Other solvolyses can be used as standards and those of 1-adamantyl species has largely supplanted the original one for reasons given later.

$$Y = m\log(k_S/k_{ss}) \qquad (48)$$

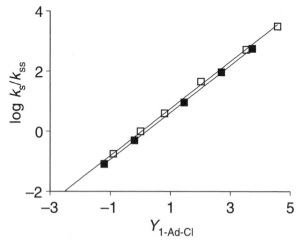

Equation (50) correlates rate constants for a number of solvolysis reactions and Figure 13 illustrates the Grunwald–Winstein plot for the solvolysis of *t*-butyl chloride against a Y value from 1-adamantyl chloride where $m = 0.78$.

$$\log(k_s/k_{ss}) = mY \qquad (50)$$

The value of m is a measure of the carbenium ion character of the transition state relative to that of the standard (1-adamantyl) reaction where m is unity (Equation 50).

Unfortunately no carbenium ion-forming equilibrium is known where it is sure that the positive charge is localised solely on the central carbon and which would be convenient to use as a reference reaction for S_N1-type processes. Further difficulties in the interpretation of the Grunwald–Winstein m values arise because the solvent itself could almost certainly be involved in the mechanism of the standard reaction. Internal return could also interfere with the value of the rate constant if decomposition of the carbenium ion became rate limiting as a result

Figure 13 *Grunwald–Winstein correlation for the solvolysis of* t-*butyl chloride* (Y)[29] *against a* Y-*value defined by the solvolysis of 1-adamantyl chloride*[30] *m = 0.78.* □, *ethanol/water 90, 80, 70, 50, 30, 0% increasing* Y; ■, *methanol/water 100, 90, 70, 50, 30% increasing* Y

of change in solvent required to estimate *m*. The first problem was recognised by Bentley and Schleyer[31] who suggested that nucleophilic assistance to dissociation occurs in the original standard reaction, as well as a complication due to an elimination reaction. The solvolysis of 1-adamantyl chloride has been employed as a standard reaction where it would be very difficult for solvent to assist the nucleophilic substitution by attack rearside to the C–Cl bond (Equation 51). Moreover, the bridgehead structure suppresses elimination. The "*Y*" values defined by this new standard depend on the nature of the leaving group (Cl–, Br– ,TsO– *etc.*).

The solvent structure can be altered by solutes and, for example, heats of solution of *t*-butyl chloride in ethanol–water mixtures are markedly non-linear in solvent composition. Enthalpies and entropies of

$$\tag{51}$$

activation for solvolysis are also non-linear in solvent composition whereas the free energy follows an almost linear function.[32] Additional terms may be added to Equation (50) (see Chapter 4) to allow for other possible modes of interaction between the forming carbenium ion and solvent.

The value of *m* (0.34)[e] for the solvolysis of ethyl bromide in 80% ethanol–water[33] compared with the value of *m* = 0.94 for the solvolysis of *t*-butyl bromide is consistent with a smaller charge separation in the transition state for the former reaction.

The cyclisation reaction (Equation 52) carried out in various solvents has a value of *m* = 0.13[e] indicating little charge expression in the transition state of the rate-limiting step compared with that of the standard solvolysis.[34]

$$\tag{52}$$

Plots of logk_S for solvolysis versus *Y* are markedly scattered in many cases but consist of a number of straight lines, one for each binary solvent mixture. A parameter, m_s, is defined which is a function of the

---

[e] Against *Y* for the solvolysis of *t*-butyl chloride.

solvolysis reaction *and* of the identity of the solvent pair; moreover, it is essential that Grunwald–Winstein methodology employs the most appropriate *Y* value (standard reaction).

2.7.2. The Kosower *Z* and Reichardt E_T Scales (Class II)

Charge transfer transitions give rise to electronic spectra which suffer medium effects because there is an alteration in charge distribution between the electronic states in the molecular system. The electronic transition involves transfer of an initial state to an excited state where the two ions have partially neutralised each other; the interaction of this state with solvent therefore provides a measure of solvent polarity. These systems suffer from the problem that the electronic process of charge transfer is very unlike that of reactions normally under investigation.

The Kosower parameter $(Z = hv)$[35] is derived from the effect of the solvent on the charge transfer spectrum of the pyridinium iodide (**5**) excitation. A high value of *Z* (the energy of the charge transfer transition) corresponds to highly polar solvents. Similar parameters have been obtained for other reporter molecules and the charge transfer system of **6** generates the Dimroth–Reichardt E_T^N scale.[36]

The values of *Z* and E_T^N values belong to a class of solvent parameters based on various standard physical processes (electromagnetic transition, dielectric constant, NMR chemical shift, *etc.*). In the context of *similarity* these parameters are not directly useful for mechanistic studies because the reference process is a physical property rather than a chemical reaction. The parameters *Z* and E_T^N are often linearly related to the Grunwald–Winstein *Y* value (Figure 14) and provide a secondary definition of Y values which are inaccessible *via* the chemical definition of those parameters.

5 (Kosower) **6** (Dimroth–Reichardt)

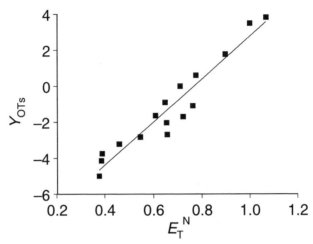

Figure 14 *Correlation between* Y_{OTs} *and* E_T^N

2.7.3. The Hansch Equation (Class II)

The favourable tendency of relatively apolar groups or molecules to associate in water results from the hydrophobic interactions between these species. The detailed understanding of the driving force of these interactions is controversial; nevertheless they constitute a very important process, featuring in enzyme catalysis, drug delivery and all manner of complexation reactions between molecules. A popular molecular interpretation of the hydrophobic interaction is based on the unusual thermodynamic parameters observed for the hydration of apolar moieties in water. The unusually large and negative entropies of dissociation of hydrated apolar species in water are often attributed to the structuring of water molecules in the hydration sphere of the solute. Association of apolar molecules is accompanied by a release of this "structured" hydration water into the bulk phase. The gain in entropy involved in the association process is thus the driving force for hydrophobic interactions. A reference model of a hydrophobic interaction is the transfer of organic solute from water to a non-aqueous phase. The non-aqueous phase could be any number of organic solvents immiscible with water but *n*-octanol has proved to be the most convenient;[37] values of the partition coefficient, P (Equation 53), are known for a very large number of solutes where $C_{n\text{-octanol}}$ and C_{water} are the equilibrium molar concentrations of the solute in *n*-octanol and water phases respectively. Hansch showed that many equilibria could be correlated with log$P$ (Equation 54) and Figure 15 illustrates this for the complexation of α-chymotrypsin with aromatic inhibitors.[38]

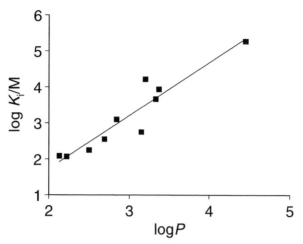

Figure 15 *Hansch plot for the complexation of α-chymotrypsin with aromatic inhibitors*[38]

$$\log P = \log C_{n\text{-octanol}}/C_{\text{water}} \tag{53}$$

$$\log K_i = p\log P + C \tag{54}$$

The $\log P$ value is an *additive* parameter[39] and this is illustrated by a plot of $\log P$ against n for $H(CH_2)_nOH$ and against n for $H(CH_2)_nPh$ (Figure 16). The linearity indicates a constant increment for the $-CH_2-$ fragment. A further parameter relates to the hydrophobicity of substituted phenyl species. In the case of substituted phenoxyacetic acids the value π is defined as $\pi = \log(P_x/P_H)$ where P_H is the partition coefficient of the parent and P_x that of the substituted species. The additivity of $\log P$ is exploited in calculations of $\log P$ for compounds (especially those of pharmacological interest) and this is discussed in more detail in Chapter 7.

2.8 OTHER STANDARD PROCESSES

2.8.1 Spectroscopic Transitions

Equation $(24)^{40}$ for the chemical shifts in ^{19}F-NMR spectroscopy is a free energy relationship where the energy of the transition in the NMR spectrum is related to the energy of the dissociation of the corresponding substituted acetic acid. Free energy relationships have been observed for a number of spectroscopic processes (IR, UV, NMR, mass

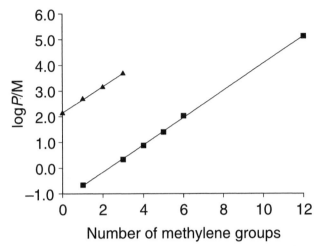

Figure 16 *Plots of logP against number of methylene groups for H(CH₂)ₙOH (■) and H(CH₂)ₙPh (▲)*

spectroscopy)[41] between the energy of the transition and some substituent constant such as Hammett's σ or σ_I.

The relationships are of substantial value in the prediction of spectral properties. Where the transition is related to a chemical process (as in charge transfer spectra) they could also be useful as standard processes for elucidating transition structures. They are also of use in studying the transmission of polar effects from substituent to the site of the energy change and providing secondary definitions of various σ parameters.

2.8.2 Acidity and Basicity Functions

The effect of a change in acidity or basicity of a solvent on the reactivity of a substrate has importance in the catalysis of many reactions.[42] The pH-dependence of a rate (a plot of log$k$ versus pH) is a free energy relationship because pH is proportional to a free energy; much useful information can be obtained about the mechanism of a reaction from its pH-profile. At high acidities the concept of pH is not useful and *acidity functions* can be determined for solvent mixtures by examining the dissociation constants of standard acids as a function of the solvent composition (Equation 55) assuming that the ratio of the activity constants (γ_A/γ_{HA}) is independent of structure for a series of structurally related acids (HA).

$$H_o = pK_a + \log[A]/[HA] = -\log h_o \tag{55}$$

The dependence of $\log k_{obs}$ on H_o for an acid-catalysed reaction is a free energy relationship[43] which can often be used to obtain information about the stoichiometry of the mechanism. Similar *basicity* functions can be written for reactions occurring at high basicity.

The free energy relationships involving concentration variables differ from the normal ones in that they give information only about stoichiometry and not about the charge structure of the transition state of a reaction. The topic is very extensive and for this reason we refer the reader to more advanced texts and references for further information.

2.8.3 Molar Refractivity

Pauling and Pressman[44] considered the molar refractivity (defined by Equation 56) as a means of assessing polarisability.

$$\text{MR} = \frac{n^2 - 1}{n^2 + 1} \cdot \frac{M}{d} \tag{56}$$

The components of MR are M = molecular weight, d = density of the liquid, and n = refractive index against the D-line of sodium. The molar refractivity is an additive constitutive property of molecules; it has been employed extensively in the past as a tool to investigate structural problems such as the constitution of dative bonds, double bonds, aromaticity and the like.[f] Values of MR have been deduced for many substituents (see Table 1 in Appendix 3) and the MR for a molecule calculated from fragmental values usually agrees well with the observed value.

The MR parameter has the dimensions of volume and can correlate with the hydrophobic substituent constant π; it can be used as a crude measure of bulk. However, the parameter has a component proportional to polarisability. These potential complications can be avoided by careful experimental design so that the chosen substituents have π values which are not linear with their MR parameters. The coefficient of MR in the free energy equation can be used to indicate a steric interaction (−ve) coefficient or an interaction involving dispersion forces (+ve coefficient).

[f] The important rôles in structure determination once played by molar refractivity (A.I. Vogel, *J. Chem. Soc.*, 1948, 1833), molar volume and the parachor do not now figure in undergraduate texts as they have been superseded by modern spectroscopic techniques (D.H. Williams and I. Fleming, *Spectroscopic Methods in Organic Chemistry*, McGraw-Hill, London, 5th edition, 1995). However, these properties are still employed, largely in pharmaceutical science, as they form part of the basis of empirical Equations relating to various medicinal and physical properties (C. Hansch and A. Leo, *Substituent Constants for Correlation Analysis in Chemistry and Biology*, J. Wiley, New York, 1979, Chapter 5, p. 44; W.E. Acree and M.H. Abraham, *Can. J. Chem.*, 2001, **79**, 1466).

Application of molar refractivity to the binding constants of methyl *N*-acylglycinate esters to α-chymotrypsin (Equation 57)[38] gives a much better correlation than correlation with log*P* or π. The results indicate that the interaction between the acylamino function and the complementary area of the active site of the α-chymotrypsin is not apolar in nature nor does it appear to be sterically demanding. The sign and magnitude of the exponent fits a mechanism whereby the substrate binds to the active site of the enzyzme through a hydrogen-bonding acceptor of the acylamido NH group.

$$\log K_m = 2.07 - 0.042 \, MR \tag{57}$$

The *parachor*,[45,46] defined in Equation (58), has the dimensions of volume and has been examined as a reference for free energy correlations in much the same way as the molar refractivity.[47]

$$\text{parachor} = \gamma^{0.25} M/(D - d) \tag{58}$$

The quantity γ is the surface tension of the liquid, M is the molecular weight and D and d are respectively the densities of liquid and vapour. The parachor is essentially a molecular volume measured at standard internal pressure and is an additive, constitutive, property because of its molar volume component.[48]

2.9 FURTHER READING

M.L. Bender, *Mechanisms of Homogeneous Catalysis: From Protons to Proteins*, Wiley-Interscience, New York, 1971.
T.W. Bentley and G. Llewellyn, Scales of Solvent Ionising Power, *Prog. Phys. Org. Chem.*, 1990, **17**, 121.
W. Blokzijl and J.B.F.N. Engberts, Hydrophobic Effects. Opinions and Facts, *Angew. Chem., Int. Ed. Engl.*, 1993, **32**, 1545.
K. Bowden, Acidity Functions for Strongly Basic Solutions, *Chem. Rev.*, 1966, **60**, 119.
T.C. Bruice and S.J. Benkovic, *Bio-organic Mechanisms*, Benjamin, New York, 1966, Vols. 1 and 2.
M. Charton, Electrical Effect Substituent Constants for Correlation Analysis, *Progr. Phys. Org. Chem.*, 1981, **13**, 119.
R.A. Cox and K. Yates, Acidity functions – an update, *Can. J. Chem.*, 1983, **61**, 2225.
R.A. Cox, Organic Reactions in Sulphuric Acid: the Excess Acidity Method, *Acc. Chem. Res.*, 1987, **20**, 27.

D.F. Ewing, Correlation of NMR Chemical Shifts with Hammett σ Values and Analogous Parameters, in *Correlation Analysis in Chemistry: Recent Advances*, N.B. Chapman and J. Shorter (eds.), Plenum Press, New York, 1978.

O. Exner, A Critical Compilation of Substituent Constants, in *Correlation Analysis in Chemistry: Recent Advances*, N.B. Chapman and J. Shorter (eds.), Plenum Press, New York, 1978.

J.F. Gal and P.C. Maria, Correlation Analysis of Acidity and Basicity, from Solution to Gas Phase, *Prog. Phys. Org. Chem.*, 1990, **17**, 159.

C.A. Grob, Polar Effects in Organic Reactions, *Angew. Chem. Int. Ed. Engl.*, 1976, **15**, 569.

H.H. Jaffé, A Re-examination of The Hammett Equation, *Chem. Rev.*, 1953, **53**, 191.

W.P. Jencks, *Catalysis in Chemistry and Enzymology*, McGraw-Hill, New York, 1969.

A.J. Leo, Calculating log P_{oct} from Structures, *Chem. Rev.*, 1993, **93**, 1281.

E.S. Lewis, Linear Free Energy Relationships, in *Investigation of Rates and Mechanisms of Reactions, Part 1*, C.F. Bernasconi (ed.), Wiley-Interscience, New York, 1986, 4th edn, p. 871.

C. Reichardt, Empirical Parameters of the Polarity of Solvents, *Angew. Chem. Int. Ed.*, 1965, **4**, 29.

C. Reichardt, *Solvent Effects in Organic Chemistry*, Verlag Chemie, Weinheim, 1979.

C. Reichardt, Solvatochromic Dyes as Solvent Polarity Indicators, *Chem. Rev.*, 1994, **94**, 2319.

C.H. Rochester, *Acidity Functions*, Academic Press, New York, 1970.

C.G. Swain and C.B. Scott, Quantitative Correlation of Relative Rates. Comparison of Hydroxide Ion with Other Nucleophilic Reagents toward Alkyl Halides, Esters, Epoxides and Acyl Halides, *J. Am. Chem. Soc.*, 1953, **75**, 141.

2.10 PROBLEMS

1 Using tables in Appendix 3 select a substituent to give a dissociation constant approximately identical for $ArNH_3^+$ and $ArSCH_2CO_2H$ in water.

2 Using data from Table 1 plot the Hammett relationship for the alkaline hydrolysis of ethyl benzoates[49] and explain the deviations from the linear regression.

Table 1

X	σ	$k_{OH}/M^{-1} sec^{-1}$	X	σ	$k_{OH}/M^{-1} sec^{-1}$
3-MeO	0.12	0.00392	4-NH$_2$	-0.66	8.64×10^{-5}
4-NMe$_2$	-0.83	6.34×10^{-5}	3-NH$_2$	-0.16	0.00166
4-Cl	0.23	0.0117	H	0.0	0.00289
3-Cl	0.37	0.0182	3-Me	-0.07	0.00169
4-Br	0.23	0.0139	4-Me	-0.17	0.00114
3-Br	0.39	0.0179	2-CN	1.06	0.122
4-I	0.18	0.0122	2-EtO	-0.01	0.00116
3-I	0.35	0.015	2-Me	0.29	3.38×10^{-4}
4-F	0.06	0.00586	2-Cl	1.26	0.0044
4-CN	0.66	0.157	2-NO$_2$	2.03	0.0169
3-CN	0.56	0.103	2-Br	1.35	0.003
3-NO$_2$	0.72	0.137	2-I	1.34	0.00163
4-NO$_2$	0.78	0.246			

3 Using the tables in Appendix 3 indicate the relative inductive electron withdrawing power of O$_2$N-, Cl-, NC-, MeOCO-, MeCO-, MeO-, and Me- compared to H in solution reactions.

4 The reaction of substituted benzyldimethylamine with methyl iodide has the following rate constants in acetonitrile and chlorobenzene at 30° (Table 2).[50] Graph the Hammett relationships and comment on the relative values of the Hammett slopes.

Table 2

X	$log(k/M^{-1} sec^{-1})$ acetonitrile	σ	$log(k/M^{-1} sec^{-1})$ chlorobenzene
3-Me	-1.19	-0.07	-2.35
H	-1.24	0.0	-2.45
3-Cl	-1.57	0.37	-3.02
3-NO$_2$	-1.87	0.72	-3.58
4-MeO	-1.06	-0.27	-2.19
4-Cl	-1.47	0.23	-2.84
4-NO$_2$	-1.92	0.78	-3.61

5 The dissociation constant for ArCO-NHOH has a Hammett equation: $pK_a = 8.78 - 1.02\Sigma\sigma$. The Hammett ρ parameter is close to that for benzoic acids. What can you deduce from this about which hydrogen is ionising?

6 Explain why the substituent effect in the dissociation of substituted 2-arylpropionic acids is less than that for the dissociation of substituted benzoic acids.

7 Table 3 lists the pK_a values of the conjugate acids of substituted ethyl benzimidate esters $(ArC(=NH_2)^+OC_2H_5)$ and rate constants for the alkaline-catalysed elimination of ethanol from the neutral species (Equation 59).[51]

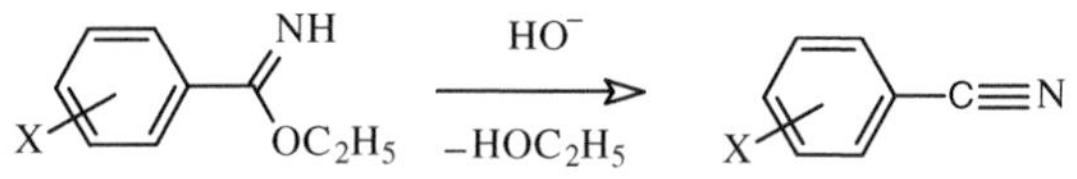

$$\tag{59}$$

Table 3

Substituent	pK_a	$10^2 k_{OH}/M^{-1} s^{-1}$
3-NO$_2$	5.30	2.1
4-NO$_2$	5.11	3.0
3-Cl	5.82	1.5
4-Cl	6.09	1.3
H	6.37	1.16
4-MeO	7.01	0.66

Draw the Hammett plots for the base-catalysed elimination (k_{OH}) and for the dissociation of the conjugate acids of the benzimidates; "map" ρ values for the overall reaction. The formation of benzonitrile from benzimidate has a ρ value of -1.7; you will need to estimate the ρ value for the dissociation of the neutral imidate.

8 Using the data of Table 4 graph the Taft plots of $\log k_{obs}/M^{-1} sec^{-1}$ for the reaction of substituted alkylamines (RNH_2) with 2,4-dinitrochlorobenzene and allyl bromide.

Assume that the polar effect is constant for these nucleophiles.

Table 4

Substituent	$-E_s$	$\log k_{obs}/M^{-1} sec^{-1}$	Substituent	$-E_s$	$\log k_{obs}/M^{-1} sec^{-1}$
2,4-Dinitrochlorobenzene			Allyl bromide		
Me	0.00	−2.5	Me	0.00	−2.081
Bu	0.39	−3.022	Et	0.07	−2.419
i-Bu	0.93	−3.168	Pr	0.36	−2.422
Pr	0.36	−3.018	i-Pr	0.47	−2.9
Et	0.07	−3.036	sec-Bu	1.13	−2.907
i-Pr	0.47	−4.0	i-Bu	0.93	−2.559
sec-Bu	1.13	−4.041	t-Bu	1.54	−3.503
t-Bu	1.54	−5.42	EtCMe$_2$	1.8	−3.569

Table 5

X	σ_m	$pK_a^H - pK_a^X$	X	σ_m	$pK_a^H - pK_a^X$
H	0.0	0.0	OH	0.12	0.370
Me	−0.07	−0.013	MeO	0.11	0.472
Et	−0.07	−0.02	CF_3	0.46	0.627
CH_2OH	0.01	0.074	CN	0.56	0.930
EtOCO	0.37	0.473	NO_2	0.72	1.058
Cl	0.37	0.739	CO_2H	0.37	0.468
Br	0.39	0.736	NMe_3^+	0.88	1.50

9 Using tables in Appendix 3, graph σ_I against σ_m for the following substituents:
Br−, Cl−, F−, H−, N_3−, Cl_3C−, Me−, NC−, MeO−, NH_2CO−, MeCO−, MeOCO−, Ph−, PhO−, $PhSO_2$−, Me_2^+NH−, F_3Ge−, Me_2S^+−, −O−, Me_3^+N−. What can you deduce from your results?

10 Using tables in Appendix 3 plot the values of σ_I versus σ^* for the substituents: C_2H_5−, H−, Me−, MeO−, Br−, Cl−, NC− and $(Me)_3N^+$−.

11 Graph values of E_s versus the Charton steric constant v_x for the substituents: C_2H_5−, H−, Me−, CH_2Br−, CH_2Cl−, CH_2I−, $CH(Me)_2$−, ^tBu−, CCl_3−, Ph−, CBr_3− and CEt_3−. (E_s and v_x values are found in tables in Appendix 3).

12 Plot the pK_a values of 4-substituted[222]octane-1-carboxylic acid given in Table 5 versus σ_m.

13 Using the data of Table 6 graph the Brønsted correlation for the mutarotation of glucose catalysed by general bases B (k_B/M^{-1} sec^{-1}).[52]

Comment briefly on the magnitude of the Brønsted β value obtained from your graph.

Table 6

Base	pK_a	$logk_B/$ $M^{-1}\,sec^{-1}$	*Base*	pK_a	$logk_B/$ $M^{-1}\,sec^{-1}$
HO−	15.96	1.80	2-Methylbenzoate ion	3.89	−3.69
Glucosate ion	12.19	−0.287	Glycollate ion	3.82	−3.64
Phenoxide ion	9.96	−1.146	Formate ion	3.68	−3.56
Histidine	6.21	−2.47	Chloroacetate ion	2.85	−4.05
α-Picoline	6.38	−3.06	Cyanoacetate ion	2.46	−4.20
Pyridine	5.13	−2.86	Sulfate ion	1.89	−4.18
Acetate ion	4.74	−3.35	Betaine	1.82	−4.35
Benzoate ion	4.19	−3.60			

Table 7

Pyridine	pK_a	$\log k/$ $M^{-1} sec^{-1}$	Amine	pK_a	$\log k/$ $M^{-1} sec^{-1}$
2,4,6-Trinitrophenyl ester			*2,4-Dinitrophenyl ester*		
4-Me$_2$N	9.87	1.114	Piperidine	10.82	2.61
4-NH$_2$	9.37	0.740	Piperazine	9.71	2.14
3,4-Me^{2+}	6.77	0.462	HN⌒NCH$_2$CH$_2$OH	9.09	1.72
4-Me	6.25	0.0414	Morpholine	8.48	1.46
3-Me	5.86	−0.102	HN⌒NCHO	7.63	0.756
H	5.37	−0.495	HN⌒$\overset{+}{N}$HCH$_2$CH$_2$OH	5.6	−0.523
3-CONH$_2$	3.43	−1.854	HN⌒$\overset{+}{N}$H$_2$	5.37	−0.658
3-Cl	2.97	−2.22			

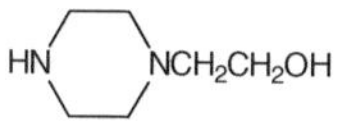

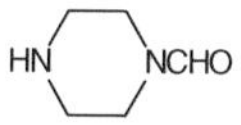

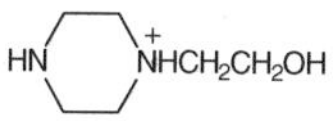

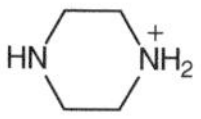

14 The second-order rate constants ($k/M^{-1}sec^{-1}$) are given in Table 7 for the reaction of substituted pyridines with 2,4,6-trinitrophenyl-*O*-ethyldithiocarbonate[53] and of alicyclic secondary amines with 2,4-dinitrophenyl-4-cyanobenzoate.[54] Plot the extended Brønsted graphs and briefly comment on the results.

15 The second-order rate constants for the reaction of α-DD-glucosyl fluoride with nucleophilic agents are given in Table 8.[55]

Graph the Swain–Scott diagram and comment on the slope of the correlation.

16 The first-order rate constants for the solvolyses of α-D-glucosyl fluoride in H$_2$O–MeOH at 30° C are shown in Table 9.[55]

Table 8

Nucleophile	n *(MeBr in H$_2$O)*	$10^6 k/M^{-1} sec^{-1}$
NO$_3^-$	1.00	0.44
MeCO$_2^-$	2.70	0.82
Cl$^-$	3.00	1.00
Br$^-$	3.90	1.26
NCS$^-$	4.77	2.10
I$^-$	5.04	2.80
S$_2$O$_3^{2-}$	6.36	3.90

Table 9

%H_2O	$10^8 k_{obs}/sec^{-1}$	$Y_{1\text{-Ad}}$
100	15 000	–
80	99	4.10
70	79	3.73
60	48	3.25
50	22	2.70
40	20	2.07
30	23	1.46
20	5.6	0.67
10	3.2	−0.20

Construct a Grunwald–Winstein plot and comment briefly on the magnitude of the slope. The $Y_{1\text{-Ad}}$ parameter is determined from the solvolysis of 1-adamantyl chloride.

17 The hydrolysis of L-*N*-acylalanine methyl esters catalysed by α-chymotrypsin have the following values of K_m.[g] Take data from Table 10.

Table 10

Acyl	π^a	$logK_m$
Furoyl	1.75	1.69
Theophenoyl	2.56	1.18
Nicotinyl	1.55	1.57
Isonicotinyl	1.66	1.46
Picolinyl	1.10	1.25
Benzoyl	2.75	0.99
2-Quinolinyl	2.45	−0.66
2-Aminobenzoyl	2.80	0.67

[a] Values from ref. 38.

Graph the Hansch plot and comment on the result in comparison with that of Figure 15.

18 Plot the ($pK_a^H − pK_a^X$) of the substituted carboxylic acids **7**:[56,57]

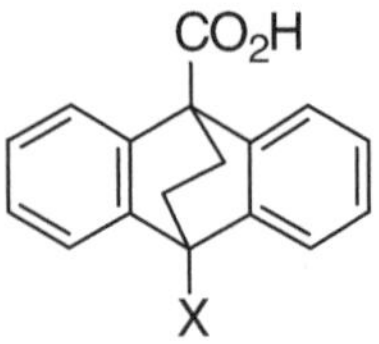

7

against the pK_a values of substituted bicyclo[2.2.2]octane-1-carboxylic acids (employ σ′ and pK_a^{HX} values given in Table 11).

Table 11

Substituent	$pK_a^H - pK_a^X$	σ'
H	0.0	0.0
Me	−0.027	−0.013
OMe	0.246	0.472
Cl	0.539	0.739
Br	0.537	0.738
NO$_2$	0.81	1.058

[g] For the purposes of this problem, K_m can be taken to be the dissociation constant of the enzyme substrate complex (ES).

Indicate how the quantity $pK_a^H - pK_a^X$ can be employed as a subdefinition of σ'.

2.11 REFERENCES

1. L.P. Hammett, Some Relations Between Reaction Rates and Equilibrium Constants, *Chem. Rev.*, 1935, **17**, 125.
2. M.J. Colthurst and A. Williams, *J. Chem. Soc., Perkin Trans. 2*, 1997, 1493.
3. R.W. Taft, Separation of Polar, Steric and Resonance Effects, in *Steric Effects in Organic Chemistry*, M.S. Newman (ed.), John Wiley, New York, 1956.
4. M.J.S. Dewar and P.J. Grisdale, Substituent effects IV. A Quantitative Theory, *J. Am. Chem. Soc.*, 1962, **84**, 3548.
5. R.W. Taft and I.C. Lewis, Evaluation of Resonance Effects on Reactivity by Application of Linear Inductive Energy Relationships. V. Concerning a σ_R Scale of Resonance Effects, *J. Am. Chem. Soc.*, 1959, **81**, 5343.
6. W.A. Pavelich and R.W. Taft, The Evaluation of Inductive and Steric Effects on Reactivity. The Methoxide Ion-catalysed Rates of Methanolysis of *1*-Menthyl Esters in Methanol, *J. Am. Chem. Soc.*, 1957, **79**, 4935.
7. M. Charton, The Quantitative Treatment of the Ortho Effect, *Progr. Phys. Org. Chem.*, 1971, **8**, 235; M. Charton, Steric Effects. I. Esterification and Acid-catalysed Hydrolysis of Esters, *J. Am. Chem. Soc.*, 1975, **97**, 1552.
8. R.W. Taft and I.C. Lewis, The General Applicability of a Fixed Scale of Inductive Effects. II. Inductive Effects of Dipolar Substituents in the Reactivities of *m*- and *p*-Substituted Derivatives of Benzene, *J. Am. Chem. Soc.*, 1958, **80**, 2436.
9. J.D. Roberts and W.J. Moreland, Electrical Effects of Substituent

Groups in Saturated Systems. Reactivities of 4-Substituted Bicyclo[2.2.2]octane-1-carboxylic Acids, *J. Am. Chem. Soc.*,1953, **75**, 2167.

10. C.A. Grob and M.G. Schlageter, The Derivation of Inductive Substituent Constants from pK_a Values of 4-Substituted Quinuclidines. Polar Effects. Part I, *Helv. Chim. Acta*, 1976, **59**, 264.

11. M. Charton, Definition of Inductive Substituent Constants, *J. Org. Chem.*, 1964, **29**, 1222.

12. R.W. Taft *et al.*, Fluorine Nuclear Magnetic Resonance Shielding in *p*-substituted Fluorobenzenes. The Influence of Structure and Solvent on Resonance Effects, *J. Am. Chem. Soc.*, 1963, **85**, 3146.

13. K. Bowden and E. Grubbs, Through-bond and Through-space Models for Interpreting Chemical Reactivity in Organic Reactions, *Chem. Soc. Rev.*, 1996, 171.

14. H. van Bekkum *et al.*, A Simple Re-evaluation of the Hammett ρσ Relation, *Recl. Trav. Pays Bas*, 1959, **78**, 815.

15. R.W. Taft, Sigma Values from Reactivities, *J. Phys. Chem.*, 1960, **64**, 1805.

16. H.H. Jaffé, A Re-examination of the Hammett Equation, *Chem. Rev.*, 1953, **53**, 191.

17. D.H. McDaniel and H.C. Brown, An Extended Table of Hammett Substituent Constants Based on the Ionization of Substituted Benzoic Acids, *J. Org. Chem.*, 1958, **23**, 420.

18. J.N. Brønsted, Acid and Base Catalysis, *Chem. Rev.*, 1928, **5**, 23.

19. M.L. Bender and A. Williams, Ketimine Intermediates in Amine-catalysed Enolisation of Acetone, *J. Am. Chem. Soc.*, 1966, **88**, 2502.

20. R.P. Bell and W.C.E. Higginson, The Catalysed Dehydration of Acetaldehyde Hydrate, and the Effect of Structure on the Velocity of Protolytic Reactions, *Proc. Roy. Soc. London*, 1949, **A197**, 141.

21. M.I. Page and A. Williams, *Organic and Bio-Organic Mechanisms*, Addison Wesley Longman, Harlow, 1997, pp. 42–43.

22. W.P. Jencks, *Catalysis in Chemistry and Enzymology*, McGraw-Hill, New York, 1969.

23. A.H.M. Renfrew *et al.*, Stepwise *versus* Concerted Mechanisms at Trigonal Carbon: Transfer of the 1,3,5-Triazinyl Group Between Aryl Oxide Ions in Aqueous Solution, *J. Am. Chem. Soc.*, 1995, **117**, 5484.

24. C.G. Swain and C.B. Scott, Quantitative Correlation of Relative Rates. Comparison of Hydroxide Ion with Other Nucleophilic Reagents toward Alkyl Halides, Esters, Epoxides and Acyl Halides, *J. Am. Chem. Soc.*, 1953, **75**, 141.

25. B.L. Knier *et al.*, Mechanism of Reactions of N-(Methoxy-methyl)-N,N-dimethylanilinium Ions with Nucleophilic Reagents, *J. Am. Chem. Soc.*, 1980, **102**, 6789.

26. C.D. Ritchie, Nucleophilic Reactivity toward Cations, *Acc. Chem. Res.*, 1972, **5**, 348.

27. C.D. Ritchie and P.O.L. Virtanen, Cation–Anion Combination Reactions. IX. A Remarkable Correlation of Nucleophilic Reactions with Cations, *J. Am. Chem. Soc.*, 1972, **94**, 4966; C.D. Ritchie, Cation–Anion Combination Reactions. CC. A Review, *Can. J. Chem.*, 1986, **64**, 2239.

28. E. Grunwald and S. Winstein, The Correlation of Solvolysis Rates, *J. Am. Chem. Soc.*, 1948, **70**, 846.

29. A.H. Fainberg and S. Winstein, Correlation of Solvolysis Rates. III. t-Butyl Chloride in a Wide Range of Solvent Mixtures, *J. Am. Chem. Soc.*, 1956, **78**, 2770.

30. T.W. Bentley and G. Carter, The S_N2–S_N1 Spectrum. 4. The S_N2 (Intermediate) Mechanisms for Solvolyses of *tert*-Butyl Chloride: A Revised *Y* Scale of Solvent Ionising Power Based on Solvolyses of 1-Adamantyl Chloride, *J. Am. Chem. Soc.*, 1982, **104**, 5741.

31. T.W. Bentley and P.v.R. Schleyer, The S_N2–S_N1 Spectrum. 1. Role of Nucleophilic Solvent Assistance and Nucleophilically Solvated Ion Pair Intermediates in Solvolyses of Primary and Secondary Arenesulfonates, *J. Am. Chem. Soc.*, 1976, **98**, 7658.

32. E.M. Arnett, Solvent Effects in Organic Chemistry. V. Molecules, Ions and Transition States in Aqueous Ethanol, *J. Am. Chem. Soc.*, 1965, **87**, 1541; 1957, **79**, 5937.

33. T.W. Bentley and G. Llewellyn, Y_X Scales of Solvent Ionising Power, *Progr. Phys. Org. Chem.*, 1990, **17**, 121.

34. F.L. Scott *et al.*, Ambident Neighbouring Groups Part V. Effects of Solvent on several O-5 Ring Closures, *J. Chem. Soc. B.*, 1971, 277.

35. E.M. Kosower, A New Measure of Solvent Polarity: Z-values, *J. Am. Chem. Soc.*, 1958, **80**, 3253.

36. C. Reichardt, *Solvent Effects in Organic Chemistry*, Verlag Chemie, Weinheim, 1979; C. Reichardt, Solvatochromic Dyes as Solvent Polarity Indicators, *Chem. Rev.*, 1994, **94**, 2319.

37. T. Fujita *et al.*, A New Substituent Constant, π, Derived from Partition Coefficients, *J. Am. Chem. Soc.*, 1964, **86**, 5175.

38. C. Hansch and E. Coats, α-Chymotrypsin: A Case Study of Substituent Constants and Regression Analysis in Enzymic Structure-Activity Relationships, *J. Pharm. Sci.*, 1970, **59**, 731.

39. C. Hansch *et al.*, On the Additive–Constitutive Character of Partition Coefficients, *J. Org. Chem.*, 1972, **37**, 3090.

40. D.F. Ewing, Correlation of NMR Chemical Shifts with Hammett σ Values and Analogous Parameters, in *Correlation Analysis in Chemistry. Recent Advances*, N. B. Chapman and J. Shorter (eds.), Plenum Press, New York, 1978.

41. R.D. Topsom, Electronic Substituent Effects in Molecular Spectroscopy, *Prog. Phys. Org. Chem.*, 1987, **16**, 193.
42. R.A. Cox and K. Yates, Acidity functions – an update, *Can. J. Chem.*, 1983, **61**, 2225.
43. A. Bagno *et al.*, Linear Free Energy Relationships for Acidic Media, *Rev. Chem. Ind.*, 1987, **7**, 313.
44. L. Pauling and D. Pressman, The Serological Properties of Simple Substances. IV, *J. Am. Chem. Soc.*, 1945, **67**, 1003.
45. S. Sugden, *The Parachor and Valency*, Routledge, London, 1930; O.R. Quayle, The Parachors of Organic Compounds, *Chem. Rev.*, 1953, **53**, 439.
46. C. Hansch *et al.*, "Aromatic" Substituent Constants for Structure–Activity Correlations, *J. Med. Chem.*, 1973, **16**, 1207.
47. J.C. McGowan, The Physical Toxicity of Chemicals. IV. Solubilities, Partition Coefficients and Physical Toxicity, *J. Appl. Chem.*, 1954, **4**, 41; D. Ahmed *et al.*, Parachors in Drug Design, *Biochem. Pharmacol.*, 1975, **24**, 1103.
48. O. Exner, Conception and Significance of the Parachor, *Nature*, 1962, **196**, 890; O. Exner, *Correlation Analysis of Chemical Data*, Plenum Press, New York, 1988, p. 171.
49. E. Tommila, Kinetic Studies of the Hydrolysis of Esters. I. Alkaline Hydrolysis of Esters of Substituted Benzoic Acids, *Ann. Acad. Sci. Fennicae*, 1941, **Ser A57**, 3.
50. T. Matsui and N. Tokura, Solvent Effects on ρ Values of the Hammett Equation. II., *Bull. Chem. Soc. Japan*, 1971, **44**, 756.
51. H.F. Gilbert and W.P. Jencks, Mechanism of the Base-catalysed Cleavage of Imido Esters to Nitriles, *J. Am. Chem. Soc.*, 1979, **101**, 5774.
52. Data from R.P. Bell, *Acid-Base Catalysis*, Clarendon Press, Oxford, 1941, p. 88.
53. E.A. Castro *et al.*, Kinetics and Mechanism of the Pyridinolysis of 2,4-Dinitrophenyl and 2,4,6-Trinitrophenyl O-Ethyl Dithiocarbonates, *J. Org. Chem.*, 1997, **62**, 126.
54. E.A. Castro *et al.*, *Int. J. Chem. Kinet.*, 1998, **30**, 267.
55. N.S. Banait and W.P. Jencks, Reactions of Anionic Nucleophiles with α-D-Glucopyranosyl Fluoride in Aqueous Solution through a Concerted, $A_N D_N$ ($S_N 2$) Mechanism, *J. Am. Chem. Soc.*, 1991, **113**, 7951.
56. H.D. Holtz and L.M. Stock, Dissociation constants for 4-substituted bicyclo[2.2.2]octane-1-carboxylic acids. Empirical and theoretical analysis, *J. Am. Chem. Soc.*, 1964, **86**, 5188.
57. C.G. Swain *et al.*, Substituent Effects in Chemical Reactivity. Improved Evaluation of Field and Resonance Components, *J. Am. Chem. Soc.*, 1983, **105**, 492.

CHAPTER 3

Effective Charge

The concept of effective charge (relative electronic charge) simplifies the interpretation of the slopes of linear free energy relationships from polar substituent effects. The slopes are related to the change in charge or dipole as reactants progress through transition structure to product. At present, polar substituent effects offer the only way in which electronic charge distribution can be investigated experimentally for transition structures. In favourable cases an effective charge map can be constructed corresponding to *all* the major bonding changes in a reaction.

Effective charge is a readily understood physical model and is a very useful concept because there are substantial changes in electrical charge on atoms in reacting bonds during a reaction. Any effect (due to variation of polar substituent) to change the charge developing in a transition structure will be converted into an energy difference resulting in a faster or slower rate.[1] This assumes that the polar substituent is located close enough on the molecule for the effect to be transmitted to the reaction site. Kirkwood and Westheimer theory[2] factors the polar substituent effect into the amount of charge variation and its distance from the substituent.

The linear free energy equations (Chapter 2) do not directly give the charge characteristics of the reaction. The following discussion shows how the polar substituent effects can be treated to obtain a quantitative measure of relative charge.

3.1 EQUILIBRIA

It is useful to discuss equilibria first, because the product and reactant molecules have known structures and they provide standards or references of charge which can be used later to determine charge in transition structures.

The effect of substituents X on the dissociation equilibrium (Equation 1) reflects the overall change in charge on the group A.[a] A plot of the logarithm of the equilibrium constant for Equation (1) against pK_a^{XAH} has unit negative slope corresponding to unit negative charge development in the product. The plot of $\log K_{eq}^{XAH}$ for Equation (2) has a slope $-\beta_{eq}$ against pK_a^{XAH} which measures the sensitivity to change in polar substituents for the equilibrium *relative* to that of the dissociation equilibrium (Equation 1). The system is exemplified by reactions given in Equations (3) and (4). Subcripts 'Nuc' and 'Lg' refer to the nucleophile and the leaving group respectively.

$$(\varepsilon_{rs} = +1) \qquad\qquad K_a^{XAH} \qquad (\varepsilon_{ps} = 0)$$
$$X\!-\!A^+\!-\!H \; + \; H_2O \; \rightleftharpoons \; X\!-\!A^0 \; + \; H_3O^+ \;\; (\text{slope} = \varepsilon_{ps} - \varepsilon_{rs} \tag{1}$$
$$= 0 - 1$$
$$= -1)$$

$$(\varepsilon_r = +\beta_{eq}) \qquad\qquad k_{Nuc} \qquad (\varepsilon_p = 0)$$
$$X\!-\!A^+\!-\!E \; + \; Nu^- \; \underset{k_{Lg}}{\overset{}{\rightleftharpoons}} \; X\!-\!A^0 \; + \; Nu\!-\!E \;\; (\text{slope} = \varepsilon_p - \varepsilon_r \tag{2}$$
$$K_{eq}^{XAH} = k_{Nuc}/k_{Lg} \qquad\qquad = 0 - \beta_{eq}$$
$$= -\beta_{eq})$$

$$(\varepsilon_{rs} = +1) \qquad K_S = K_a^{Xpy} \qquad (\varepsilon_{ps} = 0)$$

$$(\text{slope} = -1) \tag{3}$$

$$(\varepsilon_r = +1.6) \qquad\qquad k_{Nuc} \qquad (\varepsilon_p = 0)$$

$$K_{eq}^{Xpy} = k_{Nuc}/k_{Lg} \qquad (\text{slope} = -1.6) \tag{4}$$

The value of β_{eq} registers the change in relative charge on A. Since the species X–A$^+$–H may be defined as possessing 1 unit positive charge and zero in the conjugate base X–A, the charge on A in the reactant X–A$^+$–E is $+\beta_{eq}$. Since the quantities measured by the β values are not absolute charge they are given the term *effective charge*. McGowan first used the

---

[a] The effective charges, defined as if they were localised, are designated as follows: ε_{rs} and ε_{ps} are for reactant and product of the standardising equilibrium and ε_r and ε_p are for reactant and product of the reaction under investigation. Throughout the text the subscripts r, p and s refer to reactants, products and standard equilibrium respectively.

word effective charge in a slightly different context [3,4] and the concept in its present form was first applied by Jencks.[5]

The dissociation (Equation 3) is the standardising equilibrium for effective charge in reactions of substituted pyridines such as the transfer of the acetyl group (Equation 4) where E– = CH_3CO–; the charges are defined in the reference equilibrium as +1 and zero respectively on the nitrogen of the pyridinium ion and of the pyridine. The β_{eq} of –1 for the plot of $\log K_a^{Xpy}$ against pK_a^{Xpy} reflects the standard difference in charge of –1 from acid to base. The β_{eq} for the equilibrium constant of Equation (4), –1.6, indicates that the equilibrium is 1.6 times more selective than that of Equation (3); the equilibrium behaves *as if* there were a change in effective charge of 1.6 at the pyridine nitrogen relative to that of Equation (3). The negative sign indicates that the charge in the product is more negative than that in the reactant.

Charges on atoms in transition structure, reactant and product are independent of the distance of the polar substituent from the reaction centre as illustrated in Equation (5a), (see Chapter 4, Section 4.1). The greater sensitivity of the phenol dissociation compared with that of benzoic acid standard (Equation 5b) is interpreted on the basis of a smaller *distance* (*d*) between substituent and dissociation centre (compared with d_s in the benzoate standard) and to the delocalisation of charge in the benzoate ion (Equation 6, where ε_{ps} and ε_{rs} are the effective charges on product and reactant respectively).

$$\text{(}\varepsilon_r = 0\text{)} \qquad K_a^{ArOH} \qquad \text{(}\varepsilon_p = -1\text{)}$$

$$\text{X–C}_6\text{H}_4\text{–OH} \rightleftharpoons \text{X–C}_6\text{H}_4\text{–O}^- + H_3O^+ \tag{5a}$$

$$\text{(}\varepsilon_{rs} = 0\text{)} \qquad K_a^{ArCO_2H} \qquad \text{(}\varepsilon_{ps} = -1\text{)}$$

$$\text{X–C}_6\text{H}_4\text{–CO}_2\text{H} \rightleftharpoons \text{X–C}_6\text{H}_4\text{–CO}_2^- + H_3O^+$$

$$\log K_a^{ArOH}/K_a^{PhOH} = 2.23 \log K_a^{ArCOOH}/K_a^{PhCOOH} \tag{5b}$$

$$\rho_{ArOH}/\rho_{ArCOOH} = \rho_{ArOH} = \frac{\varepsilon_p - \varepsilon_r}{\varepsilon_{ps} - \varepsilon_{rs}} \cdot \frac{d_s}{d} \tag{6}$$

Equation (7) relates the effective charge change in a system under investigation ($\varepsilon_p - \varepsilon_r$) to that of the reference system ($\varepsilon_{ps} - \varepsilon_{rs}$).

$$-\beta = (\varepsilon_p - \varepsilon_r)/(\varepsilon_{ps} - \varepsilon_{rs}) \tag{7}$$

Since Equation (7) depends on $d_s = d$, it easy to see why the reference system is required to be closely similar in structure to that under investigation before the effective charges can be interpreted readily.

In the reference dissociation it is unlikely that the absolute charge on the nitrogen is a unit value. There are many studies which show that charge in ions is spread to a greater or lesser extent over the whole molecule and by solvation even in such species as ammonium ions.

In the case of the pyridines (Equations 3 and 4) $\varepsilon_p = \varepsilon_{ps}$ are the same and are designated as zero and the value of ε_{rs} can be defined as +1. Thus Equations (8a) and (8b) follow:

$$-\beta_{eq} = 1.6 = (\varepsilon_p - \varepsilon_r)/(\varepsilon_{ps} - \varepsilon_{rs}) \qquad (8a)$$

$$= (0 - \varepsilon_r)/(0 - 1) \qquad (8b)$$

therefore

$$\varepsilon_r = +1.6 \qquad (8c)$$

These results indicate that the CH_3CO- group is effectively more electron withdrawing than the $H-$ group when covalently linked to a pyridinium nitrogen.

3.1.1 Measuring Effective Charge in Equilibria

Measurement of effective charges for equilibria can be carried out directly from the equilibrium constants for a range of substituents. Explicit measurements of equilibrium constants over ranges of substituents are often difficult to obtain, although this is not the case for dissociation constants of acids because of the very sensitive nature of the analytical techniques for determining hydrogen ion concentration. Accurate determination of the substituent effect on equilibria over a wide range of substituents would require a very sensitive technique capable of analysing concentrations of species over several orders of magnitude. Although they have not been tried extensively, tools analogous to the hydrogen ion electrode, namely ion selective electrodes might be used with great effect and one such case would be that of the fluoride ion. Other electrode sensor methods are being developed and it is conceivable that methods for non-ionic species over a wide range of concentrations could also be developed.[6] However, these potentially direct methods are not yet available and the more indirect techniques are given below.

Substituent effects on equilibrium constants from kinetics derive from Equation (9b), derived from Equation (9a), which gives $\beta_{eq} = \beta_1 - \beta_{-1}$ (or $\rho_{eq} = \rho_1 - \rho_{-1}$).[b]

$$A \underset{k_{-1}}{\overset{k_1}{\rightleftharpoons}} B \qquad (9a)$$

$$K_{eq} = k_1/k_{-1} \qquad (9b)$$

The subscript of β or ρ refers to the rate constant measured.

Since the range of the equilibrium constants can be very large it is not usually a simple procedure to measure both forward and reverse rate constants under the same conditions; it must also be borne in mind that when the equilibrium constant is close to unity the expression $k_{obs} = k_1 + k_{-1}$ for the measured rate constants (k_{obs}) becomes important.[c]

The value of β_{eq} for the reaction of substituted phenyl acetates with imidazole (Equation 10) was first measured from kinetic data.[7] In this case the standard equilibrium is the dissociation of substituted phenols (Equation 11).

$$(10)$$

$$(11)$$

It is not a simple procedure to measure both k_1 and k_{-1} under the same conditions as required by application of Equation (9a) and the measurement of k_1 may be effected by use of excess imidazole to force the equilibrium of Equation (10) well to the right. Under these circumstances $k_1 = k_{obs}$ (the rate constant actually measured). The acetyl imidazole is in its *neutral* form (AcIm) at the pH of the measurements (between 5 and 8); knowledge of the dissociation constant of AcImH$^+$ (K_a^{AcImH}) and that

[b] The substituent effect in the standard dissociation equilibrium could be measured by any linear free energy equation (see Chapter 2). The Brønsted Equation is preferred and in all the arguments in this chapter we shall use β terminology. The other terminologies (such as ρ, ρ^*, ρ_I, etc.) are perfectly valid vehicles for determining effective charge.

[c] This expression also means that when the equilibrium is favourable to products the overall rate constant $k_{obs} = k_1$ and when the equilibrium is unfavourable $k_{obs} = k_{-1}$. These pitfalls only trap the unwary if k_{obs} is measured as the rate constant for *approach to equilibrium* of the system.

of the substituted phenol are required to compute k_{-1} from the return rate of the neutral AcIm to phenyl acetate (Equation 12).

$$\text{Return rate} = k_{-1}[\text{AcImH}^+][\text{ArO}^-]$$

$$= k_{-1}[\text{AcIm}][\text{H}_3\text{O}^+][\text{ArO}^-]/(K_a^{\text{AcImH}} + [\text{H}_3\text{O}^+]) \quad (12)$$

Owing to the experimental difficulties of measuring accurate rate constants for reactions with unfavourable equilibrium constants, the forward and reverse rate constants are often measured with reagents with no overlap in the ranges of substituent. In these circumstances it is necessary to have confidence that the free energy relationships are linear over the extrapolated range (line A in Figure 1) and that curvature does not exist.

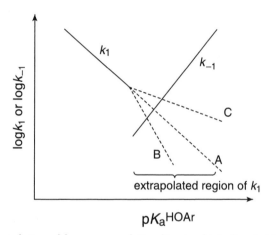

Figure 1   *Extrapolation of free energy relationships to obtain β_{eq}. A: identical mechanism and rate-limiting step; B: Change in rate-limiting step; C: Change in mechanism*

It is essential to demonstrate that both the measured forward and reverse rate constants refer to the same rate-limiting step and that there is no change in rate-limiting step in the extrapolated range. Consider the *quasi-symmetrical* reaction of phenolate ions with 4-nitrophenyl acetate (Equation 13).

$$\text{CH}_3\text{COO} - \langle \rangle - \text{NO}_2 + \ ^-\text{O} - \langle \rangle_X \underset{(k_{-1})}{\overset{K_{eq} \atop (k_1)}{\rightleftharpoons}} \text{CH}_3\text{CO-O} - \langle \rangle_X + \ ^-\text{O} - \langle \rangle - \text{NO}_2 \quad (13)$$

The return rate constant, k_{-1}, is not conveniently measured because the 4-nitrophenolate anion is too weak a nucleophile to compete with interfering reactions such as hydrolysis, and the equilibrium constants for phenols less acidic than 4-nitrophenol will be small. The strategy adopted is to measure k_{-1} for phenolate ions which are sufficiently reactive to give a measured k_{-1} value. This can be done for several phenolate ions and the values of β_{-1} thus obtained plotted against pK_a^{HOAr} to extrapolate a value for the 4-nitrophenolate ion case (Figure 2). The β_{Lg} for attack of 4-nitrophenolate ion on $ArOCOCH_3$ when combined with β_{Nuc} for attack of ArO^- on 4-nitrophenyl acetate yields the β_{eq} for the reaction $ArOCOCH_3 + Nu^- \rightarrow ArO^- + CH_3CONu$.

In the present example values of β_{Lg} (β_{-1}) are plotted as a function of the nucleophile's pK_a (pK_a^{Nuc}) and the value obeys Equation (14).

$$\beta_{Lg} = a \cdot pK_a^{Nuc} + b \tag{14}$$

The range of pK_a values employed leads to only a small extrapolation to that of 4-nitrophenol.

A related method is to determine the Equations connecting β_1 and β_{-1} with substituent for a series of related *quasi-symmetrical* reactions. For example in a nucleophilic displacement reaction ($Nu + A-Lg \rightarrow Lg + A-Nu$)

$$\beta_{Nuc} = p_{xy}pK_{Lg} + N \quad \text{and} \quad \beta_{Lg} = p_{xy}pK_{Nuc} + L$$

(see Chapter 5 for further details of p_{xy}.)
Subtracting gives

$$\beta_{eq} = \beta_{Nuc} - \beta_{Lg} = N - L$$

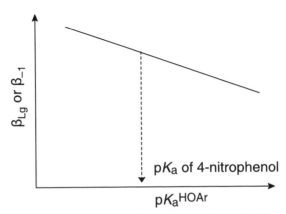

Figure 2 *Extrapolation of free energy relationships to obtain β_{Lg} for reaction between $ArOCOCH_3$ and 4-nitrophenolate ion*

Systems with overlapping values of pK_{Nuc} and pK_{Lg} give the most reliable results for β_{eq}. Effective charge data derived from β_{eq} values are given in Scheme 1 for some reactions.

An important corollary of Equations (6) and (7) is that β_{eq} for all equilibria can be calculated provided the effective charge on the donor molecule is known. Scheme 1 also provides a good basis for estimating β_{eq} for a given donor molecule. For example, it is reasonable that phenyl

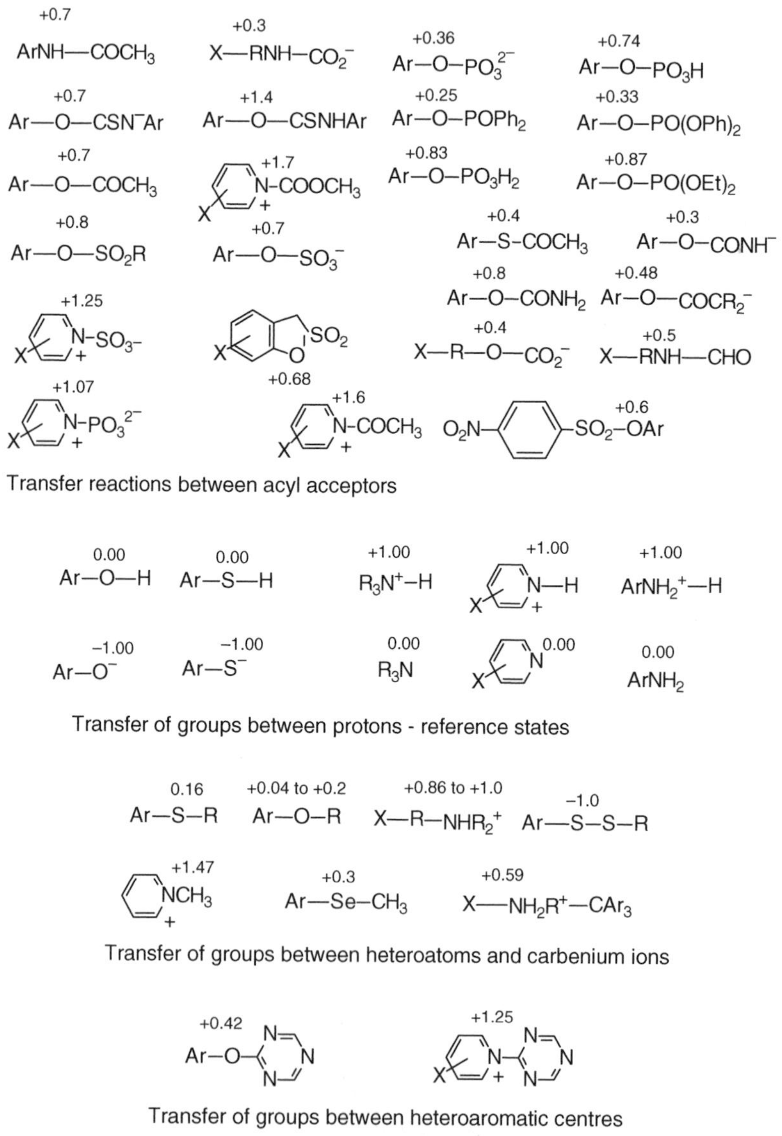

+0.7
ArNH—COCH$_3$

+0.3
X—RNH—CO$_2^-$

+0.36
Ar—O—PO$_3^{2-}$

+0.74
Ar—O—PO$_3$H

+0.7
Ar—O—CSN$^-$Ar

+1.4
Ar—O—CSNHAr

+0.25
Ar—O—POPh$_2$

+0.33
Ar—O—PO(OPh)$_2$

+0.7
Ar—O—COCH$_3$

+1.7
N-COOCH$_3$

+0.83
Ar—O—PO$_3$H$_2$

+0.87
Ar—O—PO(OEt)$_2$

+0.8
Ar—O—SO$_2$R

+0.7
Ar—O—SO$_3^-$

+0.4
Ar—S—COCH$_3$

+0.3
Ar—O—CONH$^-$

+1.25
N-SO$_3^-$

+0.68
SO$_2$

+0.8
Ar—O—CONH$_2$

+0.48
Ar—O—COCR$_2^-$

+0.4
X—R—O—CO$_2^-$

+0.5
X—RNH—CHO

+1.07
N-PO$_3^{2-}$

+1.6
N-COCH$_3$

+0.6
O$_2$N—SO$_2$—OAr

Transfer reactions between acyl acceptors

0.00
Ar—O—H

0.00
Ar—S—H

+1.00
R$_3$N$^+$—H

+1.00
N—H

+1.00
ArNH$_2^+$—H

−1.00
Ar—O$^-$

−1.00
Ar—S$^-$

0.00
R$_3$N

0.00
N

0.00
ArNH$_2$

Transfer of groups between protons - reference states

0.16
Ar—S—R

+0.04 to +0.2
Ar—O—R

+0.86 to +1.0
X—R—NHR$_2^+$

−1.0
Ar—S—S—R

+1.47
NCH$_3$

+0.3
Ar—Se—CH$_3$

+0.59
X—NH$_2$R$^+$—CAr$_3$

Transfer of groups between heteroatoms and carbenium ions

+0.42
Ar—O—⟨N=N, N⟩

+1.25
N—⟨N=N, N⟩

Transfer of groups between heteroaromatic centres

Scheme 1 *Effective charge on some reactant and product states*

esters of carboxylic acids possess an effective charge of +0.7 on the ether oxygen unless the structure of the acyl group differs grossly from that of the acetyl group; placing a formal charge on the acyl function such as in $ArOCOCR_2^-$ and $ArOCONH^-$ (Scheme 1) changes the effective charge substantially. However, a change from carbon to nitrogen (CH_3CO- to H_2NCO-) does not appear to be sufficiently gross to change the effective charge significantly (Scheme 1).

The value of β_{eq} and hence of the effective charge on atoms or groups may be determined indirectly from that for related equilibria. The β_{eq} for the equilibrium constant of Equation (15) is independent of the nucleophile. This can be shown by considering Equations (18–20) and the fact that K_2 must be independent of the aryl substituents.

$$CH_3CO-O-Ar + Nu^- \underset{}{\overset{K_{eq}}{\rightleftharpoons}} {}^--Ar + CH_3CO-Nu \tag{15}$$

$$CH_3CO-O-COCH_3 + {}^-O-Ar \overset{K_1}{\rightleftharpoons} CH_3CO_2^- + CH_3CO-O-Ar \tag{16}$$

$$CH_3CO-O-COCH_3 + Nu^- \overset{K_2}{\rightleftharpoons} CH_3CO_2^- + CH_3CO-Nu \tag{17}$$

$$K_{eq} = K_2/K_1 \tag{18}$$

$$\log K_{eq} = \log K_2 - \log K_1 \tag{19}$$

$$\beta_{eq} = d\log K_{eq}/dpK_a^{ArOH} = -d\log K_1/dpK_a^{ArOH} \tag{20}$$

The value of β_{eq} depends only on the acetyl group and the donor aryloxy group.

The relative values of the effective charges also measure polarity and those for many group transfer equilibria (Scheme 1) correlate with the electropositive character of the group being transferred.

3.1.2 A Group Transfer of Biological Interest

Phosphoryl group transfer (Equation 21) illustrates general points concerning effective charge in relation to equilibria and is a paradigm of some of the most important bio-organic processes. The Brønsted relationship is employed in this example because it relates the equilibrium in question (Equation 21) directly to the charge standardising equilibrium (Equation 22).

Charge calibrating equilibrium

$$(\varepsilon_r = +0.36) \qquad\qquad K_{eq} \qquad (\varepsilon_p = -1)$$

$$ArO\!-\!PO_3{}^{2-} + \quad Nu^- \;\rightleftharpoons\; ArO^- \quad + \quad Nu\!-\!PO_3{}^{2-}$$

$$\beta_{eq} = -1.36 \tag{21}$$

$$(\Delta\varepsilon = -1.36)$$

Charge standardising equilibrium

$$(\varepsilon_{rs} = 0) \qquad\qquad K_S = K_a^{ArOH} \qquad (\varepsilon_{ps} = \varepsilon_p = -1)$$

$$ArOH \qquad + \qquad H_2O \;\rightleftharpoons\; ArO^- + H_3O^+$$

$$\beta_S = -1 \tag{22}$$

$$(\Delta\varepsilon_S = -1)$$

The plot of $\log K_{eq}$ against $pK_a{}^{ArOH}$ is linear (Equation 23) (Figure 3)[8] and the slope, β_{eq}, compares the sensitivity of the equilibrium to change in polar substituents with that of the standard dissociation equilibrium.

$$\log K_{eq} = \beta_{eq} pK_a{}^{ArOH} + C \tag{23}$$

Since $K_s = K_a{}^{ArOH}$, Equation (24) may be deduced for the standard dissociation equilibrium (Equation 22).

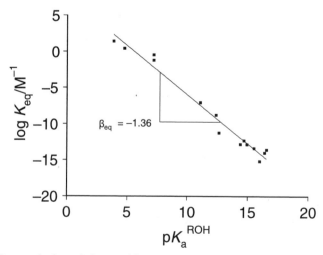

Figure 3 *Brønsted plot of the equilibrium constants for reaction of phosphate esters ($Lg\!-\!PO_3{}^{2-}$) with water (Nu^- in Equation 21)*

$$\log K_s = \beta_s p K_a^{\mathrm{ArOH}} = -1.p K_a^{\mathrm{ArOH}} \tag{24}$$

The charges on aryl oxide ion (ε_p) and phenol (ε_r) oxygen atoms are defined as -1 and zero respectively. The change in effective charge on the oxygen of the phenol is computed to be -1.36 from Equations (25) and (26).

$$\beta_{eq}/\beta_s = (\varepsilon_p - \varepsilon_r)/(\varepsilon_{ps} - \varepsilon_{rs})^d$$

$$= (-1 - \varepsilon_r)/(-1 - 0) = (\varepsilon_r + 1) = 1.36 \tag{25}$$

Thus

$$\varepsilon_r = +0.36 \tag{26}$$

Hammett ρ values could have been used in the above example instead of the Brønsted β values (see Appendix 1, Section A1.1.5) as follows: ρ_{eq} and ρ_s are respectively -3.03 and -2.23; the equilibrium (Equation 21) is thus $-3.03/-2.23 = 1.36$-fold more sensitive to substituent change than is the standard dissociation (Equation 22). The total change in effective charge is therefore -1.36 and the effective charge on the reactant is given by Equation (27) ($d_s = d$) yielding the same result as that from the Brønsted methodology (Equation 25).

$$\rho_{eq}/\rho_s = (\varepsilon_p - \varepsilon_r)/(\varepsilon_{ps} - \varepsilon_{rs}) = 1.36 \tag{27}$$

3.1.3 Additivity of Effective Charge in Reactants and Products

The aryl carboxylate ester of Scheme 2 has an effective charge on the ether oxygen of $+0.7$. Since the ester is overall neutral this positive charge is balanced by -0.7 units of effective charge on the acyl group assumed to be located on the acyl oxygen atom.

Similar arguments can be applied to the other species in Scheme 2; in some cases it is reasonable to assume that the balance of effective charge is spread evenly over the oxygen atoms in the acyl function.

3.2 RATES

Knowledge of the transition structure of a reaction is the goal of mechanistic studies but it presents an immediate problem. The transition

---

^{d} Derived from Equation (6) with $d_s = d$.

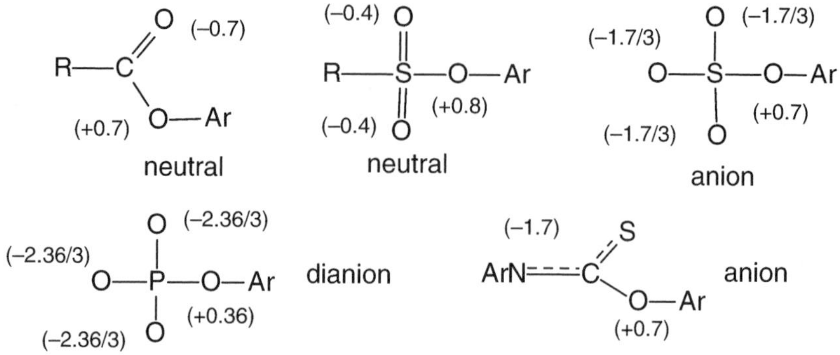

Scheme 2 *Additivity of effective charge for some acyl groups*

structure is an average structure which is taken up by molecular *systems* as they pass from reactant to product: it cannot be studied in the same way as for a molecule because it is not a discrete species and cannot be *isolated* even in principle (see Chapter 1). It is impossible to measure attributes of the transition structure in the same way that can be done for regular collections of molecules. Since the transition state can be considered *as if* it were an "equilibrium" state it is possible to define its effective charges in the same way as those just considered for reactant and product molecules in equilibrium reactions. Equation (28) has "rate constants" for breakdown of the transition state "species" (represented as "‡") forward (k_+) and return (k_-) which are essentially invariant because they register the collapse of the transition structure. These "rate constants" are independent of substituent changes and are therefore associated with zero β values. The "equilibrium constants" for formation of the transition state (k_1/k_-) and for its breakdown to products (k_+/k_{-1}) vary only according to changes in k_1 and k_{-1}.

$$\text{Reactant} \underset{k_-}{\overset{k_1}{\rightleftharpoons}} \text{Transition state} \underset{k_{-1}}{\overset{k_+}{\rightleftharpoons}} \text{Product} \qquad (28)$$
$$\text{(r)} \hspace{5cm} \ddagger \hspace{4cm} \text{(p)}$$

The polar effects on the rate constants k_1 and k_{-1} (corresponding to β_1 and β_{-1} respectively) therefore measure changes in effective charge from reactant or product to the transition structure.

The effective charge in a hypothetical general base-catalysed reaction (Equations 29 and 30) is given by Equations (31) and (32) ($d_s = d$, $\varepsilon_p = \varepsilon_{ps} = 0$ and $\varepsilon_r = \varepsilon_{rs} = -1$).

$$\overset{(\varepsilon_{rs}=-1)}{RCO_2^-} + H_3O^+ \underset{\text{Standard equilibrium}}{\overset{K_s = K_a^{RCO_2H}}{\rightleftharpoons}} \overset{(\varepsilon_{ps}=0)}{RCO_2H} + H_2O \qquad (29)$$

$$\underset{\substack{\text{General base-catalysed}\\\text{reaction}}}{\overset{(\varepsilon_r=-1)}{RCO_2^-} + \overset{(\varepsilon_+)}{H\text{–}S} \rightarrow \Big|\; RCO_2^{\delta-} \text{----} H \text{--} S^{\delta-} \Big|^{\ddagger} \rightarrow \overset{(\varepsilon_p=0)}{RCO_2H} + S^-} \qquad (30)$$

$$\beta = \{(\varepsilon_{\ddagger} - \varepsilon_r)/(\varepsilon_{ps} - \varepsilon_{rs})\} = \{\varepsilon_{\ddagger} - (-1)\}/\{0 - (-1)\} \qquad (31)$$

$$= (\varepsilon_{\ddagger} + 1)/1$$

Thus

$$\varepsilon_{\ddagger} = \beta - 1 \qquad (32)$$

3.2.1 Additivity of Effective Charge in Transition Structures

It is a reasonable assumption that the total effective charge in a system does not change as the reaction progresses. Thus the effective charge of 0.8 on the ether oxygen of the carbamate ester (Scheme 3) is balanced by a charge of -0.8 on the NH_2CO- residue because the ester is electrically neutral. The reactant state comprising ester and hydroxide ion has an overall effective charge of -1 due to the formal charge on the hydroxide ion. Scheme 3 maps the changes in effective charge during the alkaline hydrolysis of aryl carbamates. Once the overall change in effective charge is known, in this case -1.8 units, the effective charges through the reaction can be calculated by addition or subtraction of β. Thus the change in charge on the ether oxygen through dissociation of the NH_2CO- group is -0.50 and the charge on the ether oxygen in the conjugate base is $+0.3$. The carbamate ester accepts the additional negative charge of the hydroxide ion nucleophile in the transition structure, which has an overall *change* in effective charge on the leaving oxygen of $-0.34 - 0.8 = -1.14$.

3.2.2 Effective Charge Maps for More Than Two Bond Changes

Effective charges measured for a reaction involving two major bond changes give a much better description of the electronic charge distribution in the transition structure than that for a single bonding change. The ring opening of sultones by aryl oxide ion (Scheme 4) has an effective charge development of 0.81 units on the attacking oxygen closely balanced by a change of -0.86 units on the sultone oxygen.[10] The effective charge distribution is illustrated in the transition structure (Scheme 5) which is written assuming that charge additivity holds. The lack of substantial excess charge ($\Delta\varepsilon = +0.05$) (distributed to the sulfone group)

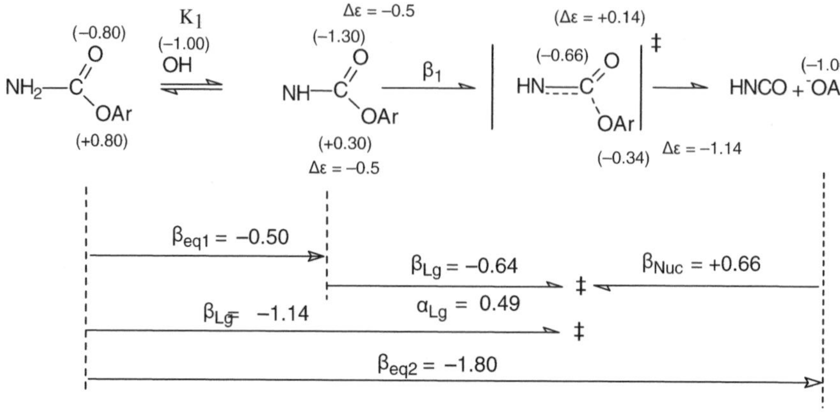

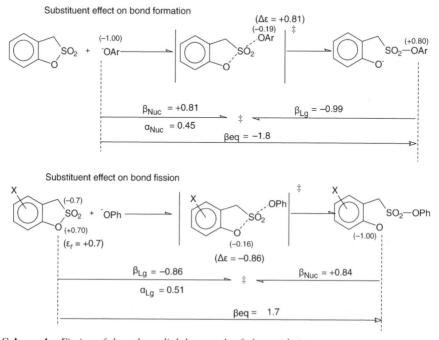

Scheme 3 *Carbamate hydrolysis in alkali. Subscript* Lg *refers to substituent change in the leaving group (−OAr); subscript* Nuc *refers to substituent change in the nucleophile.* Δε *refers to* change *in effective charge*

is consistent with a *synchronous concerted*[11] displacement mechanism where change in charge due to bond fission is balanced by an equal extent of change in charge due to bond formation.

Scheme 4 *Fission of the sultone link by attack of phenoxide ions*

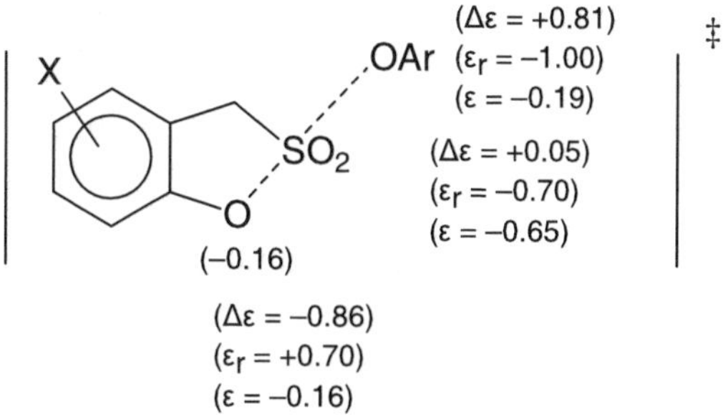

Scheme 5 *Effective charge distribution (ε) in the transition structure of the sultone ring opening reaction deduced from substituent effects on both forming and breaking bonds. Values of Δε are written relative to respective atoms in the reactant structure (ε$_r$)*

3.3 FURTHER READING

W.P. Jencks, Are Structure–Activity Correlations Useful? *Bull. Soc. Chim., France*, 1988, 218.

W.P. Jencks, General Acid–Base Catalysis of Complex Reactions in Water, *Chem. Rev.*, 1972, **72**, 706.

W.P. Jencks, *Catalysis in Chemistry and Enzymology*, McGraw-Hill, New York, 1969.

W.P. Jencks, Structure-Reactivity Correlations and General Acid–Base Catalysis in Enzymic Transacylation Reactions, *Cold Spring Harbor Symp. Quant. Biol*, 1971, **36**, 1.

A. Williams, Effective Charge and Transition-state Structure in Solution, *Adv. Phys. Org. Chem.*, 1991, **27**, 1.

A. Williams, Diagnosis of Concerted Organic Mechanisms, *Chem. Soc. Rev.*, 1994, 93.

3.4 PROBLEMS

Effective Charge Maps – Equilibria

1 Using data from Scheme 1 calculate effective charges for reactants and products in the reaction of Ar'S$^-$ with ArOCOCH$_3$ and hence derive β_{eq} values for the equilibrium for variation of both nucleophile and leaving group.

2 The values of β_{Lg} for reaction of phenolate ions with substituted phenyl esters of diphenylphosphoric acid ((PhO)$_2$PO–OAr) are –0.52, –0.68 and –0.81 respectively for phenolate ion, 4-acetylphenolate ion and 4-formylphenolate ion nucleophiles.[12] Attack of phenolate ions on 4-nitrophenyldiphenylphosphate ester has $\beta_{Nuc} = 0.53$. Given that the pK_a values of 4-nitro-, 4-formyl-, 4-acetyl- and parent phenols are 7.14, 7.66,

8.05 and 9.99 respectively, determine β_{eq} for the transfer of the diphenyl-phosphoryl group between phenyl esters and nucleophiles.

3 The ρ value for the dissociation of ArXOH is on average near unity.[e] However, the dissociation of $ArB(OH)_2$ has $\rho = 2.15$. Suggest an explanation.[13]

4 Using Hammett Equations recorded in Appendix 4 calculate the σ value where the pK_a of a substituted anilinium ion equals that of a substituted phenoxyacetic acid. Suggest a substituent to provide this σ value and calculate pK_a at the equality.

5 Given that the ρ_{eq} value for the hydrolysis of substituted aryl acetates is 3.8 and (ρ_s) for dissociation of phenols is 2.23 calculate ε_r for aryl acetate. Use this value to derive β_{eq} for the reaction of substituted aryl acetate esters (Scheme 6) to form the aryloxide ion (see Appendix 1, Section A1.1.5). Note that $\varepsilon_p = \varepsilon_{ps}$ and the parameters β_{eq} and ρ_{eq} for a dissociation reaction have opposite signs.

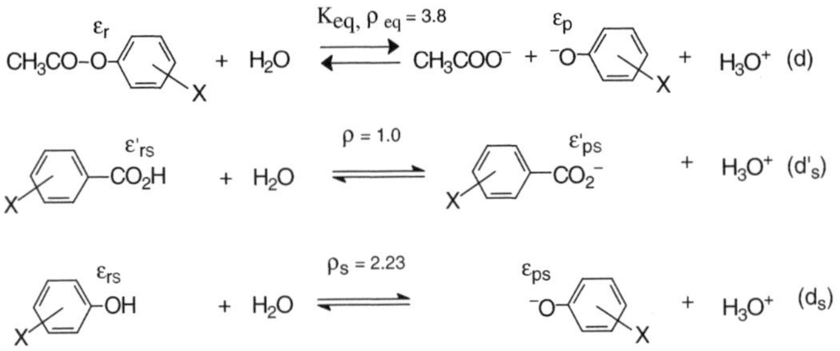

Scheme 6

6 Estimate β_{eq} for the reaction:

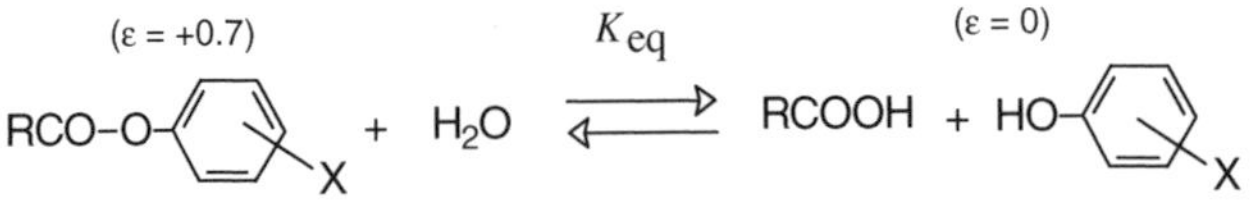

$$(33)$$

where $R = CH_3-$, C_2H_5- or $CH(CH_3)_2-$

7 Estimate β_{eq} for the reaction:

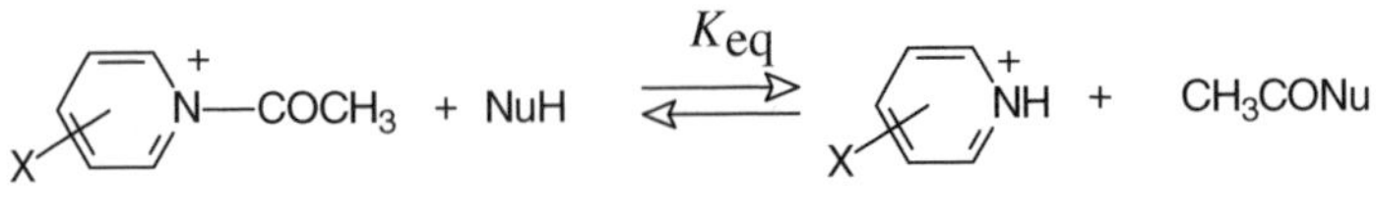

$$(34)$$

[e] The values of ρ range from 0.7 and 1.4 (Appendix 4, Table 1).

8   Using Scheme 1 and a ρ_s value of -2.24 for pK_a of ArSH, estimate β_{eq} and ρ_{eq} for the reaction:

$$ArS^- + RI \rightleftharpoons ArSR + I^- \qquad\qquad 35$$

9   Using data from Scheme 1 estimate the effective charge on the pyridine nitrogen in reactants and products in the reaction of substituted pyridines with 4-nitrophenyl acetate; calculate β_{eq} for the reaction.

10   Use the data from Scheme 1 to draw the map for the effective charges on the aryl oxygen in reactants and products in the formation of aryl acetate esters from phenolate ions and (**1**).

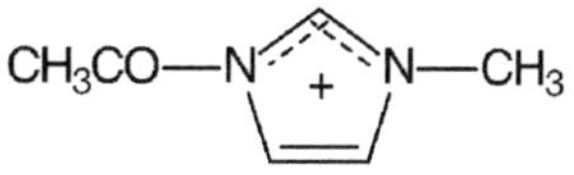

1

11   Using data from Scheme 1 map the effective charges on pyridine nitrogen and on aryl oxygen in reactants and products for the phosphorylation of aryl oxide ions by N-phosphopyridines (Equation 36). Comment on the different values for the changes in effective charge on nitrogen and oxygen atoms.

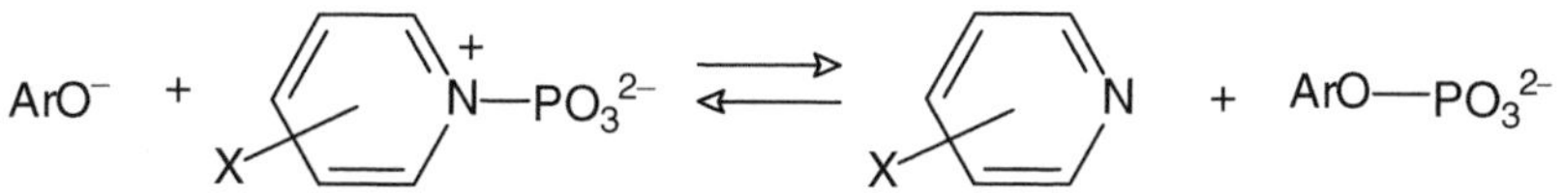

$$ArO^- \; + \quad \rightleftharpoons \quad + \; ArO-PO_3^{2-} \qquad (36)$$

Effective Charge Maps – Rates

12   Use ρ values to determine Leffler's α_{Nuc} for the cyclisation of phenyl-substituted benzoylglycinates in alkali *via* their conjugate bases

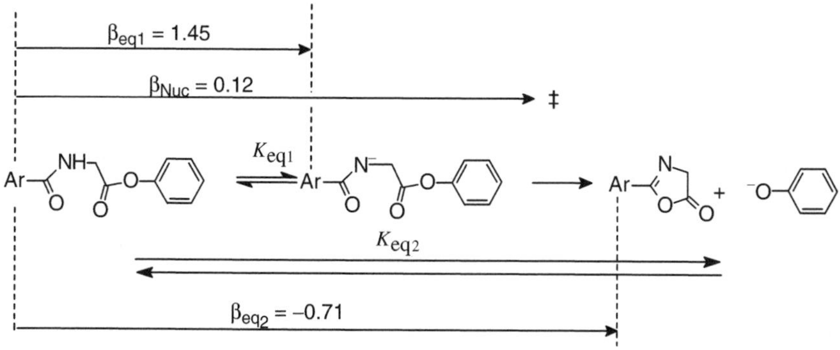

Scheme 7

(Scheme 7). Assume that ρ_{eq} for the overall reaction is -0.71, $\rho_{Nuc} = 0.12$ for the second-order rate constant for reaction of ester with hydroxide ion and $\rho = 1.45$ for the dissociation of the $-NH-$ group.[14]

13 The cyclisation reaction of Problem 12 can be studied by investigating the effect of varying the substituent on the phenoxy leaving group (Scheme 8). Given the Brønsted coefficients $\beta_{eq1} = -0.07$, $\beta_{eq2} = -1.86$ and $\beta_{Lg} = -1.05$ calculate α_{Lg} for the first-order rate constant for cyclisation from the anion.

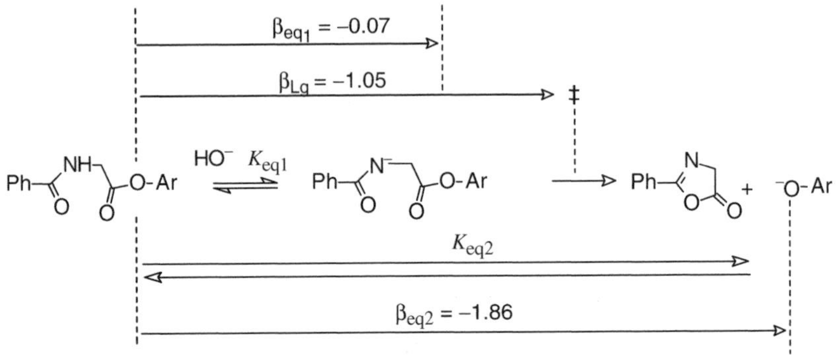

Scheme 8

Comment on your result for α_{Lg} compared with α_{Nuc} obtained from Problem 12.

14 Construct two effective charge maps (one for leaving group variation and one for variation of the nucleophile) for the *concerted* transfer of SO_3^- between pyridine leaving groups (donors) and phenoxide ion nucleophiles (acceptors) (Equation 37). Assume that β_{Nuc} and β_{Lg} are respectively 0.23 and -1.0 and $\beta_{eq} = 1.25$ for varying X substituents and $\beta_{eq} = 1.74$ for varying substituents in ArO^-.

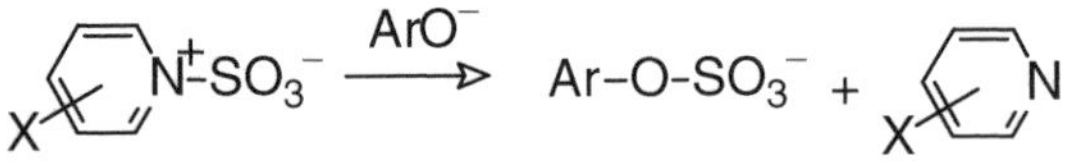

$$\tag{37}$$

Estimate values of α_{Nuc} and α_{Lg} for phenoxide attacking groups and pyridine leaving groups and comment on the relative values of these parameters.[15]

15 The displacement of substituted phenoxide ions from aryl monophosphate monoanions by nicotinamide (Equation 38) has a β_{Lg} of -0.95. Reaction of substituted pyridines with 2,4-dinitrophenylphosphate monoanion has a β_{Nuc} of 0.56. Using effective charge data from Scheme 1 construct a combined effective charge map for the reaction assuming a concerted displacement mechanism and that the dissociation of $XpyPO_3H^-$ has a $\beta_{eq} = -0.16$.[16]

$$\text{(Xpy)} \quad + \quad Ar\text{--}O\text{--}PO_3H^- \rightleftharpoons \quad \text{(XpyPO}_3\text{H}^-) \quad + \; ^-OAr \tag{38}$$

3.5 REFERENCES

1. J. Hine, Polar Effects on Rates and Equilibria (III), *J. Am. Chem. Soc.*, 1960, **82**, 4877.
2. J.G. Kirkwood and F.H. Westheimer, The Electrostatic Influence of Substituents in the Dissociation Constants of Acids, *J. Chem. Phys.*, 1938, **6**, 506, 513.
3. J.C. McGowan, Relationship Between Certain Dissociation Constants, *Chem. Ind.*, 1948, 632.
4. J.C. McGowan, Inductive Effects and Rate and Equilibrium Constants if Chemical Reactions, *J. Appl. Chem.*, 1960, **10**, 312.
5. W.P. Jencks and M. Gilchrist, Nonlinear Structure-Reactivity Correlations. The Reactivity of Nucleophilic Reagents Toward Esters, *J. Am. Chem. Soc.*, 1968, **90**, 2622; A.R. Fersht and W.P. Jencks, Reactions of Neclophilic Reagents into Acylating Agents of Extreme Reactivity and Unreactivity. Correlation of β values for Attacking and Leaving Group Variation, *J. Am. Chem. Soc.*, 1971, **92**, 5442; W.P. Jencks, B. Schaffhausen, K. Tornheim and H. White, Free Energies of Acetyl Transfer from Ring Substituted Acetanilides, *J. Am. Chem. Soc.*, 1971, **92**, 3917.
6. R.W. Cattrall, *Chemical Sensors*, Oxford University Press, Oxford, 1977; T.E. Edmonds, *Chemical Sensors*, Chapman & Hall, London, 1988; I. Janata, *Principles of Chemical Sensors*, Plenum Press, New York, 1989.
7. J. Gerstein and W.P. Jencks, Equilibria and Rates for Acetyl Transfer among Substituted Phenyl Acetates, Acetylimidazole, O-Acetylhydroxamic Acid and Thiol Esters, *J. Am. Chem. Soc.*, 1964, **86**, 4655.
8. N. Bourne and A. Williams, Effective Charge in Phosphoryl $(-PO_3^{2-})$ Group Transfer from an Oxygen Donor, *J. Org. Chem.*, 1984, **49**, 1200.
9. H. Al-Rawi and A. Williams, Elimination–Addition Mechanisms of Acyl Group Transfer: the Hydrolysis and Synthesis of Carbamates *J. Am. Chem. Soc.*, 1977, **99**, 2671.

10. T. Deacon *et al.*, Reactions of Nucleophiles wth Strained Cyclic Sulphonate Esters: Bronsted Relationships for Rate and Equilibrium Constants for Variation of Phenolate Anion Nucleophile and Leaving Group, *J. Am. Chem. Soc.*, 1978, **100**, 2525.
11. A. Williams, *Concerted Organic and Bio-Organic Mechanisms*, CRC Press, Boca Raton, FL, 2000.
12. S.A. Ba-Saif *et al.*, Dependence of Transition State Structure on Nucleophile in the Reaction of Aryl Oxide Anions with Aryl Diphenylphosphate Esters, *J. Chem. Soc., Perkin Trans. 2*, 1991, 1653.
13. H.H. Jaffé, A Re-Examination of the Hammett Equation, *Chem. Rev.*, 1953, **53**, 191.
14. T.P. Curran *et al.*, Structure-Reactivity Studies on the Equilibrium Reaction Between Phenolate Ions and 2-Aryloxazolin-5-ones: Data Consistent with a Concerted Acyl-Group Transfer Mechanism, *J. Am. Chem. Soc.*, 1980, **102**, 6828.
15. A.R. Hopkins *et al.*, Sulphonate Group $(-So_3)^-$ Transfer Between Nitrogen and Oxygen: Evidence Consistent with an Open 'Exploded' Transition State, *J. Am. Chem. Soc.*, 1983, **105**, 6062.
16. A.J. Kirby and A.G. Varvoglis, The Reactivity of Phosphate Esters: Reactions of Mono Esters with Nucleophiles. Nucleophilicity Independent of Basicity in a Bimolecular Substitution Reaction, *J. Chem. Soc. B*, 1968, 135.

Multiple Pathways to the Reaction Centre

Polar substituent effects can be transmitted to a reaction centre *via* three main routes:[1] *sigma-inductive* (I) by successive polarisation of intervening sigma bonds; *field* (F) where the electric field of the polar substituent influences the reaction centre across space; and *resonance* (R) through polarisation of a conjugated π-system. The inductive route of electronic transmission through sigma bonds and the field effect are not easy to distinguish and "F" is conventionally taken to mean the combined inductive and field effects.[1,2] The contribution of "through-bond" inductive transmission is thought in any case to be small.[3]

The resonance pathway (R) requires π-bonding for transmission of the polar effect and is comprised of at least three distinct processes:[1]

- π-inductive effect where the electrostatic charge at a conjugated atom adjacent to the substituent polarises the corresponding π-electron system;
- polarisation of the the π-electron system by resonance.
- mutual conjugation between substituent and reaction centre through the intervening conjugated system.

The substituent can also interact with the reaction centre by virtue of its bulk (S) as discussed in Chapter 2. In addition to effects caused by substituents the reaction centre is also subject to interaction with solvent.[4]

4.1 STRENGTH OF THE INTERACTION

The strength of the interaction *via* field and inductive transmission is dependent on the distance between the substituent and the reaction centre.[5,a] The attenuation may be exemplified by the transmission of a given polar effect through increasing lengths of aliphatic chain. The Hammett ρ values for dissociation of phenylacetic and phenylpropionic acids decrease with the increasing number of intervening methylene groups. The polar effect falls roughly as the power of the number of intervening atoms as in Equation (1a) where "f" is the attenuation factor for one group (see Table 1), n is the number of intervening groups and ρ_o is the ρ value when no group intervenes.

$$\rho = \rho_o(f)^n \tag{1a}$$

$$\sigma = \sigma_o (f)^n \tag{1b}$$

The effect of attenuation could also be written as Equation (1b) where σ is the value of $\log(K/K_H)$ for the dissociation of the acid with n intervening methylene groups and σ_o is that for the substituents in benzoic acid.[6]

The values of ρ are not the same (*c.f.* Figure 2, Chapter 2) for the various dissociation equilibria and this means that the effect of the substituent on the charge is dependent on the transmission pathway because the charge differences must be the same. It is therefore very important to employ the appropriate standard dissociation equilibrium for comparative purposes between systems which have different transmission modes and efficiencies if effective charge is to be determined. Moreover it is necessary to use the same solvent in the reference system. The concept of attenuation can also be applied to the inductive effect of substituents through the methylene group for substituent constants other

Table 1 *Attenuation factors, f, for Hammett ρ values transmitted through various groups. All groups except the methylene and ethylene also involve some form of resonance transmission.*

$N{=}N$	(cyclopropane)	(cyclopropene)	(phenylene)	$-NH-$	$-S-$	$-O-$
0.57	0.39	0.28	0.34	0.69	0.68	0.64

$-CH_2-$	$-CH_2-CH_2-$	$-CH{=}CH-$	$-C{\equiv}C-$	$-NH_2^+-$	$-Se-$	$-SO_2-$
0.47	0.22	0.47	0.39	0.27	0.74	0.53

Data taken from A.F. Hegarty & P. Tuohey, Hydrolysis of Azo Esters *via* Azoformate Intermediates, *J. Chem. Soc., Perkin Trans. 2*, 1980, 1238; A. Williams, Free Energy Correlations and Reaction Mechanisms, in M. I. Page, ed., Chemistry of Enzyme Action, Elsevier, Amsterdam, 1984, p. 145.

---

a   See also Chapter 3, Equation (6).

than Hammett's σ; the attenuation constant for Taft's σ^* constants is similar to that for the Hammett σ values (see Equation 14, Chapter 2).

Equation 6 of Chapter 3 can be applied to the dissociation of benzoic and phenylacetic acids (Scheme 1) and since $\varepsilon_p' = \varepsilon_p = -1$ and $\varepsilon_r' = \varepsilon_r = 0$ the value of d' is $d/0.47$; this is qualitatively acceptable for phenylacetic acid owing to the increase in distance between reaction centre and substituent (d') compared with that (d) in benzoic acid.

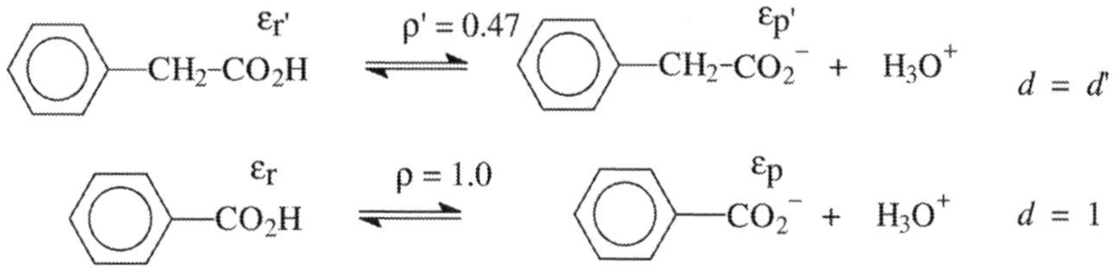

Scheme 1 *Dissociation of phenylacetic and benzoic acids having the same effective charge*

$$\rho'/\rho = \frac{\varepsilon_p' - \varepsilon_r'}{\varepsilon_p - \varepsilon_r)} \cdot \frac{d}{d'} = 0.47/1 \tag{2}$$

There is a strong empirical correlation between the Hammett ρ for dissociation of a proton and the number, i, of atoms between the ionisable hydrogen and the aromatic carbon (Equation 3).[7,8]

$$\rho = (2.4)^{2-i} \tag{3}$$

Thus the pK_a values of substituted phenylacetic acids ($i = 3$) have a calculated ρ value of $(2.4)^{-1} = 0.42$ and the pK_a values of substituted phenols ($i = 1$) have a calculated ρ value of $(2.4)^1 = 2.4$. Both these values come close to those of the observed ones (see Table 1 in Appendix 4).

The ρ^* parameters for aliphatic systems are related by Equation (4) to the number of atoms intervening in the saturated chain between the substituent and the atom bearing the dissociating proton ($R-(C)_i-XH$).

$$\rho^* \sim 0.8(2.0)^{2-i} \tag{4}$$

The predicted values for substituted acetic (RCH_2CO_2H, $i = 2$) and formic acids (RCO_2H, $i = 1$) are 0.8 and 1.6 respectively, agreeing with the observed values of 0.67 and 1.62.[9] The intervention of a double or triple bond decreases the effective chain length, increasing ρ^*, and a double bond can be regarded effectively as having an i value of 1. An equation similar to Equation (4) can be advanced for σ_I values.

4.2 ADDITIVITY OF INDUCTIVE SUBSTITUENT EFFECTS

The effects of multiple substituents in a benzene ring are additive leading to a global σ value of $\Sigma\sigma$,[10] and the reactivity can be expressed as in Equation (5).

$$\log(k/k_o) = \rho \, \Sigma\sigma \tag{5}$$

Additivity is only valid for *meta* and *para* substituents (X) in structure **1** and also for substituents X in different benzene rings *equally* disposed in structure **2** relative to the reaction centre Y. The combined σ value for **1** is $(\sigma_{X_1} + \sigma_{X_2} + \sigma_{X_3})$ and that for (**2**) is $(\sigma_{X_1} + \sigma_{X_2} + \sigma_{X_3} + \sigma_{X_4} + \sigma_{X_5} + \sigma_{X_6} + \sigma_{X_7} + \sigma_{X_8} + \sigma_{X_9})$.

Additivity also applies to σ^* and σ_I values as well as to the pK_a for phenols, and Scheme 2 illustrates a comparison of calculated and observed values for some examples. The concept of additivity can be applied to the calculation of pK_a values which is very useful for conjugate acids where, perhaps owing to their instability, measured values are inaccessible.

Additivity is a very useful property, as multiple substitution can substantially increase the range of σ and pK_a (and hence range of polarity) over the relatively limited range for single substituents. The phenomenon of additivity of physical properties is important in many other quantitative structure–activity relationships. Such quantities as Molecular Volume, Molecular Refractivity, the parachor and the Hansch logP values (see Chapter 2) can be computed with reasonable accuracy from the structure of a drug candidate by summing fragmental constants.

	σ^*exp	σ^*calc
—C(CH$_3$)$_3$	–0.3	$3 \times -0.1 = -0.3$
—CH(Ph)(OH)	+0.765	$0.215 + 0.555 = +0.77$

		p$K_a$exp	p$K_a$calc
O$_2$N—⬡—OH (with O$_2$N, Cl)		5.45	$7.14 - (9.95 - 8.48) = 5.67$
⬡—CO$_2$H (with O$_2$N)		2.82 (σexp = 1.38)	$4.2 - (4.2 - 3.49) \times 2 = 2.78$ (σcalc = 1.42)
CH$_3$O ⬡—NH$_3^+$ (with CH$_3$O)		3.82	$4.62 - (4.62 - 4.2) \times 2 = 3.78$
CH$_3$ ⬡N—NH$^+$ (with CH$_3$)		6.14	$5.17 - (5.17 - 5.68) \times 2 = 6.19$

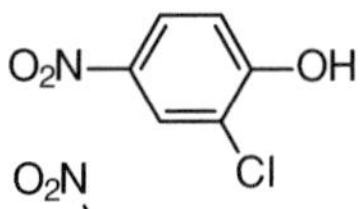

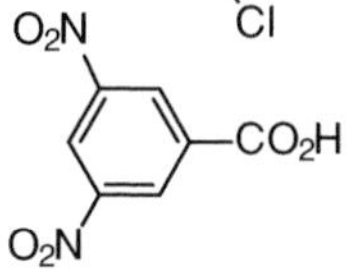

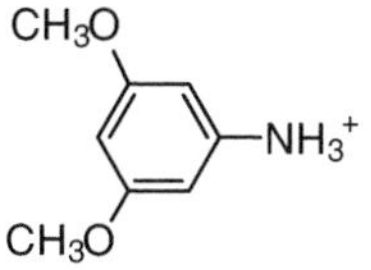

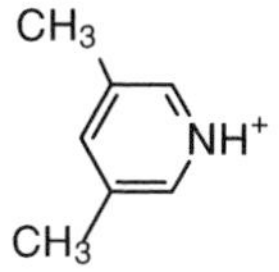

Scheme 2 *Comparison of observed and calculated values of pK_a and σ^* for multiple substituted species. See Table 1 in Appendix 3 for σ^* and σ values. The following pK_a values are employed: benzoic acid, 4.20; 4-nitrophenol, 7.14; 2-chlorophenol, 8.48; phenol, 9.95; 3-nitrobenzoic acid, 3.49; aniline, 4.62; 3-methoxyaniline, 4.2; 4-methylpyridine, 6.02; pyridine, 5.17; 3-methylpyridine, 5.68. pK_a values are calculated by adding incremental pK_a values to that of the parent*

4.3 MULTIPLE INTERACTIONS

The effect of a substituent on a rate or equilibrium is likely to be transmitted by multiple interaction paths to the reaction centre. The success of single interaction parameters such as that of the Hammett Equation is largely due to the experimental conditions where one effect is isolated or the different transmission routes through the benzene nucleus are utilised to the same extent with each substituent. The Pavelich–Taft Equation (Chapter 2, Equation 15) recognises steric and polar effects as contributing separately, and the two parameter equation is successful when steric requirements are significant. The addition of extra independent variables to an equation will of necessity provide a better fit to data even though the quantities that they represent may not be related to the system under investigation. This does not invalidate the use of multi-parameter equations in prediction of physical quantities but a proliferation of parameters is only valid if there are good theoretical reasons why each should be included. However, the fundamental understanding of a system becomes tenuous as the number of fitting parameters increases.

4.3.1 The Jaffé Relationship[11,12]

Transmission of the polar effect from substituent (X) to reaction centre (R_x) in benzenoid systems involves more than one route. The *meta* substituent in **3** transmits its effect to the *ipso* position[b] *via* the 2-position or *via* the 4, 5 and 6 positions. The overall effect at R_x is due to the combined transmission routes and this is constant provided the only route to R_x is through the *ipso* position.

In some aromatic systems, particularly in fused benzenoid structures, the two transmission routes can be separated experimentally. The function Z in structure **4** lies on the transmission track to R_x and the substituent X is *para* to Z but *meta* to R_x. Which value is to be used, σ_m or σ_p? Structures **5** and **6** illustrate this problem for reaction at R_x in substituted benzothiazoles and **7** and **8** for the hydrolysis of salicylate esters which includes transmission *via* the carboxylate anion.

A more explicit separation of transmission routes is illustrated in the dissociation of *N*-phenyl benzene sulphonamides (**9**) where the effect of the substituent in the aniline residue Y has a manifestly different path from that of X in the benzene sulfonyl residue.

---

[b] In this case the *ipso* position is that on the aromatic ring adjacent to the reaction centre, R_x.

The similarity coefficient for each substituent effect can be solved separately in this case by keeping either X or Y constant and varying the other substituent.

The relative effectiveness of the two transmission routes in systems such as those discussed in structures **5** to **9** can be determined by use of the Jaffé Equation (6).

$$\log(k/k_H) = \rho_A\sigma_A + \rho_B\sigma_B \qquad (6)$$

The equation predicts the substituent effect if it is transmitted to the reaction centre by two pathways (A and B), for example by the substituent having an effect on both the nucleophile and the electrophile in an intramolecular reaction.

The intramolecular acid-catalysed hydrolysis (Equation 7)[13] provides a good example of the application of the Jaffé equation.

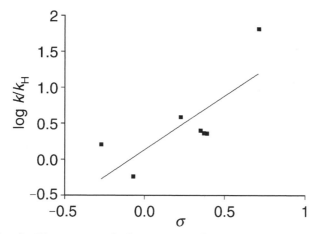

$$(7)$$

A plot of $\log(k/k_H)$ versus σ (relative to either A or B) gives a very scattered plot (Figure 1) indicating simply that there is more than one discrete path between substituent and reaction site.

Equation (6)[14] may be rearranged to give Equation (8) and the plot of $[\log(k_X/k_H]/\sigma_A$ versus σ_B/σ_A gives a straight line with slope ρ_B and an intercept of ρ_A.

Figure 1 *Simple Hammett graph for a two pathway system for the reaction of Equation (7)*

$$\frac{\log(k/k_{\rm H})}{\sigma_{\rm A}} = \rho_{\rm A} + \frac{\log(k/l_{\rm H})}{\sigma_{\rm A}} \tag{8}$$

The form of this equation often leads to a poor correlation owing to error magnification in the arithmetical manipulation (Figure 2).

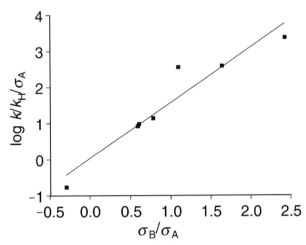

Figure 2 *Graphical analysis of Jaffé equation by use of Equation (8)*

Data fitting computer programs are available (see Appendix 1, Section A1.1.4) which are preferable to the graphical method to fit Equation (6) to the data.[c] If $\sigma_{\rm A}$ and $\sigma_{\rm B}$ correlate with each other the correlation

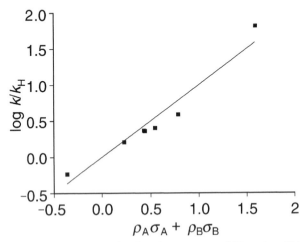

Figure 3 *Graphical display of results for the reaction of Equation (7) from multiple regression of the Jaffé Equation (6)*

[c] A problem is included where the student can try out the two approaches, and it is good practice, whatever the fitting technique, to have some form of graphical demonstration of the fit.

becomes unreliable and the criterion for significance is that the correlation coefficient between the σ_m and σ_p values employed in the relationship should be less than 0.90.[14]

Although the advent of high-speed computers has enabled data to be directly fitted to the Jaffé Equation, the results cannot be directly illustrated. Instead the results for ρ_A and ρ_B are determined statistically and the goodness of fit is demonstrated graphically by plotting $\log(k/k_H)$ (observed) against its calculated value $(\rho_A\sigma_A + \rho_B\sigma_B)$ (Figure 3).

4.3.2 Resonance Effects

Equations (9a) and (9b) can be assumed to govern the resonance and inductive components of the polar substituent effect.

$$\log(k_X/k_H) = R + I \tag{9a}$$

$$\sigma = \sigma_R + \sigma_I \tag{9b}$$

The resonance interaction is likely to be small in the dissociation of substituted benzoic acids owing to cancellation of the similar energies of the resonance contributions in acid and conjugate base (Scheme 3) and is not possible for *meta* substituents.

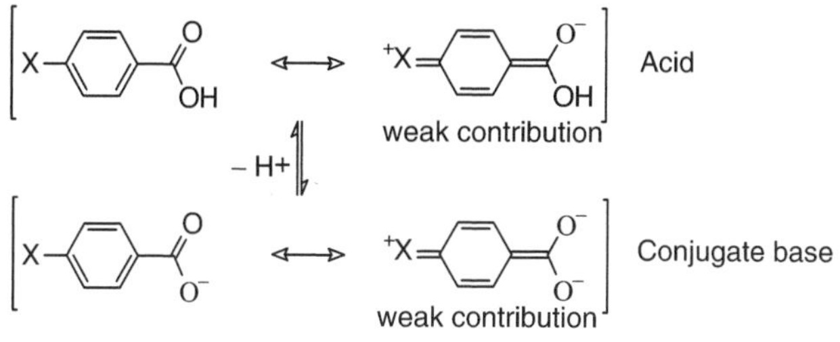

Scheme 3 *Cancellation of the effect of resonance contributions in the dissociation of benzoic acids*

Dissociation of phenols or anilinium ions, however, can involve substantial resonance stabilisation of product base but not of the reactant acid (Schemes 4 and 5) and the relevant pK_a values do not correlate well with Hammett σ values as illustrated in Figure 4.[11]

The *meta* substituent constant σ_m has no significant resonance interaction: thus $\sigma_R = 0$ and $\sigma_I \sim \sigma_m$. A parameter (σ_R) may be defined as σ^- by Equation (10).

$$\sigma^- = (pK_a^{\,PhOH} - pK_a^{\,ArOH})/2.23 \tag{10}$$

The value of 2.23 is the ρ value for the dissociation of *meta* substituted

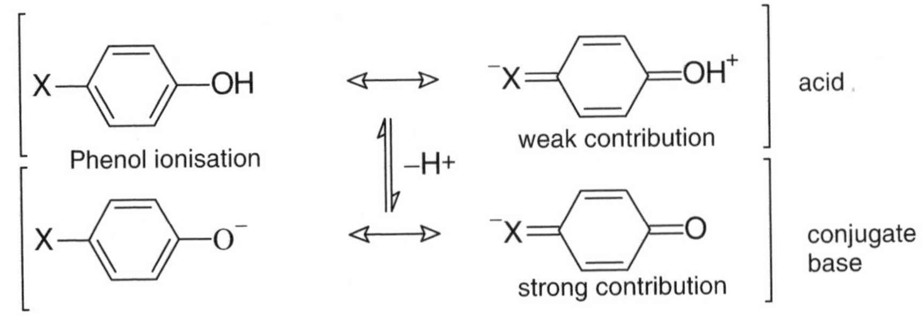

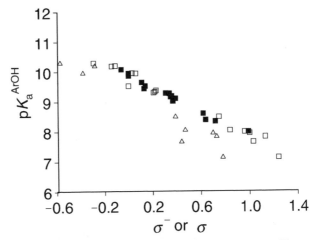

Scheme 4   *Resonance transmission of charge withdrawal in phenolate ions*

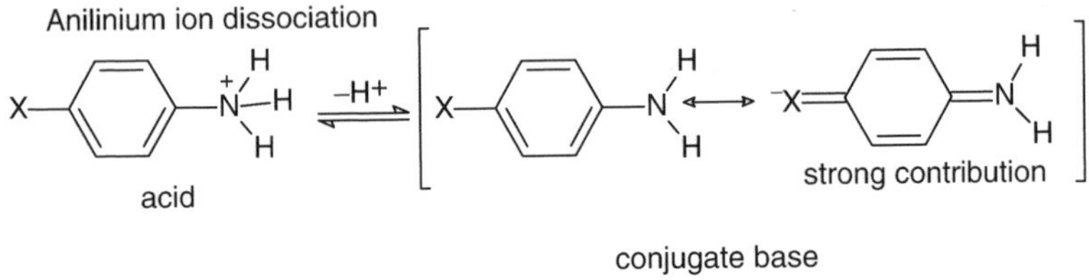

Scheme 5   *Resonance transmission of charge withdrawal in anilines*

Figure 4   *Dissociation of substituted phenols.* ■, meta *substituents;* □, para *substituents against* σ^-; △, para *substituents against* σ

phenols against Hammett σ_m constants where no resonance is expected, and is employed as a normalising factor.

A small amount of resonance transmission is suffered by *meta* substituents and is due to the π-inductive effect where electrostatic charge at a conjugated atom adjacent to the substituent polarises the corresponding π-electron system. For our purposes this contribution is taken to be negligible

Poor Hammett σ correlations are often obtained for reactions in which a positively charged centre is formed that can resonate with an electron *donating* substituent, because of the additional resonance transmission for these substituents.[15] This phenomenon is illustrated by the solvolysis of substituted 2-phenyl-2-chloropropane (Figure 5). Scheme 6 shows how a transition structure may obtain extra stabilisation by resonance interaction between carbenium ion and 4-methoxy and 4-dimethylamino groups. In these cases there is no effect on the energy of the reactant molecules.

The parameter (σ^+) may be defined by Equation (11) derived from the solvolysis of ArCMe$_2$Cl in 90% acetone/water at 25°.[15]

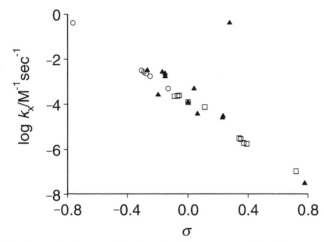

Figure 5 *Solvolysis of substituted 2-phenyl-2-chloropropane.* □, σ *for* meta *substituents;* ▲, σ *for* para *substituents;* ○, σ$^+$ *for* para *substituents*

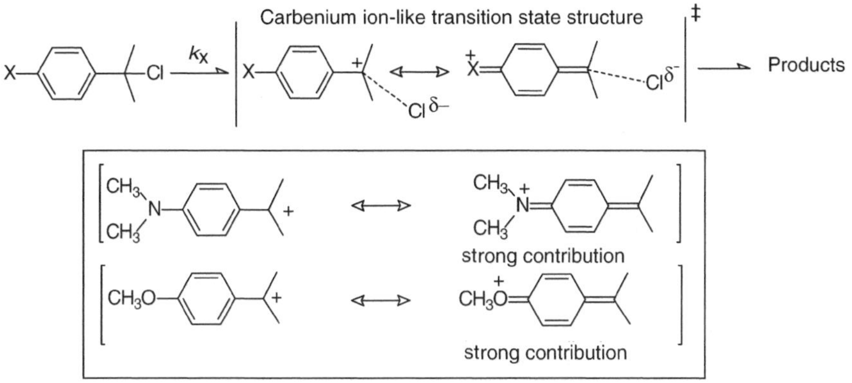

Scheme 6 *Resonance interactions with electron donating substituents*

$$\sigma^+ = (\log k_H - \log k_x)/(-4.54) \tag{11}$$

Values of σ^+ are tabulated in Appendix 3, Table 1 for a selection of substituents; the quantity -4.54 is the ρ value for the standard solvolysis reaction for *meta* substituents. The σ^+ parameters are defined by use of rate constants from the standard reaction because it is difficult to measure equilibrium constants for formation of a localised carbenium ion for a substantial number of substituents.[16]

The resonance interaction will only be substantial when the substituent itself is able to participate in resonance. Thus nitro, cyano and keto functions withdraw electrons by resonance and methoxy and amino substituents donate electrons by resonance. Reference to tables of σ^+ and σ^- values in Appendix 3 shows that *para* substituents that would not normally undergo resonance have values slightly differing from those of the Hammett σ series.

4.3.3 The Yukawa–Tsuno Equation

Maximum resonance interaction is likely to occur in the transition structure for the solvolysis reaction of Scheme 6 when the transition structure fully resembles a carbenium ion. In reactions where the transition structure does not completely resemble the reference state the substituent will not exert its full potential in resonance transmission with the reaction centre. The "r" parameter of Yukawa and Tsuno[17,18] (Equation 12) provides a measure of the extent of the resonance interaction for a reaction centre which builds up positive charge.

$$\log(k/k_H) = \rho[\sigma + r(\sigma^+ - \sigma)] \tag{12}$$

Where resonance occurs between substituent and a reaction centre suffering negative charge build-up, Equation (13) may be employed.[19]

$$\log(k/k_H) = \rho[\sigma + r(\sigma^- - \sigma)] \tag{13}$$

An application of Equation (13) is illustrated in Figure 6 for the alkaline hydrolysis of substituted phenyl carbamates. Here the points for resonating *para* substituents lie above the line. When σ^- is used for these points instead of σ the data obey a linear equation and $r = 1$.

Young and Jencks proposed a modified form of the Yukawa–Tsuno Equation (14) which has the advantage over Equation (12) in that $\rho^r(\sigma^+ - \sigma)$ directly measures the proportion of free energy change caused by electron donation via a resonance route.[21]

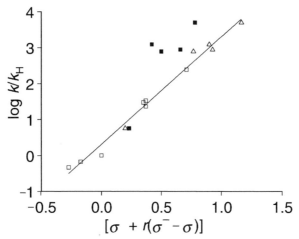

Figure 6 *Alkaline hydrolysis of substituted phenyl carbamates ($\rho = 3.35$, $r = 0.785$).*[20]
■, para *substituents uncorrected;* □, meta *substituents;* △, para *substituents (corrected)*

$$\log k/k_o = \rho_o\sigma + \rho^r(\sigma^+ - \sigma) \tag{14}$$

A similar argument applies to the case where electron withdrawal involves a resonance pathway.[d]

The development of the resonance interaction as measured by ρ^r can be compared with the value of ρ^r_{eq} determined for the equilibrium reaction. Such a case is offered in the addition of bisulfite ion to acetophenones (Scheme 7) where both equilibrium and rate constants are available.

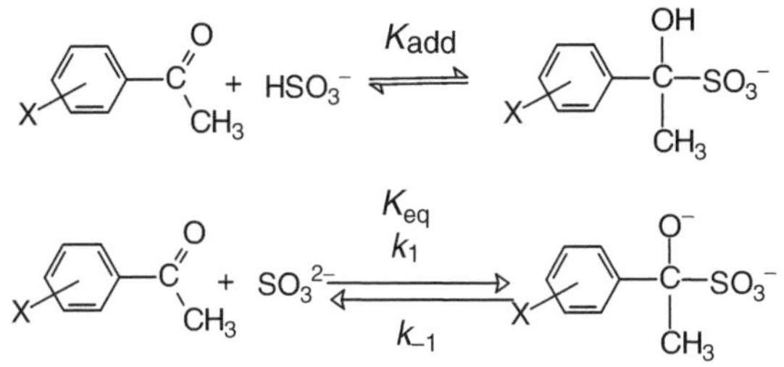

Scheme 7 *Addition of bisulfite ion to acetophenones*

The equilibrium formation of monoanion is given by Equations (15–17):

$$K_{add} = [>C(OH)SO_3^-]/[>C{=}O][HSO_3^-] \tag{15}$$

---

[d] The r value of the Yukawa–Tsuno equation compares the energy of the resonance route with that in the standard (where r is defined as unity).

$$= [>C(O^-)SO_3^-]K_1/[>C=O][SO_3^{2-}]K_2 \qquad (16)$$

$$= K_{eq}K_1/K_2 \qquad (17)$$

where K_1 and K_2 are the dissociation constants respectively of HSO_3^- and $>C(OH)SO_3^-$ and $K_{eq} = [>C(O^-)SO_3^-]/[>C=O][\ SO_3^-]$. The values of K_{add} have a ρ^r_{eq} of 0.95 due to stabilisation of the acetophenone carbonyl group through electron donation by resonance. Since K_1 is substituent independent and there is no change in conjugation in the equilibrium K_2, the resonance contribution to the substituent effect on K_{eq} may be taken as essentially the same as that on K_{add}. The value of ρ^r for k_1 is 0.45 so that the destruction of resonance in the transition structure is some 47% of that in the product $(45 \times 1/0.95)$.

In the case of the alkaline hydrolysis of aryl carbamates the ρ^r is 2.64 ± 0.44 which is, within experimental error, the same as the ρ^r for the dissociation of substituted phenols (2.23); this is consistent with complete expression of the resonance interaction between developing oxyanion and the substituent.

The Yukawa–Tsuno and Young–Jencks equations have the advantage that they utilise both *meta* and *para* substituents in the same equation unlike the Dewar–Grisdale and Swain–Lupton equations (Sections 4.3.4.1 and 4.3.4.2).

4.3.4 Other Two-parameter Equations

Several two-parameter equations have been suggested based on the additivity of I and R effects, each being the combination of various components as described earlier. Taft[22] originally introduced this approach for the electronic effects of benzene and defined σ_R values employing values of σ_I (from the aliphatic series – see Chapter 2) and Hammett's σ for *para* substituents (Equation 18).

$$\sigma_p = \sigma_I + \sigma_R \qquad (18)$$

The parameters σ_R defined by Equation (18) are named according to whether σ_p is σ^+ or σ^- and are σ_R^+, or σ_R^-.[23]

4.3.4.1 The Dewar–Grisdale Equation An approach to field and resonance transmission of the polar effect essentially the same as that of Taft's group was suggested by Dewar and Grisdale.[1] The localised and delocalised polar effects are represented by F and M which are taken to be additive in any given σ value. The value σ_{ij} is related to the sum of field effect (F) between atoms i and j and the combined π-inductive-resonance (M) effect of the substituent (Equations 19 and 20).

$$\sigma_{ij} = F/r_{ij} + M.q_{ij} \qquad (19)$$

$$\sigma_{ij} = F'/r_{ij} - M'.\pi_{ij} \qquad (20)$$

The distance between the atoms i and j is r_{ij}, the formal charge induced at atom j by $^-CH_2-$ placed at i is q_{ij}, and π_{ij} is the atom–atom polarisability of atoms i–j. Since π_{ij} and q_{ij} can be calculated for benzene, application of equations (19) and (20) in combination with known values of Hammett's σ_{ij} gives a series of F, F', M and M' parameters.

The approach is based on a theoretical model for the transmission of the polar effect but its further extension is beyond the scope of this book. Future models of the polar transmission effect will undoubtedly be firmly based in MO theory.[24,25,26]

4.3.4.2 The Swain–Lupton Equation. A more generally applicable set of Dewar–Grisdale-type parameters was developed by Swain and Lupton.[27] The large number of σ parameters (derived from different standard reactions) may be simplified by attributing to each substituent two standard components: a field constant F (combining through-space and through-bond transmission interaction) and a resonance constant R.[e] The substituent effect is given by Equation (21) where the reaction constants f and r (analogous to ρ selectivities) are defined in Equation (22).

$$\log k_X/k_H = fF + rR \qquad (21)$$

$$\sigma_{RX} = f_{RX}F + r_{RX}R \qquad (22)$$

The F and R parameters may be determined using the Roberts and Moreland σ' values which are assumed to measure the field effect (F) and are related by Equation (23) from which a and b are calculated from a basis set of σ_m and σ_p values (originally 14 substituents).

$$\sigma' = a\sigma_m + b\sigma_p = F \qquad (23)$$

Values of F for other substituents (not possessing σ' values) are calculated using Equation (23) from known Hammett σ_m and σ_p parameters.

---

[e] *F* and *R* come close to a universal set of parameters measuring polarity of a given substituent.

Setting the value of $R = 0$ for the substituent $-NMe_3^+$ enables c to be calculated in equation (24) and hence R can be calculated from F and σ_p values.

$$\sigma_p = cF + R \qquad (24)$$

An estimation of r, R, f and F parameters[28] may be derived from Equation (25) by a non-linear least squares procedure to fit all the parameters r_i, f_i, R_j and F_j.

$$\log(k_{ij}/k_{iH}) = r_i R_j + f_i F_j + h_i \qquad (25)$$

The parameters F and R depend on the substituent j; f, r, and h depend on the *identity of the series*, i. The procedure utilises an iterative calculation to converge to a best fit between predicted and observed values of $\log k_{ij}$; the term h_i is added to avoid k_{iH} for the unsubstituted compound having infinite weight. The parameters are transformed so that $F = R = 0$ for the hydrogen substituent, $F = R = 1$ for $-NO_2$ and $r = 0$ for the $\log(K_a/K_a'')$ values (σ'') of *trans*-4-substituted cyclohexanecarboxylic acids in water and $R = 0$ for the $-NMe_3^+$ substituent.

The Swain–Lupton, Taft and Dewar–Grisdale equations all require that the substituent in the reaction under investigation remains at the same position relative to the reaction centre. Thus *meta* and *para* series have to be treated separately unlike the Yukawa–Tsuno method where both *meta* and *para* substituents may be taken together. The Swain–Lupton treatment may be illustrated by the reaction of chloride ion with substituted benzenediazonium chlorides (Figure 7, Equation 26).[29]

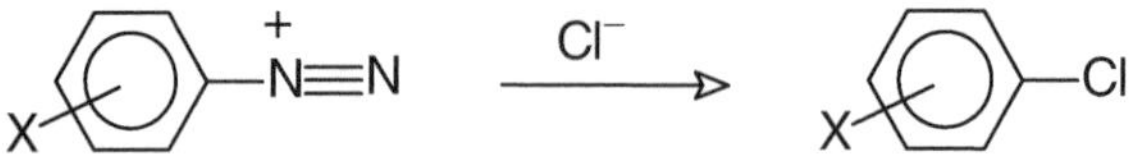

$$\qquad (26)$$

The mechanism of the reaction involves a transition structure resembling that of a phenyl cation.[29] The following reaction parameters were obtained for $\log(k/k_H)$ for *meta* and *para* substituents: $f_m = -2.74$ and $f_p = -2.60$; $r_m = -3.18$ and $r_p = +5.08$, Figure 8.

The relative importance of field and resonance effects (%R) may be judged by use of the equation

$$\%R = 100|r|/(|r| + |f|)^{28}$$

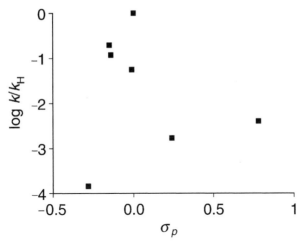

Figure 7 *Reaction of chloride ion with substituted benzenediazonium ions. Hammett plot for* para *substituents*

This value is tabulated for various standard reactions (Appendix 4, Table 7) and for the displacement at the benzenediazonium ion the *para* substituents exhibit resonance stabilisation of the reactant state. The benzenediazonium ion reaction has $\%R = 54$ for the *meta* and $\%R = 66$ for the *para* substituents, consistent with increased resonance interaction in the *para* case. Values of the reaction parameters, f and r, are given in Appendix 4, Table 7.

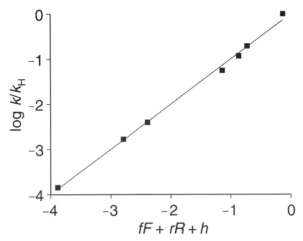

Figure 8 *Swain–Lupton plot. The data are taken from reference 29 and the Swain–Lupton regression carried out using* F *and* R *constants from the Appendix (not those of reference 28)*

4.3.4.3 The Edwards Equation for Nucleophilic Aliphatic Substitution.
Reactions of nucleophiles of similar structure (such as all phenolate ions) exhibit good extended Brønsted equations (Chapter 2) but there is poor correlation when the nucleophiles have widely differing structures (Figure 9).[30,31,32] In this case the simple dissociation equilibrium is a poor model of the nucleophilic process.

The Edwards equation sums component processes modelling the transfer of the pair of bonding electrons from substrate to product. The component model processes are proton transfer to the nucleophile (the proton affinity) (Equation 27) and the polarisability of the nucleophile as judged by its standard electrode potential (E°) (Equation 28).

$$Nu^- + H^+ \underset{K_a}{\rightleftharpoons} NuH \tag{27}$$

$$Nu^- \overset{E^\circ}{\rightleftharpoons} Nu + e^- \tag{28}$$

The Edwards equation (Equation 29) sets $E_N = E^\circ + 2.60$ and $H_N = pK_a^{NuH} + 1.74$ and has been successful in correlating a wide range of data for displacements at tetrahedral carbon but a major disadvantage is that E_N values are available for very few reagents.

$$\log(k_{Nu}/k_{water}) = aE_N + bH_N \tag{29}$$

Figure 10 illustrates the Edwards equation for the reaction of nucleophiles with methyl iodide.

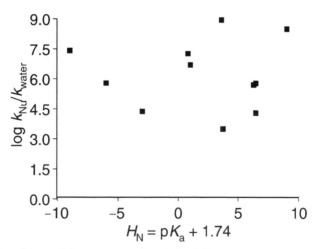

Figure 9   *Correlation of the reactivity of methyl iodide and the pK$_a$ of the conjugate acid of the nucleophile of widely disparate structures*

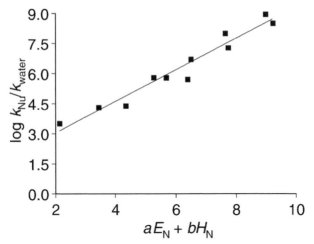

Figure 10 *Edwards correlation applied to the data of Figure 9; the line is calculated from Equation (29) with a = 3.54 and b = 0.0132.*[f]

The Edwards equation may be modified using molecular refractivity as a measure of polarisability instead of ionisation potentials (Equation 30).[31]

$$\log(k_{Nu}/k_{water}) = aP + bH_N \qquad (30)$$

The parameter H_N is the same as in Equation (29) and P is $\log(R/R_{H_2O})$ where R is the molecular refractivity. There are even fewer values of R compared with E_N so that Equation (30) loses much of its interest as a correlating tool. Moreover, as we saw in Chapter 2, molecular refractivity possesses a substantial molar volume component.

4.3.4.4 The Mayr Equation. Mayr and his co-workers have developed Equation (31)[33,34] which correlates the reactivity of carbocations and related electrophiles with π nucleophiles.

$$\log k = s(N + E) \qquad (31)$$

Equation (31) has an *electrophilic* (E) and a *nucleophilic* (N) component and the value, s, is a nucleophile specific parameter. This equation has a close family relationship with the Ritchie and Swain–Scott equations (Chapter 2) and the Edwards equation (29). The equation successfully correlates rate constants for a wide range of disparate structures and

---

[f] The relative values of a and b indicate that the oxidation equilibrium (Equation 28) is a good model of the nucleophilic attack; indeed a plot of log k_{Nu}/k_{water} against E_N indicates a good linear correlation even without the introduction of H_N.

reactions and has considerable predictive capacity suitable for the design of synthetic routes.

4.4 MULTI-PARAMETER SOLVENT EQUATIONS

4.4.1 Extension of the Grunwald–Winstein Equation

The success of the single parameter Grunwald–Winstein equation is largely due to the limited range of solvent change such as variation of composition of mixtures. When different solvent types are employed extra terms are needed in the equation to fit the data. Solvolysis reactions in aliphatic nucleophilic substitution involve nucleophilic attack of the solvent and it is unlikely that solvents of different structure would have similar nucleophilicities excepting those in a series of mixed solvents such as ethanol–water (Chapter 2). The simplest treatment involves dividing the solvent action into nucleophilic and electrophilic components as shown in Equation (32).[35,36]

$$\log(k/k_o) = lN + mY \qquad (32)$$

The parameter N is defined as a function of the solvolysis of methyl tosylate with l set at 1.00 and m at 0.3 (Equation 33).

$$N = [\log(k/k_o) - mY]/l \qquad (33)$$

The parameter Y is defined by the solvolysis of 2-adamantyl tosylate and Figure 11 illustrates the fit to Equation (32) for the solvolysis of iso-propyl tosylate. The additional term introduces a better correlation and indicates that the solvolysis has substantial nucleophilic assistance ($l = 0.4$) in the transition structure as would be expected.

4.4.2 The Swain Equation

Swain advanced Equation (34)[28], analogous to that in the Swain–Lupton treatment of polar effects (Equation 21).

$$\log(k_X/k_o) = aA + bB + h \qquad (34)$$

The terms a and b are reaction parameters and A and B are dependent only on the solvent. A data set of 61 solvents and 77 reactions and processes were fitted to Equation (34) using 1080 data points, yielding a set of 154 reaction parameters (a and b) and 122 solvent parameters (A and B). The solvent parameter scale was calculated using the following defined boundary conditions: $A = B = 1$ for water solvent, $A = B = 0$ for

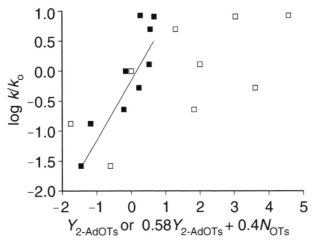

Figure 11 *Extended Grunwald–Winstein plot for the solvolysis of isopropyl tosylate.*[37] $\square$, *plot of the data against* $Y_{2\text{-AdOTs}}$; $\blacksquare$, *improved plot when bilinear equation is fitted with* $l = 0.4$ *and* $m = 0.58$

n-heptane, $A = 0$ for hexamethylphosphoramide and $B = 0$ for trifluoro-acetic acid.

The A and B terms of Equation (34) are not based on any model reaction; the meaning of A and B comes from the boundary con-ditions, which assign A to be an anion-solvating parameter and B to be cation solvating. The 1080 data points used in determining the Swain parameters exhibit excellent internal consistency and the final fit has a correlation coefficient of 0.9906; the individual reactions are fit with correlation coefficients greater than 0.9750. The high correlation coefficients indicate that explicit incorporation of solvation parameters related directly to molecular processes is not necessary as far as fit is concerned. Terms due to anion-solvating power, hydrogen-bonding acidity and electrophilicity are adequately represented as a single parameter (A) (the *acity*) for each solvent. Likewise cation solvation, hydrogen-bonding basicity and nucleophilicity are represented by the *basity* parameter (B). Although providing an excellent fit, the Swain approach does not employ a *standard reaction* so that the method has little significance for elucidating transition structure.

The two-parameter equations possess a family resemblance. The original Swain–Scott equation (Chapter 2)[38] was written as a special case of the *general* equation[g] $\log k_{\text{Nuc}}/k_o = ns + es'$ where the parameters n and s have already been defined. The term es' is a function of the

---

[g] The polar displacement reaction involves attack of nucleophile (N) and electrophile (E) on the substrate (S).

electrophilicity of the solvent and for a given solvent this is a constant in the Swain–Scott equation. Both the Swain Equation (34) and the extended Grunwald–Winstein Equation (32) are similar in that the nucleophilic and solvent interactions are treated as additive quantities. If ns is negligible compared to *es'* and the electrophile is the solvent the Equation reduces to the Grunwald–Winstein special case ($\log(k/k_o) = es' \equiv Ym$).

4.4.3 Extended Hansch Equations

Biological properties such as the reactivity of enzymes, toxicity or drug action usually involve a combination of fundamental effects that give rise to equations which include more than one parameter. The Hansch equation is usually modified by the *addition* of terms such as steric parameters (E_s) or polar substituent constants (σ). Examples are found in the catalytic action of α-chymotrypsin on 4-nitrophenyl esters of aliphatic acids (Equation 35).

$$\log[(k_{cat}/K_m)^X/(k_{cat}/K_m)^H] = \delta E_s + p\log P \qquad (35)$$

Inhibition of α-chymotrypsin by substituted phenoxyacetophenones (ArOCH$_2$COPh) includes a polar term in its Hansch equation (Equation 36).

$$\log[(K_i)^H/(K_i)^X] = \rho\,\sigma_x + p\log P \qquad (36)$$

4.4.4 Solvent Equations Based on Macroscopic Quantities

The Swain Equation (34)[28] provides very accurate predictive power for a limited range of solvents and processes based on a statistical analysis of a five-parameter equation. The origins of the Swain parameters are not explicit, although the *acity* (A) and *basity* (B) coefficients are related to electrophilic and nucleophilic processes respectively. Many fundamental processes have been implicated in solvent effects and the equation of Koppel and Palm (Equation 37)[39] incorporates the major factors thought to be involved.

$$\log(k/k_o) = gf(\varepsilon) + pf(n) + cE + bB \qquad (37)$$

The exponents *g*, *p*, *c* and *b* are parameters of the reaction or process being affected by change in solvent. The term f(ε) is a function of the dielectric constant, usually ($\varepsilon - 1$)/($2\varepsilon + 1$), f(n) is a refractive index function $(n^2 - 1)/(n^2 + 2)$ and E and B are measures of electrophilic and nucleophilic solvating ability respectively. The solvolysis of *t*-butyl chloride in various solvents can be fitted to Equation (37) with the exponents given in Equation (38).

$$\log k/\text{sec}^{-1} = -19.5 + 5.67f(\varepsilon) + 17.27f(n) + 0.379E \tag{38}$$

Taft advanced Equation (39), analogous to that of Kopple and Palm, where π^* is an average solvation parameter obtained from solvato-chromic shifts and α is a hydrogen-bond acidity scale.

$$\log k/\text{sec}^{-1} = -15.06 - 6.94\pi^* - 5.25\alpha \tag{39}$$

4.5 FURTHER READING

M.H. Abraham *et al.*, Solvent Effects in Organic Chemistry – Recent Developments, *Can. J. Chem.*, 1988, **66**, 2673.

M.H. Abraham, Scales of Solute Hydrogen-bonding: Their Construction and Application to Physicochemical and Biochemical Processes, *Chem. Soc. Rev.*, 1993, 73.

T.W. Bentley and G. Llewellyn, Y_X Scales of Solvent Ionising Power, *Progr. Phys. Org. Chem.*, 1990, **17**, 121.

N.B. Chapman and J. Shorter (eds.), *Correlation Analysis in Chemistry*, Plenum Press, New York, 1978.

N.B. Chapman and J. Shorter (eds.), *Advances in Linear Free Energy Relationships*, Plenum Press, New York, 1972.

J.O. Edwards and R.G. Pearson, The Factors Determining Nucleophilic Reactivities, *J. Am. Chem. Soc.*, 1962, **84**, 16.

I.A. Koppel and V.A. Palm, The Influence of Solvent on Organic Reactivity, in *Advances in Linear Free Energy Relationships*, N.B. Chapman and J. Shorter (eds.), Plenum Press, New York, 1972, Chapter 5, p. 203.

Y. Yukawa and Y. Tsuno, Resonance Effect in Hammett Relationship. III. The Modified Hammett Relationship for Electrophilic Reactions, *Bull. Soc. Chem. Japan*, 1959, **32**, 971.

4.6 PROBLEMS

1 Calculate (see Scheme 2) the $\Sigma\sigma^*$ value for the following, $-\text{CH}(\text{C}_6\text{H}_5)_2$, $-\text{CH}_2\text{C}(\text{CH}_3)_3$, $-\text{CH}(\text{OH})\text{C}_6\text{H}_5$, $-\text{CHCl}_2$, $-\text{CCl}_3$. Use Table 1 in Appendix 3 to obtain σ^* for the fragments.

2 Calculate the pK_a values of the following acids making use of the additivity principle (see Scheme 2): 2,4-dichlorophenol, 3,4-dinitro-benzoic acid, 3,4-dibromo-anilinium ion and 3,4-dimethyl-pyridinium ion. Use the following pK_a data: 2-chlorophenol, 8.48; 4-chlorophenol, 9.38; phenol, 9.95; benzoic acid, 4.20; anilinium ion, 4.62; 3-bromoanilinium ion, 3.51; 4-bromoanilinium ion, 3.51; pyridinium ion, 5.17; 3-methylpyridinium ion, 5.68 and 4-methylpyridinium ion, 6.02. Calculate the overall σ value for 2,4-dinitrobenzoic acid from its calculated pK_a.

3 The hydrolyses of substituted aspirins (**10**)[14]

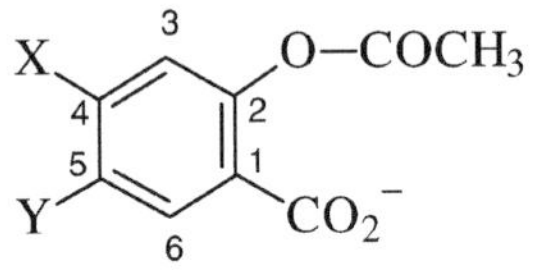

10

have the rate parameters given in Table 2.

Try a graphical method and a regression analysis method for solving the Jaffé bilinear Equation governing these data. Discuss the values of ρ_1 and ρ_2 which you obtain in terms of a map of effective charges.

Table 2

Substituent	$10^5 k/sec^{-1}$	σ_1	σ_2
Parent	1.108	0	0
4-Cl	1.955	0.23	0.37
4-Br	1.923	0.23	0.39
4-I	1.770	0.18	0.35
4-NO$_2$	2.107	0.78	0.72
4-MeO	2.087	−0.27	0.11
5-Cl	1.235	0.37	0.23
5-Br	1.292	0.39	0.23
5-I	1.283	0.35	0.18
5-NO$_2$	2.339	0.72	0.78
5-MeO	0.952	0.11	−0.27

4 Substituted *N*-phenylbenzene sulphonanilides (**9**) have the following pK_a values in aqueous ethanol at 20°C (Table 3).[40]

Table 3

X	Y	σ_X	pK_a^{XY}	σ_Y
H	H	0.0	9.1	0.0
H	4-F	0.0	8.9	0.06
H	3-Br	0.0	8.25	0.39
3-NO$_2$	H	0.72	7.93	0.0
3-NO$_2$	4-CH$_3$O	0.72	8.44	-0.27
3-NO$_2$	4-Br	0.72	7.42	0.23
4-Cl	4-CH$_3$O	0.23	9.19	-0.27
4-Cl	4-Cl	0.23	8.30	0.23
4-Cl	3-NO$_2$	0.23	7.19	0.72
4-F	4-CH$_3$O	0.06	9.32	-0.27
4-F	3-NO$_2$	0.06	7.27	0.72
4-CH$_3$	4-CH$_3$	-0.17	9.65	-0.17
4-CH$_3$	4-Cl	-0.17	8.78	0.23
4-NH$_2$	3-CH$_3$	-0.66	10.44	0.07
4-Cl	4-NO$_2$	0.23	6.24	0.78
3-NO$_2$	4-NO$_2$	0.72	5.51	0.78

Construct a Jaffé plot and determine ρ_X and ρ_Y assuming that they are independent of the substituents Y and X respectively. Comment on the fact that σ^- values are not required to make the data fit the correlation for the 4-nitro substituents.

5   Plot the pK_a of substituted anilines (given in Table 4) against the corresponding Hammett σ and σ^- values. Explain your results in terms of a separation of inductive and resonance transmission of the polar effect.

Table 4

para substituent	pK_a	σ^-	σ_p	meta substituent	pK_a	σ_m
H	4.58	0.0	0.0	H	4.58	0.0
(CH$_3$)$_3$N$^+$	2.51	0.96	0.82	(CH$_3$)$_3$N$^+$	2.26	0.88
CH$_3$CO	2.19	0.84	0.50	CH$_3$CO	3.58	0.38
C$_2$H$_5$OCO	2.3	0.75	0.45	CH$_3$SO$_2$	2.68	0.60
Br	3.91	0.22	0.23	Br	3.51	0.39
F	4.52	0.06	0.06	F	3.38	0.34
CH$_3$	5.07	-0.14	-0.17	CH$_3$	4.67	-0.07
CH$_3$O	5.29	-0.11	-0.27	CH$_3$O	4.2	0.11
NO$_2$	0.98	1.24	0.78	NO$_2$	2.45	0.72
CHO	1.76	1.03	0.44	OH	4.17	0.12
CN	1.74	1.0	0.66	CN	2.75	0.56
I	3.78	0.21	0.18	I	3.61	0.35
Cl	3.81	0.23	0.23	Cl	3.32	0.37
NH$_2$	6.08	-0.29	-0.66	NH$_2$	4.88	-0.16
SO$_3^-$	3.32	0.37	0.09	SO$_3^-$	3.8	0.05
CH$_3$SO$_2$	1.48	1.13	0.72			

6   The values of pK_R^+ for Equation (40) are given in Table 5.[41]

$$H_3O^+ + \underset{\underset{Ar}{|}}{\overset{\overset{Ar}{|}}{Ar-COH}} \quad \underset{K_R^+}{\rightleftharpoons} \quad \underset{\underset{Ar}{\textstyle\backslash}}{\overset{\overset{Ar}{\textstyle/}}{Ar-C^+}} + \quad 2\,H_2O \qquad (40)$$

Plot a Hammett graph employing both σ and σ^+ parameters and comment on your results.

Table 5

Substituents	pK_R^+	σ	σ^+
3,3′,3″-(CH$_3$)$_3$	−6.35	−0.07	
Parent	−6.63	0	
3,3′,3″-Cl$_3$	−11.03	0.37	
4,4′,4″-(NO$_2$)$_3$	−16.27	0.78	
4,4′,4″-[N(CH$_3$)$_2$]$_3$	9.36	−0.83	−1.7
4,4′,4″-(NH$_2$)$_3$	7.57	−0.66	−1.3
4,4′,4″-(CH$_3$O)$_3$	0.82	−0.27	−0.78
4,4′,4″-(CH$_3$)$_3$	−3.56	−0.17	−0.31
4,4′,4″-Cl$_3$	−7.74	0.23	0.11

7   Fit the data in Table 6 for the solvolysis of *meta*-substituted[42] phenyldimethylmethyl chlorides according to the Swain–Lupton equation using values of F and R tabulated by Leo and Hansch (Appendix 3, Table 1).

Table 6

Substituent	$\log(k/k_H)$ para	$\log(k/k_H)$ meta	F	R
NH$_2$	−1.3	−0.16	0.08	−0.74
Br	0.15	0.405	0.45	−0.22
(CH$_3$)$_3$C	−0.256	−0.059	−0.02	−0.18
Cl	0.114	0.399	0.42	−0.19
CN	0.659	0.562	0.51	0.15
F		0.3452	0.45	−0.39
H	0.00	0.00	0.00	0.00
I		0.359	0.42	−0.24
CH$_3$O		0.047	0.29	−0.56
NO$_2$	0.79	0.674	0.65	0.13
C$_6$H$_5$	−0.179	0.109	0.12	−0.13
F$_3$C	0.612	0.52	0.38	0.16
Me$_3$N$^+$	0.408		0.86	−0.04
HO	−0.92		0.33	−0.70
C$_6$H$_5$O	−0.50		0.37	−0.40
Me	−0.311		0.01	−0.18

8 Fit the data in Table 7 for the *para*-substituted analogues of Problem 7 using the *F* and *R* parameters tabulated by Leo and Hansch (Appendix 3, Table 1).

Compare the results *r* and *f* of the *para* case with those of the *meta* system using the %*R* methodology.

Table 7

para substituent	$log(k/k_H)$	*F*	*R*
NH$_2$	−1.3	0.08	−0.74
Br	0.15	0.45	−0.22
(CH$_3$)$_3$C	−0.256	−0.02	−0.18
Cl	0.114	0.42	−0.19
CN	0.659	0.51	0.15
H	0.00	0.00	0.00
HO	−0.92	0.33	−0.70
CH$_3$	−0.311	0.01	−0.18
NO$_2$	0.79	0.65	0.13
C$_6$H$_5$O	−0.50	0.37	−0.40
C$_6$H$_5$	−0.179	0.12	−0.13
F$_3$C	0.612	0.38	0.16
(CH$_3$)$_3$N$^+$	0.408	0.86	−0.04

9 Construct a Hammett and a Yukawa–Tsuno graph for the alkaline hydrolysis of substituted phenyl triethylsilyl ethers from the data in Table 8.[43]

Comment briefly on the results regarding the timing of bond fission and formation in the transition structure.

Table 8

Substituent X	$log(k_X/k_H)$	σ	σ^-
H	0.00	0	0
3-CH$_3$	−0.093	−0.07	−0.07
3-COCH$_3$	1.223	0.38	0.38
3-CN	2.13	0.56	0.56
3-NO$_2$	2.332	0.71	0.72
4-N=NC$_6$H$_5$	1.903	0.31	0.69
4-CO$_2$C$_2$H$_5$	2.02	0.45	0.64
4-COC$_6$H$_5$	2.488	0.43	0.88
4-COCH$_3$	2.291	0.50	0.84
4-CN	2.862	0.66	0.88
4-CHO	2.914	0.43	1.03

10 Calculate the σ^* values for:

$-CH_2CH_2Cl$, $-CH_2CH_2CF_3$, $-CH_2CH_2CH_2CF_3$, $-CH_2C_6H_5$,
$-CH_2CH_2C_6H_5$ and $-CH_2CH_2CH_2C_6H_5$

given that the attenuation factor for $-CH_2-$ in an aliphatic system (Equation 14, Chapter 2) is 0.41 and taking σ^* for $-C_6H_5$, $-CH_2Cl$ and $-CH_2CF_3$ from tables in Appendix 3.

11 Calculate the ρ value for ionisation of **11**

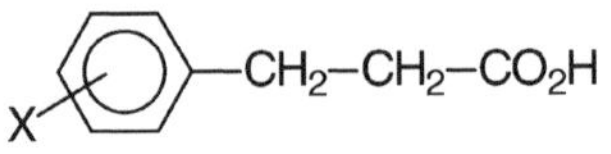

11

using the ρ for the dissociation of substituted benzoic acids (1.0) and an attenuation factor of 0.47 (Table 1).

12 Use the Edwards equation to correlate the data of Table 9 for the reaction of nucleophiles with ethyl tosylate in aqueous solution at 25°C (Data and parameters from reference 44).

Table 9

Nucleophile	$logk/M^{-1} sec^{-1}$	E_N	H_N
H_2O	-6.936	0.00	0.00
NaOH	-3.900	1.65	17.48
$Na_2S_2O_3$	-3.066	2.52	3.60
Na_2SO_3	-2.947	2.57	9.00
NaN_3	-3.824	1.58	6.46
NaSH	-2.783	3.08	14.66
KI	-4.187	2.06	-9.00
KSCN	-4.167	1.83	1.00
$NaNO_2$	-3.810	1.73	5.09

13 Indicate $\Sigma\sigma$ for structure **12**, assuming that reaction occurs at R_x:

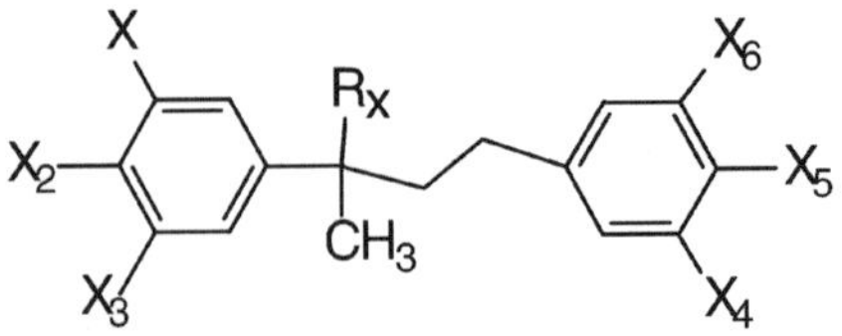

12

14   Calculate the β_{eq} for the dissociation of the amido NH in Equation (41) given that the dissociation of phenols has a β_{eq} value of -1.0 and assuming a attenuation factor of 0.4 per atom.

$$H_2O + PhCONHCH_2CO\text{-}O\text{-}\underset{X}{\langle\rangle} \overset{K_a}{\rightleftharpoons} H_3O^+ + PhCON^-CH_2CO\text{-}O\text{-}\underset{X}{\langle\rangle} \quad (41)$$

15   Calculate the β_{eq} for dissociation of the *N*-phosphoryl-pyridine, using as reference the dissociation of substituted pyridinium ions and an attenuation factor of 0.4 per atom.

$$\underset{X}{\langle\rangle}N^+\text{-}P(\text{=}O)(O^-)\text{-}OH + H_2O \overset{K_a}{\rightleftharpoons} \underset{X}{\langle\rangle}N^+\text{-}P(\text{=}O)(O^-)\text{-}O^- + H_3O^+ \quad (42)$$

16   The ρ_{eq} value for the dissociation of the amide NH in the following reaction is +1.45.

$$H_2O + \underset{X}{\langle\rangle}\text{-}CONHCH_2COOPh \overset{K_a}{\rightleftharpoons} H_3O^+ + \underset{X}{\langle\rangle}\text{-}CON^-CH_2COOPh \quad (43)$$

Indicate why this equilibrium could be more sensitive to substituents than in the similar dissociation reaction of benzoic acids.

$$H_2O + \underset{X}{\langle\rangle}\text{-}CO_2H \overset{K_a}{\rightleftharpoons} H_3O^+ + \underset{X}{\langle\rangle}\text{-}CO_2^- \quad (44)$$

4.7 REFERENCES

1. M.J.S. Dewar and P.J. Grisdale, Substituent Effects I, *J. Am. Chem. Soc.*, 1962, **84**, 3539.
2. C.G. Swain and E.C. Lupton, Field and Resonance Components of Substituent Effects, *J. Am. Chem. Soc.*, 1968, **90**, 4328.
3. K. Bowden and E.J. Grubbs, Through-Bond and Through-Space Models for Interpreting Chemical Reactivity in Organic Reactions, *Chem. Soc. Rev.*, 1996, 171.
4. M.H. Abraham *et al.*, Solvent Effects in Organic Chemistry – Recent Developments, *Can. J. Chem.*, 1988, **66**, 2673.
5. J.G. Kirkwood and F.H. Westheimer, The Electrostatic Influence of Substituents in the Dissociation Constants of Acids, *J. Chem. Phys.*, 1938, **6**, 506, 513.

6. J.E Leffler and E. Grunwald, *Rates and Equilibria of Organic Reactions*, John Wiley, New York, 1963, p. 224.

7. G.B. Barlin and D.D. Perrin, Prediction of the Strengths of Organic Acids, *Quart. Rev. Chem. Soc.*, 1966, **20**, 75.

8. R.W. Taft and I.C. Lewis, The General Applicability of the Fixed Scale of Inductive Effects II., *J. Am. Chem. Soc.*, 1958, **80**, 2436.

9. D.D. Perrin *et al.*, *pK$_a$ Prediction for Organic Acids and Bases*, Chapman & Hall, London, 1981, p. 88.

10. J. Shorter and F.J. Stubbs, The Additive Effect of Substituents on the Strength of Benzoic Acid, *J. Chem. Soc., Perkin Trans. 2*, 1949, 1180.

11. H.H. Jaffé, A Re-examination of the Hammett Equation, *Chem. Rev.*, 1953, **53**, 250.

12. H.H. Jaffé, Some Extensions of the Hammett Equation, *Science*, 1953, **118**, 246; H.H. Jaffé, Application of the Hammett Equation to Fused Ring Systems, *J. Am. Chem. Soc.*, 1954, **76**, 4261.

13. A.R. Hopkins *et al.*, Electrophilic Catalysis of Sulphate ($-SO_3^-$) Group Transfer: Hydrolysis of Salicyl Sulphate Assisted by Intramolecular Hydrogen Bonding, *J. Chem. Soc., Perkin Trans. 2*, 1983, 1279.

14. A.R. Fersht and A.J. Kirby, Structure and Mechanism in Intramolecular Catalysis. The Hydrolysis of Substituted Aspirins, *J. Am. Chem. Soc.*, 1967, **89**, 4853.

15. H.C. Brown and Y. Okamoto, Electrophilic Substituent Constants, *J. Am. Chem. Soc.*, 1958, **80**, 4979.

16. N.C. Deno and W.L. Evans, Carbonium Ions VI σ^+-Parameters, *J. Am. Chem. Soc.*, 1957, **79**, 5804.

17. Y. Yukawa and Y. Tsuno, Resonance Effect in Hammett Relationship III The Modified Hammett Relationship for Electrophilic Reactions, *Bull. Chem. Soc. Japan*, 1959, **32**, 971.

18. J. Shorter, Multiparameter Extensions of the Hammett Equation, in *Correlation Analysis in Chemistry*, N.B. Chapman and J. Shorter (eds.), Plenum Press, New York, 1978, Chapter 4, p. 119.

19. J.J. Ryan and A.A. Humffray, Rate Correlation Involving Linear Combination of Substituent Parameters. I. Hydrolysis of Aryl Acetates, *J. Chem. Soc., B*, 1966, 842.

20. H. Al-Rawi and A. Williams, Elimination–Addition Mechanisms of Acyl Group Transfer: the Hydrolysis and Synthesis of Carbamates, *J. Am. Chem. Soc.*, 1977, **99**, 2671.

21. P.R. Young and W.P. Jencks, Separation of Polar and Resonance Substituent Effects in the Reaction of Acetophenones with Bisulfite and of Benzyl Halides with Nucleophiles, *J. Am. Chem. Soc.*, 1979, **101**, 3288.

22. R.W. Taft and I.C. Lewis, The General Applicability of a Fixed Scale of Inductive Effects. II. Inductive Effects of Dipolar Substituents in the Reactivities of *m*- and *p*-Substituted Derivatives of Benzene, *J. Am. Chem. Soc.*, 1958, **80**, 2436.

23. O. Exner, A Critical Compilation of Substituent Constants, in *Correlation Analysis in Chemistry*, N.B. Chapman and J. Shorter (eds.), Plenum Press, New York, 1978, Chapter 10, p. 439.

24. M. Godfrey, Theoretical Models for Interpreting Linear Correlations in Organic Chemistry, in *Correlation Analysis in Chemistry*, N.B. Chapman and J. Shorter (eds.), Plenum Press, New York, 1978, Chapter 3, p. 85.

25. A. Warshel, *Computer Modelling of Chemical Reactions in Enzymes and Solutions*, Wiley-Interscience, New York, 1991, pp. 92–96.

26. M.J.S. Dewar and R.C. Dougherty, *The PMO Theory of Organic Chemistry*, Plenum Press, New York, 1975, Chapter 4, pp. 131–195.

27. C.G. Swain and E.C. Lupton, Field and Resonance Components of Substituent Effects, *J. Am. Chem. Soc.*, 1968, **90**, 4328.

28. C.G. Swain *et al.*, Substituent Effects in Chemical Reactivity. Improved Evaluation of Field and Resonance Components, *J. Am. Chem. Soc.*, 1983, **105**, 492.

29. C.G. Swain *et al.*, Evidence for Phenyl Cation as an Intermediate in Reactions of Benzenediazonium Salts in Solution, *J. Am. Chem. Soc.*, 1975, **97**, 783.

30. J.O. Edwards, Correlation of Relative Rates and Equilibria with a Double Basicity Scale, *J. Am. Chem. Soc.*, 1954, **76**, 1540.

31. J.O. Edwards, Polarisability, Basicity and Nucleophilic Character, *J. Am. Chem. Soc.*, 1956, **78**, 1819.

32. C. Duboc, The Correlation of Nucleophilicity, in *Correlation Analysis in Chemistry*, N.B. Chapman and J. Shorter (eds.), Plenum Press, New York, 1978, Chapter 7, p. 313.

33. H. Mayr and M. Patz, Scales of Nucleophilicity and Electrophilicity: A System for Ordering Polar Organic and Organometallic Reactions, *Angew. Chem. Int. Ed.*, 1994, **33**, 938.

34. H. Mayr *et al.*, *J. Phys. Org. Chem.*, 1998, **11**, 642.

35. S. Winstein *et al.*, Correlation of Solvolysis Rates. VIII. Benzhydryl Chloride and Bromide, Comparison of *m*.Y and Swain's Correlations, *J. Am. Chem. Soc.*, 1957, **79**, 4146.

36. T.W. Bentley and G. Llewellyn, Y_X Scales of Solvent Ionising Power, *Progr. Phys. Org. Chem.*, 1990, **17**, 121.

37. T.W. Bentley and P.v.R. Schleyer, The SN1-SN2 spectrum. 1. Role of Nucleophilic Solvent Assistance and Nucleophilically Solvated Ion

Pair Intermediates in Solvolyses of Primary and Secondary Arenesulfonates, *J. Am. Chem. Soc.*, 1976, **98**, 7658.

38. C.G. Swain and C.B. Scott, Quantitative Correlation of Relative Rates. Comparison of Hydroxide Ion with Other Nucleophilic Reagents Toward Alkyl Halides, Esters, Epoxides and Acyl Halides, *J. Am. Chem. Soc.*, 1953, **75**, 141.

39. I.A. Koppel and V.A. Palm, The Influence of the Solvent on Organic Reactivity, in *Advances in Linear Free Energy Relationships*, N.B. Chapman and J. Shorter (eds.), Plenum Press, New York, 1972, Chapter 5, pp. 203–280.

40. G. Dauphin and A. Kargomard, Étude de la Dissociation Acide de Quelques Sulfonamides, *Bull. Soc. Chim. France*, 1961, 486.

41. N.C. Deno and A. Schriesheim, Carbonium Ions. II. Linear Free Energy Relationships in Arylcarbonium Ion Equilibria, *J. Am. Chem. Soc.*, 1955, **77**, 3051.

42. Data from reference 27.

43. A.A. Humffray and J.J. Ryan, Rate Correlation Involving Linear Combination of Substituent Parameters. IV. Base-Catalysed Hydrolysis of Triethylphenoxy-silanes, *J. Chem. Soc., B*, 1969, 1138.

44. R.E. Davis *et al.*, The Oxibase Scale and Displacement Reactions. XVII. The Reaction of Nucleophiles with Ethyl Tosylate and the Extension of the Oxibase Scale, *J. Am. Chem. Soc.*, 1969, **91**, 91.

45. A.F. Hegarty and P. Tuohey, Hydrolysis of Azo Esters *via* Azoformate Intermediates, *J. Chem. Soc., Perkin Trans. 2*, 1980, 1238; A. Williams, Free Energy Correlations and Reaction Mechanisms, in M.I. Page, ed., *Chemistry of Enzyme Action*, Elsevier, Amsterdam, 1984, p. 145.

CHAPTER 5

Coupling Between Bonds in Transition Structures of Displacement Reactions

5.1 CROSS AND SELF INTERACTION BETWEEN TWO OR MORE SUBSTITUENT EFFECTS

As well as reporting variation in charge at a reaction centre the substituent variation can itself alter the transition structure. In principle, the slope of a free energy relationship varies as a function of the substituent. There are two types of variation of slope: (1) *cross interaction* where the slope measured by one substituent change is altered by a substituent elsewhere in the reacting system and (2) *self interaction* where the substituent being altered causes a variation in electronic structure (Marcus Curvature, Chapter 6). Cross- and self-coupling[a] effects were recognised by Miller and other workers[1,2,3,4] and have been quantified in terms of the shape of the free energy surface in the region of the transition structure.[2] Extensive experimental work has been carried out[3] on cross and self coupling in displacement reactions under a large range of conditions.

Let us consider a general two-bond reaction with variation in substituents x and y each of which will affect the reactivity (Scheme 1).

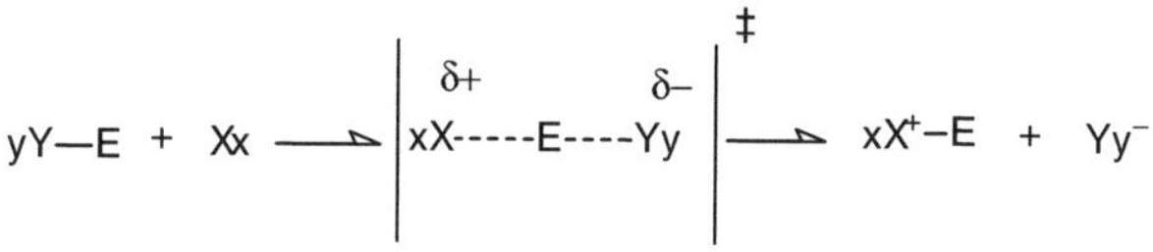

$$yY-E \ + \ Xx \ \longrightarrow \ \left| \ \overset{\delta+}{xX}\text{-----}E\text{----}\overset{\delta-}{Yy} \ \right|^{\ddagger} \ \longrightarrow \ xX^+-E \ + \ Yy^-$$

Scheme 1   *General two-bond displacement reaction with substituents x and y on leaving and entering groups*

---

[a] The terms *coupling* and *interaction* are regarded as synonymous.

In order to simplify the arguments we shall employ ρ and σ parameters although any other similarity and substituent constant, such as β and pK_a, could be used in the equations. The rate constant can be expressed as a two-dimensional quadratic Equation (1) (where ρ_x° or ρ_y° are invariant or standard values).

$$\log(k_{xy}/k_{oo}) = \rho_x^\circ \sigma_x + \rho_y^\circ \sigma_y + p_{xy}\sigma_x\sigma_y + 0.5p_x\sigma_x\sigma_x + 0.5p_y\sigma_y\sigma_y \tag{1}$$

Partial differentiation of $\log(k_{xy}/k_{oo})$ with respect to σ_x (keeping y constant) gives ρ_x; likewise $\partial\log k_{xy}/\partial\sigma_y$ is ρ_y.[b]

$$\rho_x = \partial\log k_{xy}/\partial\sigma_x = \rho_x^\circ + p_{xy}\sigma_y + p_x\sigma_x$$

and

$$\rho_y = \partial\log k_{xy}/\partial\sigma_y = \rho_y^\circ + p_{xy}\sigma_x + p_y\sigma_y$$

Partial differentiation of these two equations gives:[c]

$$p_{xy} = \partial^2\log k_{xy}/\partial\sigma_x\partial\sigma_y = \partial\rho_x/\partial\sigma_y = \partial\rho_y/\partial\sigma_x \tag{2}$$

$$p_x = \partial^2\log k_{xy}/\partial\sigma_x\partial\sigma_x = \partial\rho_x/\partial\sigma_x \tag{3}$$

$$p_y = \partial^2\log k_{xy}/\partial\sigma_y\partial\sigma_y = \partial\rho_y/\partial\sigma_y \tag{4}$$

The magnitude of the cross-interaction coefficient, p_{xy}, is dependent on the electronic coupling between the two bonds undergoing change.[d]

The arguments can be extended to more than two substituent variations and the next level of complexity is where a substituent on the fragment "E" of Scheme 1 would be varied. Such variation is exemplified in displacement reactions at centres adjacent to substituted aromatic rings. Solvent effects can also be included in equations similar to (1) but these are not considered here.

The dual interaction routes of Scheme 1 are analogous to those discussed in Chapter 4. The equations of Chapter 4 are essentially special cases of Equation (1) where cross-interaction (p_{xy}) and self-interaction (p_x or p_y) coefficients are absent. The chemical difference between the two conditions is that in Scheme 1 the two interactions are not mutually exclusive whereas the multiple interactions of Chapter 4 (such as in

[b] k_{oo} is invariant in x or y.
[c] Take care because ρ_x and p_x and ρ_y and p_y are easily confused!
[d] The coupling of motions based on IR data provides little evidence that motions beyond 3–4 bonded atoms are coupled.

reactions conforming to the Jaffé relationship) are not affected by each other and are additive. The reason for this is that the effects discussed in Chapter 4 involve a single part of a reaction centre whereas the substituent variations in Scheme 1 affect two parts of the reaction centre (such as bond formation and bond fission).

The cross-interaction coefficient, p_{xy}, can be determined from the experimental data k_{xy} by measuring ρ_x as a function of σ_y making use of Equation (5). Alternatively p_{xy} can be determined from a plot of ρ_y against σ_x.

$$\rho_x = p_{xy}.\sigma_y + C \qquad (5)$$

Multiple linear regression analysis of Equation (1) can also be used and for this k_{xy} is determined with as many different combinations of σ_x and σ_y as is possible.[5] The self-interaction coefficients (p_x or p_y) can be measured by fitting the Brønsted or Hammett data to a binomial expression ($\log k = a + bx + cx^2$) by regular statistical software packages or by a program based on the statistical equation in Appendix 1 (Section A1.1.4.4).

5.1.1 Hammond and Cordes–Thornton Coefficients

The acronym *Bema Hapothle*[6] describes the effects of structure and energy change on the shape of the energy surface at the transition structure for a reaction.[e] The first derivative of $\log k_{xy}$ (for example ρ or β) with respect to a measure of substituent polarity characterises the transition structure of a reaction in terms of an effective charge. The second derivative of $\log k_{xy}$ ($\partial\beta/\partial pK_a$, etc) measures the *variation* in effective charge as a function of substituent polarity. The first derivative positions the transition structure on the energy surface of the More O'Ferrall–Jencks diagram (see Appendix 1, Section A1.2) of the reaction and the second derivatives (p_x, p_y or p_{xy}) measure the shape of the surface at that location.

The existence of a detectable change in p_x or p_y (Hammond co-efficient)[2] depends on the sharpness of the curvature of the reaction coordinate at the *saddle point*;[f] "broad" curvature gives a large change in

---

[e] The term Bema Hapothle is useful as a shorthand reference to effects on the transition structure; it is derived from the names of some of the principal workers who have contributed to this field namely *Bell, Marcus, Hammond, Polanyi, Thornton* and *Leffler*.

[f] The term *saddle point* is synonymous with *transition structure*. The saddle point is distinguished by its possessing convex curvature along and concave curvature perpendicular to the reaction coordinate. Discussions of reaction surfaces often use the term *saddle point* interchangeably with *transition structure*.

position of the transition structure as a function of substituent whereas "sharp" curvature will give only a little change. The existence of a non-zero value of p_x or p_y means that the free energy relationships have curvature. A positive p_x or p_y indicates concave curvature (viewed from above) and negative values indicate convex curvature.

The *Cordes*[4]*–Thornton*[7] coefficient (p_{xy}) is the second derivative describing change in ρ_x with respect to σ_y (Equation 2).[2] It is equally permissible to use β and pK_a as well as combinations of these in the definition of the Cordes–Thornton coefficient (for example p_{xy} could be $\partial\rho_x/\partial pK_a^y$, $\partial\beta_x/\partial pK_a^y$ or $\partial\beta_x/\partial\sigma_y$). The student should also be aware that other similarity coefficients and polarity constants such as those registered in Chapter 2 can also be employed in the definition of p_{xy}. A non-zero value of p_{xy} does not necessarily indicate curvature in the free energy relationships.

5.2 SHAPES OF ENERGY SURFACES FROM COUPLING COEFFICIENTS

The arguments in this section are largely due to Jencks[2] and Thornton.[7] The extent of structural change perpendicular to the reaction coordinate at the transition structure (registered by p_{xy}) depends on the sharpness of the curvature at the saddle point in the More O'Ferrall–Jencks map. The effect of structural change on the position of the transition structure perpendicular to the reaction coordinate moves it towards the corner of the diagram where the energy *decreases*. Movement of the saddle point parallel to the reaction coordinate is towards a corner registering an *increase* in energy.[g]

The Hammond and Cordes–Thornton coefficients can be utilised to characterise transition states in terms of structure reactivity surfaces as follows. The Hammond coefficient is defined so that it is positive when an increase in energy of the products (relative to reactants) accompanies an increase in the first derivative. This describes the generalisation (Hammond effect) that endergonic reactions tend to have product-like transition structures. A positive Hammond coefficient is expected for a *fundamental* process where structural change affects the energy of one end of a reaction coordinate, as a consequence of the maximum (negative curvature) at the saddle-point in the direction along the reaction coordinate.

---

[g] See Appendix 1 (Section A1.2) for graphical demonstrations of movement perpendicular and parallel to the reaction coordinate at the transition structure in surfaces approximating to parabolas.

Equations which are used to model the shape of the surface at the transition structure differ from that used to develop the Marcus Theory (Chapter 6) where the cross-over point of the two inverted parabolas is sharp and cannot have a smooth surface. Equations with smooth turning points are required for modelling the region of the surface close to the transition structure.

A two-dimensional free energy reaction diagram may be described intuitively for a fundamental reaction (Figure 1). The curve at the saddle point and for most of the reaction coordinate can be modelled by the paraboloid Equation (6) but the equation fails to predict the minima in potential energy at the reactant and product positions. The value of β_{Nuc} will increase as pK_a^{Nuc} increases and Equation (6) will have a as negative. The p_x or p_y value for this change is defined as $-\partial\beta_{Nuc}/\partial pK_a^{Nuc}$.

$$G_{edge} = ax^2 + bx + c \tag{6}$$

The paraboloid representation can be extended to model the three-dimensional x,y-surface (Equation 7) (which is analogous to Equation 1).

$$G_{surface} = ax^2 + by^2 + cxy + dx + ey + f \tag{7}$$

The system may be represented as in Figure 2 where three different graphical systems illustrate the surface. The quadratic Equation (7) fails at the corners of the whole energy surface diagram (Figure 2). However, the equation provides a reasonably good approximation to the surface in the region of the saddle point as a function of the energies at the edges and corners. Equations more consistent with chemical intuition have been described[8,9,10,11] using higher orders and one of these, the quartic

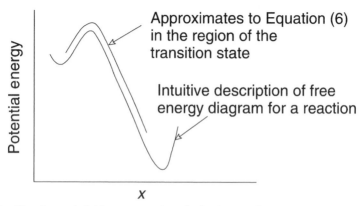

Figure 1 *Simple paraboloid representation of a fundamental process*

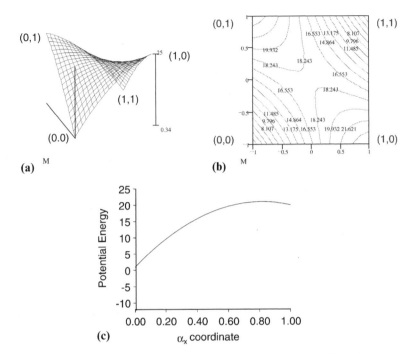

Figure 2 *Structure–reactivity surface (a), contour (b) and edge profile (c) as represented by the quadratic Equation (7); the edge profile refers to the x coordinate*

Equation (8), predicts the turning points required by chemical intuition for the edges, corners and transition structure. The edge Equations (8) and (9) and the surface Equation (10) for the quartic model are illustrated in Figure 3.

$$G_{edge} = ax^4 + bx^3 + cx^2 + d \tag{8}$$

$$= x^4/4 - (\alpha + 1)x^3/3 + \alpha x^2/2 + d \tag{9}$$

$$G_{surface} = a_1x^2 + a_2y^2 + a_3x^3 + a_4y^3 + a_5x^4 +$$
$$a_6y^4 + a_7x^2y^3 + a_8x^3y^2 + a_9x^3y^3 + a_{10}x^2y^2 \tag{10}$$

Both quadratic and quartic equations for a fundamental process satisfactorily model the Hammond effect whereby the transition structure moves towards that of the reactant as the fundamental process becomes more exergonic.

The quadratic formula (Equation 7) is employed[2] to deduce a relationship between p_x, p_y and p_{xy} and the shape of the surface at the transition structure because the calculations can be readily achieved without the complicated algebra and calculus required for the quartic Equation (10);

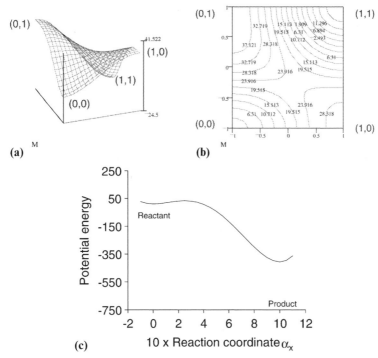

Figure 3 *Structure–reactivity surface (a), contour (b) and edge profile (c) as represented by the quartic Equations (8–10)*

moreover, the equation has exact solutions. The resulting cross-inter-action equations characterise the shape of the surface at the transition structure provided it is understood that the formula breaks down at the edges and corners of the More O'Ferrall–Jencks diagram.

The terms a and b of Equation (7) define the curvature at the saddle point parallel to the x and y axes respectively and will be positive for curvature representing a minimum and negative for curvature at a maximum. Terms e and d represent energies of top and right-hand edges respectively of the diagram. The sign of the curvature parameter, c, is negative for reaction coordinates through saddle points in the lower left and top right quadrants of the diagram; $4ab - c^2$ is negative for surfaces that describe a saddle point.

Coordinates of the transition structure (Equations 11a and 11b, $x^{\ddagger}$, $y^{\ddagger}$) are determined by setting $\partial \log k / \partial x$ and $\partial \log k / \partial y$ to zero and are:

$$x^{\ddagger} = (ce - 2abd)/(4ab - c^2) = \text{Leffler's } \alpha_x \qquad (11a)$$

$$y^{\ddagger} = (cd - 2ae)/(4ab - c^2) = \text{Leffler's } \alpha_y \qquad (11b)$$

The Hammond and Cordes–Thornton coefficients are related to the parameters *a*, *b* and *c* by Equations (12–14).[2]

$$p_x = \partial x^\ddagger / \partial d = -2b/(4ab - c^2) \tag{12}$$

$$p_y = \partial y^\ddagger / \partial e = -2a/(4ab - c^2) \tag{13}$$

$$p_{xy} = \partial x^\ddagger / \partial e = \partial y^\ddagger / \partial d = c/(4ab - c^2) \tag{14}$$

Equations (12–14) can be solved to obtain the curvatures *a*, *b* and *c* in terms of Hammond and Cordes–Thornton coefficients (Equations 15–17).

$$a = -p_y/[2(p_x p_y - p_{xy}^2] \tag{15}$$

$$b = -p_x/[2(p_x p_y - p_{xy}^2] \tag{16}$$

$$c = p_{xy}/(p_x p_y - p_{xy}^2) \tag{17}$$

The parameters *a*, *b* and *c* measure respectively the horizontal (*x*), vertical (*y*) and diagonal curvature at the saddle point.

The curvature of the surface and the orientation of the saddle-point can be deduced from the angles of the two *level lines* (which join points of the same energy as that of the transition structure – see Figure 4) through the saddle point. The coordinates g_1, g_2, of a point on one of the level lines are relative to the coordinates of the transition structure, $x^\ddagger, y^\ddagger$. The coordinates of the point on the level line relative to the 0,0 corner of the More O'Ferrall–Jencks diagram are $(x^\ddagger + g_1), (y^\ddagger + g_2)$. The slopes of the level lines g_1/g_2 and g_2/g_1 relative to the *y* and *x* axes respectively are given by equations (18) and (19).

$$g_1/g_2 = [-c \pm (c^2 - 4ab)^{0.5}]/2a \tag{18}$$

$$g_2/g_1 = [-c \pm (c^2 - 4ab)^{0.5}]/2b \tag{19}$$

The orientation of the reaction coordinate at the saddle point can be deduced from the angles of the level lines; they also indicate the ratios of the curvatures parallel and perpendicular to the reaction coordinate. The extreme cases of the curvature at the saddle point are illustrated in Figure 4: (a) a narrow ridge over which the reaction coordinate passes and (b) a narrow pass which the reaction coordinate traverses. Change in transition structure (caused by energy changes at the edges and corners of the diagram) depends on the angle of the reaction coordinate and the

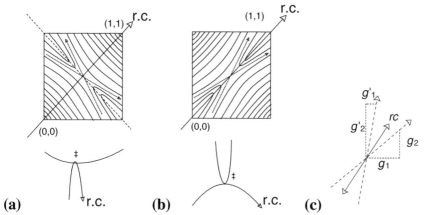

Figure 4 *Angle of level lines and the shape of the saddle point. (a) Narrow ridge perpendicular to reaction coordinate; (b) narrow pass parallel to reaction coordinate; (c) combination of level lines (dashed) give a vector (rc, the direction of the reaction coordinate at the saddle point). Product is at (1,1) and reactant at (0,0); reaction coordinate is r.c.*

curvature at the saddle point. Tilting the surface will cause the position of the transition structure to move a large distance when the curvature is small (perpendicular to the reaction coordinate in Figure 4a and parallel in 4b) and to change very little when the curvature is steep (parallel and perpendicular to the reaction coordinate in Figure 4a and 4b respectively).[2,7] When the two curvatures of the saddle point are significantly different the resultant change in position of the transition structure (r.c. in Figure 4c) is not easily predicted.

Substantial cross interaction[h] between effects on forming and breaking bonds (p_{xy}) is consistent with a concerted mechanism. A stepwise process, where there is no substantial energy transfer between bond formation and bond fission, would give only small cross-interaction coefficients.

The following examples illustrate the application of cross- and self-interaction coefficients in the study of surface shapes at the saddle points of a selection of reactions.

5.2.1 General Acid-catalysed Addition of Thiol Anions to Acetaldehyde

This system has $p_x = 0$, $p_y = \partial\beta_{Nuc}/-\partial pK_{Nuc} = 0.089$ and $p_{xy} = 0.026$.[12] Substitution into Equations (15–17) gives $a = 66$, $b = 0$, $c = -39$ and g_1/g_2 for level lines (Equations 18 and 19) are then 0 and 0.59, corresponding

---

[h] Sometimes called cross coupling.

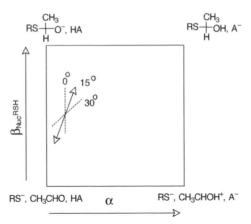

Figure 5   *General acid-catalysed reaction of thiolate anion with acetaldehyde*

to the angles 0° and 30° against the y axis. The saddle point is described in Figure 5 with the reaction coordinate bisecting the level lines at an angle of 15° from the y axis. The zero angle of one of the level lines indicates no curvature along the y axis ($b = 0$, Equation 12) through the saddle point. This is consistent with the observed absence of an effect of increasing acid strength on α ($p_x = 0$).

5.2.2 Reaction of Amines with Phosphate Monoesters

This reaction exhibits low β_{Nuc} and large negative β_{Lg} values for attacking amine and leaving phenolate ions respectively.[13] The attacking amines and leaving phenolate ions in this reaction exhibit low β_{Nuc} and large negative β_{Lg} values which position the transition structure in the SE (1,0) corner of the More O'Ferrall–Jencks diagram (Figure 6). The values of p_{xy} and p_x

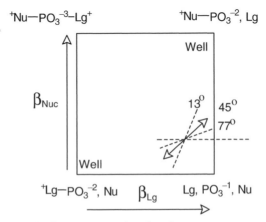

Figure 6   *Nucleophilic displacement at a phosphoryl centre*

and p_y for pyridinolysis of N-phosphopyridinium species are 0.014, 0.006 and 0.006 respectively. These give rise to curvatures of $a = 18.8$, $b = 18.8$ and $c = -87.5$ and slopes $g_1/g_2 = 0.23$ and 4.44 for the two level lines at the saddle point. These values correspond to angles of 13° and 77° relative to the y axis and by geometry the reaction coordinate at the saddle point is 45° to the y axis. The reaction coordinate therefore goes across a saddle point where there is steeper upward curvature perpendicular to the reaction coordinate than downward curvature along the reaction path (corresponding to Figure 4b). The relatively small curvature *along* the reaction coordinate is responsible for the predominant Hammond effect on reactivity of change in structure of the nucleophile or leaving group.

5.2.3 Elimination Reactions of 2-Arylethylquinuclidinium Ions

This reaction (Figure 7) undergoes a change in mechanism from stepwise E1cb for the 2-(4-nitrophenyl)ethyl group to a concerted E2 mechanism for derivatives with substituents which provide least stabilisation to a carbanion intermediate. The mechanism for the 4-nitro species has a transition structure, along the y axis, between reactants and carbanion intermediate (Figure 7). The sensitivity of the base-catalysed reaction to leaving group (β_{Lg}) is independent of the pK_a of the base catalyst over 12 pK_a units, consistent with a Cordes–Thornton coefficient (p_{xy}) of zero. The curvatures a, b and c for the other 2-aryl species such as 4-cyano and 4-acetylphenyl (which have less-stabilised carbanions at 0,1 than that of the 4-nitrophenyl group) are respectively 0, 51 and -55.[14] The curvatures give rise to level lines which have slopes g_1/g_2 relative to the y axis of 1.078 and 0 corresponding to angles of 47° and 0°. A reaction coordinate

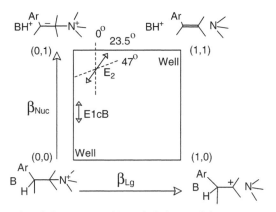

Figure 7 *Base-catalysed elimination of 2-arylethylquinuclidinium ions*

bisecting these level lines lies at an angle of 23.5° relative to the y axis as shown in the figure.

5.3 MAGNITUDE AND SIGN OF CORDES–THORNTON COEFFICIENTS

The methodology of Section 5.2 can also be applied when there is no interaction between the bonding changes, as might occur in a stepwise displacement process, except through the group being transferred. Values of p_{xy} are normally small for such systems but large values of p_{xy} should be observed for stepwise or concerted front-side nucleophilic displacements when there is a propensity for substantial through-space coupling between forming and breaking bonds. Nucleophilic aliphatic displacements by anilines of the sulfonate group from substituted benzyl sulfonate esters have relatively small p_{xy} values ($\sim$−0.11) consistent with a classical *in-line* mechanism (Scheme 2).

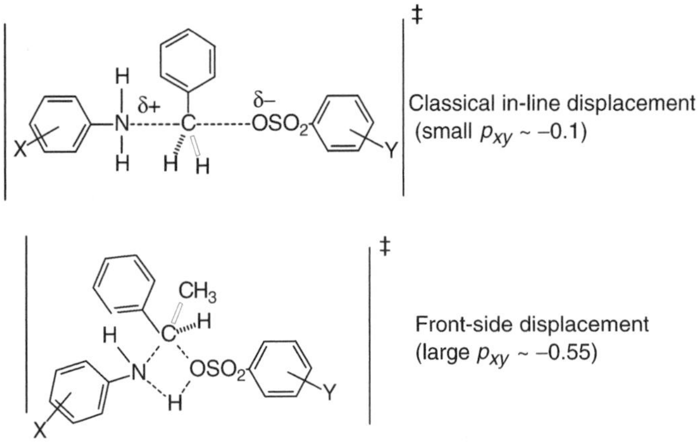

Scheme 2   *Front-side and in-line nucleophilic displacement diagnosed by the magnitude of* p_{xy} [15,16]

The displacement reaction at the substituted 1-phenylethyl centre has a substantial p_{xy} value of −0.55 which indicates strong interaction between leaving group and nucleophile as in a four-centre transition structure (Scheme 2). The additional methyl group would assist the mechanistic change from in-line to front-side owing to its stabilisation effect on an incipient carbenium ion.

   Lee showed that the sign of the Cordes–Thornton coefficient indicates the movement of the transition structure in the More O'Ferrall–Jencks map as a function of the change in energies of the corners.[3,17] Table 1 correlates the changes in ρ_x and ρ_y to be expected for change in σ_y and σ_x

Table 1 *Correlation of $\partial p_x/\partial\sigma_y$ and $\partial p_y/\partial\sigma_x$ as a function of p_{xy}*

Sign of $\partial\sigma_y$	Sign of ∂p_x	Effect of increase in nucleophilicity of X land eaving ability of Y (vectors of Figure 8 in parentheses)
A, positive p_{xy}		
−ve	−ve	earlier bond fission (*a*)
+ve	+ve	earlier bond formation (*b*)
B, negative p_{xy}		
−ve	+ve	later bond fission (*c*)
−ve	+ve	later bond formation (*d*)

as a function of p_{xy}. In the map (Figure 8) for the general displacement of X by Y, increasing nucleophilicity ($\partial\sigma_y$ is negative) or increasing leaving ability of X ($\partial\sigma_x$ is positive) moves the transition structure to the SW corner (0,0) for a positive p_{xy} and to the NE (1,1) for a negative p_{xy}.

5.4 IDENTITY REACTIONS

Identity reactions (defined as reactions giving product identical with the reactant) are very useful for discussions of *concerted* nucleophilic displacements because they require that the transition structure lies on the tightness diagonal shown in the More O'Ferrall–Jencks reaction map

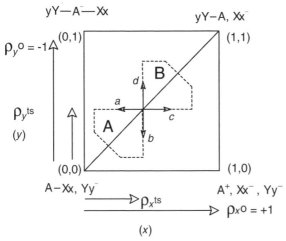

Figure 8 *More O'Ferrall–Jencks map for a general displacement reaction illustrating change in transition structure for the conditions of Table 1. The identities of the vectors a, b, c and d are shown in the table. Quadrants A and B represent +ve and −ve p_{xy} values respectively*

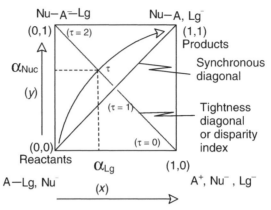

Figure 9   *Identity displacement reaction. The map could equally well be used to describe a radical or electrophilic displacement. τ may be obtained from α_{Lg} or α_{Nuc} by simple geometry (see Equation 25)*

(Figure 9).[18,19,20] The tightness diagonal is sometimes called the disparity index[21] and it essentially maps change of the transition structure on the reaction surface *perpendicular* to the reaction coordinate.

The sum of the bond orders, η, of forming and breaking bonds in the transition structure is defined as τ. The values are defined according to Equations (20), (21) and (22) assuming that η is a linear function of bond length.[18,20]

$$\tau = \eta_y + \eta_x \tag{20}$$

$$\eta_y = [(r_y)_{RC} - r_y]/[(r_y)_{RC} - 1] \tag{21}$$

$$r_y = y/y_{product} \tag{22}$$

The parameter r_y is the bond length, y, at any point in the reaction map divided by the bond length in the product ($r_{y(product)}$); $(r_y)_{RC}$ is the value of r_y in the reaction complex. The terms r_x, $(r_x)_{RC}$ and η_x for the bond fission component, x, are similarly defined. Since τ can be estimated from Leffler's α parameters (see later) it provides a useful experimental index of tightness in a reaction where substantial change is suffered by two bonds.

Identity rate constants (k_{ii}) may be studied by use of isotope replacement techniques[22] but need specialised skill and equipment for their accurate measurement. Free energy relationships for quasi-symmetrical reactions (Equation 23)[i] offer a simple approach to calculating identity

---

[i]   A quasi-symmetrical reaction has entering and leaving groups of similar structure. Thus Lg or Nu in Equation (23) could be all phenolate ions or all pyridines.

rate constants with reasonable accuracy. The More O'Ferrall–Jencks map of an identity displacement (Figure 9) has the advantage that it can be constructed without recourse to the *explicit* measurement of identity rate constants and the equations which enable this to be done are given below. The quantity K_{ij} is the equilibrium constant of a system with varying substituent, i, against a standard substituent, j (thus $K_{ii} = 1$).

$$A-Lg_i + Nu_j^- \xrightarrow{K_{ij}} Nu_j-A + Lg_i^- \tag{23}$$

$$\delta = \partial \log k_{ii}/\partial \log K_{ij} = \tau - 1 = \beta_{ii}/\beta_{eq} \tag{24}$$

Where β_{ii} is $\partial \log k_{ii}/\partial pK_a$ where the pK_a is for the conjugate acid Lg_iH and $\beta_{eq} = \partial \log k_{ij}/\partial pK_a$.

The quantity δ may be defined by Equation (24) and used as a scale[19] for τ where the values of $\tau = 2$, 1 and 0 refer respectively to NW, central and SE positions on the tightness diagonal, corresponding to 0,1; 0.5,0.5; 1,0 coordinates respectively). The value of τ may be determined from a *single* Brønsted experiment either from β_{Lg} or from β_{Nuc} provided β_{eq} is known so that the Leffler α_{Lg} (or α_{Nuc}) can be calculated (Figure 9) (Equation 25)[j] (α_{Nuc} and α_{Lg} have opposite signs).

$$\tau = 2\alpha_{Nuc} = 2(1 + \alpha_{Lg}) \tag{25}$$

The Marcus Equations (26) and (27) for forward and reverse reactions can be used to derive the relationships between β_f, β_r, and β_{eq} and β_{ii} (Equations 28 and 29)[22,k] where pK_a is that of the conjugate acid Lg_iH and ω is a work term (p 126).

$$\Delta G^{\ddagger}_{ij} = \omega^R + \gamma + \Delta G^{\circ}_{ij}/2 + \Delta G^{\circ}_{ij}{}^2/16\gamma \text{ (forward reaction)} \tag{26}$$

$$\gamma = 0.5(\Delta G^{\ddagger}_{ii} + \Delta G^{\ddagger}_{jj}) \tag{27}$$

$$\beta_{eq} = \beta_f - \beta_r \tag{28}$$

$$\beta_{ii} = \partial \log k_{ii}/\partial pK_a = \beta_f + \beta_r \tag{29}$$

The coefficients β_f and β_r correspond to β_{Nuc} and β_{Lg} in a nucleophilic displacement reaction and their respective dependence on pK_{Lg} and

---

[j] See Problem 2 for the derivation of Equation (25).
[k] See Problem 1 for the derivation of Equations (28) and (29) from Equations (26) and (27) and Chapter 6.

pK_{Nuc} (p_{xy}) can be determined readily. The slope (β_{ii}) is related to β_f (β_{Nuc}) and β_r (β_{Lg}) by Equation (29). The values of β_{Lg} and β_{Nuc} can vary linearly with pK_{Nuc} and pK_{Lg} respectively (Equations 30 and 31) and combining with Equation (29), and integration, yields Equation (32) and Equation (33), the Lewis–Kreevoy correlation.[23]

$$\beta_{\mathrm{Lg}} = p_{xy}\, pK_{\mathrm{Nuc}} + L \tag{30}$$

$$\beta_{\mathrm{Nuc}} = p_{xy}\, pK_{\mathrm{Lg}} + N \tag{31}$$

$$\beta_{ii} = 2p_{xy}\, pK_a + (L + N) = \partial \log k_{ii}/\partial pK_a \tag{32}$$

$$\log k_{ii} = p_{xy}\, pK_a^{\,2} + (L + N)pK_a + C \tag{33}$$

The parameter p_{xy} determined experimentally from Equation (30) is usually not identical to that from Equation (31), because of the small number of data points generally available and the intervention of microscopic medium effects (Chapter 6). Data for equations (30) and (31) can be combined to solve for p_{xy}, L and N by a *global* equation (Appendix 1, Section A1.1.4.3).

Equations (30) and (31) for the reactions of phenolate ions with phenyl acetates have parameters $p_{xy} = 0.16$, $N = -0.21$ and $L = -1.91$ and substituting in Equation (33) gives a parabola (Figure 10) which only requires vertical adjustment (C) to fit the experimental data for $\log k_{ii}$. The More O'Ferrall–Jencks diagram (Figure 11) describes the identity transition structure as a function of increasing basicity of the nucleophile (and leaving group). Transition structures for increasing nucleophile basicity are at the intersection of reaction coordinates and the tightness

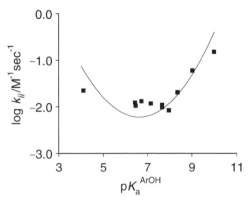

Figure 10 *Lewis–Kreevoy correlation for* k_{ii} *(the identity rate constant) for the phenolysis of substituted phenyl acetates;[24] the line is a parabola drawn using Equation (33)*

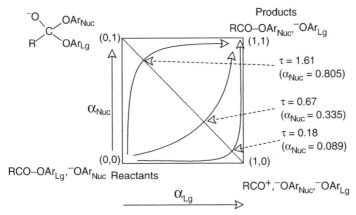

Figure 11 *More O'Ferrall–Jencks diagram for the identity reactions of substituted phenolate ions with substituted phenyl esters (data from reference 24). See text for the identities of the three identity reactions*

diagonal: $t = 0.67$ for 2,4-dinitrophenyl acetate; $t = 1.61$ for phenyl acetate and $\tau = 0.18$ for 2,4-dinitrophenyl–4-methoxy-2,6-dimethylbenzoate.

The cross-interaction coefficient, p_{xy}, cannot be strictly invariant over a range of substituents because $p_{xy} = -1/(6\Delta G^{\ddagger}_{o})$.[25] Over the relatively small ranges of substituent variation for reactions so far investigated the value of p_{xy} appears to be constant.

5.5 FURTHER READING

W.J. Albery and M.M. Kreevoy, Methyl Transfer Reactions, *Adv. Phys. Org. Chem.*, 1978, **16**, 87.

J.P. Guthrie, Correlation and Prediction of Rate Constants for Organic Reactions, *Can. J. Chem.*, 1996, **74**, 1283.

D.A. Jencks and W.P. Jencks, On the Characterisation of Transition States by Structure–Reactivity Coefficients, *J. Am. Chem. Soc.*, 1977, **99**, 7948.

W.P. Jencks, When is an Intermediate Not an Intermediate? Enforced Mechanisms of General Acid–Base Catalysed, Carbocation, Carbanion, and Ligand Exchange Reactions, *Accs. Chem. Res.*, 1980, **13**, 161.

W.P. Jencks, A Primer for the Bema Hapothle. An Empirical Approach to the Characterisation of Changing Transition State Structures, *Chem. Rev.*, 1985, **85**, 511.

W.P. Jencks, Are Structure–Reactivity Correlations Useful? *Bull. Soc. Chim. de France*, 1988, 219.

I. Lee, Characterisation of Transition States for Reactions in Solution by Cross-Interaction Constants, *Chem. Soc. Rev.*, 1990, **19**, 317.

I. Lee, Cross-Interaction Constants and Transition State Structure, *Adv. Phys. Org. Chem.*, 1992, **27**, 57.

E.S. Lewis, Rate-Equilibrium LFER Characterisation of Transition States. The Interpretation of α, *J. Phys. Org. Chem.*, 1990, **3**, 1.
A. Williams, Bonding in Phosphoryl ($-PO_3^{2-}$) and Sulphuryl ($-SO_3^-$) Group Transfer between Nitrogen Nucleophiles as Determined from Rate Constants for Identity Reactions, *J. Am. Chem. Soc.*, 1985, **107**, 6335.

5.6 PROBLEMS

1   Derive Equations (28) and (29) using the Marcus Equations (26) and (27).
2   Derive Equation (25) using Equations (28) and (29) $\alpha_{Nuc} = \beta_{Nuc}/\beta_{eq}$ and $\alpha_{Lg} = \beta_{Lg}/\beta_{eq}$ where $\beta_{Lg} = \beta_r$ and $\beta_N = \beta_f$. Also deduce that $\delta = \beta_{ii}/\beta_{eq}$.
3   Calculate the identity rate constant, k_{ii}, for the reaction of 4-nitrophenolate ion with 4-nitrophenyl acetate given that the Brønsted equation for the attack of substituted phenolate ions on the ester is:

$$\log k_{ArO^-} = 0.75 pK_a^{ArOH} - 7.28$$

(pK_a of 4-nitrophenol = 7.14).
4   The reaction of substituted phenolate ions with substituted phenoxytriazines has the Brønsted data given in Table 2.

Table 2

pK_{Nuc}	β_{Lg}	pK_{Lg}	β_{Nuc}
7.80	−0.534	7.14	0.951
8.32	−0.537	7.75	0.860
8.87	−0.527	8.19	0.910
7.66	−0.503	8.99	0.936
9.81	−0.405	5.28	0.708
		6.80	0.768
		6.68	0.820
		7.66	0.830

Use a least squares program such as that derived from statistical equations in Appendix 1, Section A1.1.4.3) to fit the data to Equations (30) and (31) and deduce β_{eq}.
5   Estimate β_{eq} for nucleophilic attack of substituted phenoxide ions on diphenylphosphate esters of phenols (($PhO)_2PO-OAr$) given that the following equations hold:

$$\beta_{Lg} = 0.115\, pK_{Nuc} - 1.63 \text{ and } \beta_{Nuc} = 0.115\, pK_{Lg} - 0.38$$

6 Use Equation (32) to derive the following equation for an identity reaction:

$$d\tau/dpK_a = 2p_{xy}/\beta_{eq}$$

7 Use Equation (32) and $\tau - 1 = \beta_{ii}/\beta_{eq}$ to deduce:

$$\tau = 1 + 2(p_{xy}pK_a + B)/\beta_{eq}$$

8 By fitting the data of Table 3[26] to the binomial expression

$$\log k = -0.5p_x pK_a^2 + \beta_o pK_a + C$$

estimate p_x for tritium abstraction from fluorene derivatives. (The data of Table 3 refer to detritiation in 95%D$_2$O–5%Me$_2$SO at 28°C). Use curve-fitting software or a program from the statistical equations in the Appendix 1, Section A1.1.4. The data should be corrected statistically.

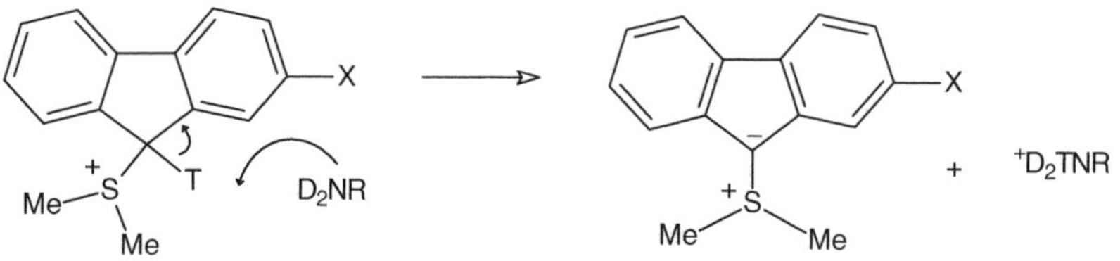

(34)

Table 3

Primary amine	pK_a^{RNH3}	$\log k_B/M^{-1}sec^{-1}$
D$_2$NC(O)NDND$_2$	4.36	−2.678
CH$_3$OND$_2$	5.22	−2.000
NCCH$_2$ND$_2$	6.1	−1.201
CF$_3$CH$_2$ND$_2$	6.43	−1.108
NCCH$_2$CH$_2$ND$_2$	8.640	0.733
ClCH$_2$CH$_2$ND$_2$	9.36	1.114
DOCH$_2$CH$_2$ND$_2$	10.45	1.813
CH$_3$CH$_2$ND$_2$	11.6	2.362

9 Estimate p_{xy} for the reactions of substituted anilines (YC$_6$H$_4$NH$_2$) with substituted benzenesulphonyl chlorides (XC$_6$H$_4$SO$_2$Cl) using Hammett ρ values given in Table 4.[27]

Table 4

Substituent Y	σ_Y	ρ_X
3,4-Me$_2$	−0.24	1.14
4-Me	−0.17	1.11
3-Me	−0.07	1.02
Parent	0.0	1.01
4-Cl	0.23	0.78
3-Cl	0.37	0.71
3-NO$_2$	0.72	0.44

10 The simplest variation of ρ as a function of σ is the linear dependence[28]

$$\rho = \rho_o + 2m\sigma$$

Use the relationship to derive the quadratic equation:

$$\log k_X/k_H = \rho_o\sigma + m\sigma^2$$

5.7 REFERENCES

1. S.I. Miller, Multiple Variation in Structure–Reactivity Correlations, *J. Am. Chem. Soc.*, 1959, **81**, 101.
2. D.A. Jencks and W.P. Jencks, On the Characterisation of Transition States by Structure–Reactivity Coefficients, *J. Am. Chem. Soc.*, 1977, **99**, 7948.
3. I. Lee, Cross-interaction Constants and Transition State Structure, *Adv. Phys. Org. Chem.*, 1992, **27**, 57.
4. E.H. Cordes and W.P. Jencks, General Acid Catalysis of Semicarbazone Formation, *J. Am. Chem. Soc.*, 1962, **84**, 4319; L. do Amaral *et al.*, Some Aspects of Mechanism and Catalysis for Carbonyl Addition Reactions, *J. Am. Chem. Soc.*, 1966, **88**, 2225.
5. J. Shorter, *Correlation Analysis of Organic Reactivity*, Research Studies Press, Chichester, 1982, Chapter 2.
6. W.P. Jencks, A Primer for the Bema Hapothle. An Empirical Approach to the Characterisation of Changing Transition State Structures, *Chem. Rev.*, 1985, **85**, 511.
7. E.R. Thornton, A Simple Theory for Predicting the Effects of Substituent Changes on Transition-State Geometry, *J. Am. Chem. Soc.*, 1967, **89**, 2915.
8. J.P. Guthrie, Concerted Mechanism for Alcoholysis of Esters: An Examination of the Requirements, *J. Am. Chem. Soc.*, 1991, **113**, 3941.

9. J.P. Guthrie, Concertedness and E2 Elimination Reactions: Prediction of Transition State Position using Two-Dimensional Reaction Surfaces based on Quadratic and Quartic Approximations, *Can. J. Chem.*, 1990, **68**, 1643.

10. B.M. Dunn, Pathways of Proton Transfer in Acetal Hydrolysis, *Int. J. Chem. Kinetics*, 1974, **6**, 143.

11. W.J. le Noble *et al.*, A Simple, Empirical Function Describing the Reaction Profile, and Some Applications, *J. Org. Chem.*, 1977, **42**, 338.

12. H.F. Gilbert and W.P. Jencks, Mechanisms for Enforced General Acid Catalysis of the Addition of Thiol Anions to Acetaldehyde, *J. Am. Chem. Soc.*, 1977, **99**, 7931.

13. M.T. Skoog and W.P. Jencks, Reactions of Pyridines and Primary Amines with N-Phosphorylated Pyridines, *J. Am. Chem. Soc.*, 1984, **106**, 7597.

14. J.R. Gandler and W.P. Jencks, General Base Catalysis, Structure–Reactivity Interactions, and Merging of Mechanisms for Elimination Reactions of (2-Arylethyl)quinuclidinium Ions, *J. Am. Chem. Soc.*, 1982, **104**, 1937.

15. I. Lee, Nucleophilic Substitution Reactions of Benzyl Benzenesulfonates with Anilines in MeOH-MeCN Mixtures-I. Effects of Solvent and Substituent on the Transition-state Structure, *Tetrahedron*, 1985, **41**, 2635.

16. I. Lee, Nucleophilic Substitution Reactions of 1-Phenylethyl Benzenesulfonates with Anilines in Methanol-Acetonitrile, *J. Org. Chem.*, 1988, **53**, 2678.

17. I. Lee, Characterisation of Transition States for Reactions in Solution by Cross-Interaction Constants, *Chem. Soc. Rev.*, 1990, **19**, 317.

18. A. Williams, Bonding in Phosphoryl ($-PO_3^{2-}$) and Sulphuryl ($-SO_3^-$) Group Transfer between Nitrogen Nucleophiles as Determined from Rate Constants for Identity Reactions, *J. Am. Chem. Soc.*, 1985, **107**, 6335.

19. M.M. Kreevoy and I.S.H. Lee, Marcus Theory of a Perpendicular Effect on α for Hydride Transfer between NAD^+ Analogues, *J. Am. Chem. Soc.*, 1984, **106**, 2550.

20. W.J. Albery and M.M. Kreevoy, Methyl Transfer Reactions, *Adv. Phys. Org. Chem.*, 1978, **16**, 87.

21. E. Grunwald, Structure-Energy Relations, Reaction Mechanisms, and Disoparity of Progress of Concerted Reaction Events, *J. Am. Chem. Soc.*, 1985, **107**, 125.

22. E.S. Lewis and D.D. Hu, Methyl Transfers. 8. The Marcus Equation and Transfers between Arenesulphonates, *J. Am. Chem. Soc.*, 1984, **106**, 3292.

23. S.A. Ba-Saif *et al.*, Concerted Acetyl Group Transfer between Substituted Phenolate Ion Nucleophiles: Variation of Transition State Structure as a Function of Substituent, *J. Am. Chem. Soc.*, 1989, **111**, 2647.

24. C.J. Murray and W.P. Jencks, Proton Abstraction from Dimethyl(2-substituted-9-fluorenyl)sulphonium Ions. Evidence for Changes in Transition State Structure, *J. Am. Chem. Soc.*, 1990, **112**, 1880.

25. O. Rogne, The Kinetics and Mechanism of the Reactions of Aromatic Sulphonyl Chlorides with Anilines in Methanol: Brønsted and Hammett Correlations, *J. Chem. Soc. B*, 1971, 1855.

26. M. O'Brien and R.A. More O'Ferrall, Application of a Quadratic Free Energy Relationship to Non-Additive Substituent Effects, *J. Chem. Soc., Perkin Trans. 2*, 1978, 1045.

CHAPTER 6

Anomalies, Special Cases and Non-linearity

6.1 FREE ENERGY RELATIONSHIPS ARE NOT ALWAYS LINEAR

Even when they were discovered by Brønsted some 80 years ago[1] it was predicted that linear free energy relationships would become non-linear in extremes of reactivity. This was confirmed experimentally by Eigen[2] for proton transfer from donor acid (AH) to acceptor base (B) over a range of pK_a values. The simplest types of non-linear free energy relationship are due to a change in mechanism or rate-limiting step leading to intersecting linear plots (Chapter 7). Such non-linear correlations conform to equations where the rate constant at the break point is either *twice* or *half* that of the intersection point depending on whether there is a change respectively in mechanism or rate-limiting step. A changing transition structure brought about by the effect of the substituent on the system also induces curvature in free energy relationships.

Free energy relationships are predominantly linear, even slightly curved plots are rare, and there are several cases where the free energy relationships are linear over *very* large ranges of rate constant.[3] The existence of non-systematic scatter of data points[4] (see Section 6.4) can make it difficult to demonstrate curvature (or its absence) even when the data points are for structurally similar reagents.

In principle, free energy relationships should be curved even when there is no change in mechanism or in rate-limiting step; this is due to a change in shape of the energy surface at the saddle-point or transition structure. Curvature is often very small because only small energy changes are involved in changing a substituent. The lack of curvature in cases where there is a large change in energy can be traced to compensatory factors.

Curvature is most reliably detected in proton transfer reactions because of the minimal microscopic medium effects (see later) for these reactions. Proton transfer between electronegative atoms A and B usually involves diffusion steps (k_1 and k_3) and a chemical proton transfer (k_2) within the encounter complex (Equation 1 and Figure 1).

$$AH + B \underset{k_{-1}}{\overset{k_1}{\rightleftarrows}} [AH.B] \underset{k_{-2}}{\overset{k_2}{\rightleftarrows}} [A^-.HB^+] \underset{k_{-3}}{\overset{k_3}{\rightleftarrows}} A^- + HB^+ \tag{1}$$

Equation (1) gives rise to the rate laws (Equation 2) for forward (k_f) and reverse (k_r) reactions ($D = k_{-1}(k_{-2} + k_3) + k_2k_3$, see Problem 8 for a derivation):

$$k_f = k_1k_2k_3/D \quad \text{and} \quad k_r = k_{-1}k_{-2}k_{-3}/D \tag{2}$$

When a rate constant for proton transfer within the encounter complex is larger than that of complex formation or decomposition (*i.e.* $k_2 > k_{-1}$ or $k_{-2} > k_3$) Equations (2) reduce to Equations (3):

$$k_f = k_1/(1 + K_2k_{-1}/k_3) \quad \text{and} \quad k_r = k_{-3}/(1 + k_3/k_{-1}K_2) \tag{3}$$

where $K_2 = k_{-2}/k_2$. A non-linear Brønsted dependence is predicted (the "Eigen" plot, curve A in Figure 2) resulting from two intersecting straight lines and a break-point at $K_2 = 1$ consistent with a change in rate-limiting step at the pK_a of HB^+ equal to that of AH ($\Delta pK_a = 0$, $pK_a^{HB} = pK_a^{AH}$).

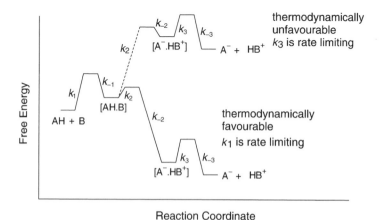

Figure 7 *Proton transfer from nitroalkanes to base. See text for details of the structure at (1,0).*

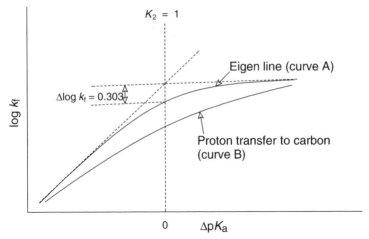

Figure 2 *Brønsted relationship for proton transfer between acids and bases. Curve A is for proton transfer between electronegative heteroatoms.[a] Curve B is for proton transfer between heteroatoms and carbon*

The rate-limiting step in proton transfer between electronegative atoms is either k_1, the diffusion-controlled encounter of the acid–base pair for thermodynamically favourable transfers or k_3, the diffusion-controlled dissociation of the acid–base pair, for thermodynamically unfavourable transfers (Figure 1). The rate-limiting step is rarely the proton transfer within the encounter complex. Eigen plots are predominantly observed in reactions in which proton transfer occurs to or from heteroatoms in reactive intermediates.

6.1.1 Marcus Curvature[b]

Proton transfer reactions involving carbon do not usually exhibit sharp breaks in the Brønsted plot in the region of $\Delta pK = 0$. Curve B of Figure 2 does *not* have a sharp break between the linear segments although the mechanism follows Equation (1). The reason for this is that the k_{-2} and k_2 steps within the encounter complex are small compared with that for diffusion (the k_{-1} and k_3 steps). Curvature results from the reaction which occurs *within* the encounter complex (the k_2 step in Equation 1) and the free energy diagram for this result is illustrated in Figure 3.

---

[a] The Eigen equations are: $\log k_f = \log k_1 - \log(1+10^{\{pK(AH) - pK(HB)\}})$ and
$\log k_r = \log k_{-3} - \log(1+10^{\{pK(HB) - pK(AH)\}})$ where k_{-1} is assumed to equal k_3.
[b] An excellent and up-to-date treatment of Marcus theory is given in Chapters 8 and 9 of E.F. Caldin, *The Mechanisms of Fast Reactions in Solution*, IOS Press, Amsterdam, 2001.

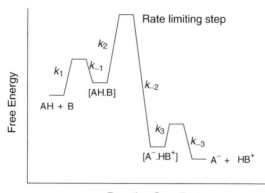

Figure 3   *A proton transfer reaction between a heteroatom (B) and a carbon acid (AH). The rate-limiting step can be the proton transfer (k₂) in contrast with the case in Figure (1)*

Curvature can result from non-linearity of the change in the shape of potential energy surfaces in the region where they intersect. A reasonable assumption (over a small range of pK_a variation) is that the two intersecting curves are parabolic[5] and that the change in entropy within the series is constant. The equations for free energy may therefore be written for the reactant parabola (Equation 4) and product parabola (Equation 5).[c]

$$y = ax^2 \ (= \Delta G) \tag{4}$$

$$y = a(x - b)^2 - c \tag{5}$$

The transition structure resides at the intersecting point of the two parabolas where $y = \Delta G^{\ddagger}$ (see Figure 4), and equations (4) and (5) can be solved for x by combining them to give Equation (6).

$$ax^2 = a(x - b)^2 - c \tag{6}$$

Thus

$$x = (ab^2 - c)/2ba \tag{7}$$

Substituting in Equation (4) gives y at the saddle-point ($\Delta G^{\ddagger}$, Equation 8):

---

[c]   This approach is a mathematical alternative to the qualitative arguments given in Chapter 1 and in Appendix 1 (Sections A1.1.2 and A1.1.3) for the deduction of Class I free energy relationships and the variation of the selectivity coefficient, a, with ΔG_o.

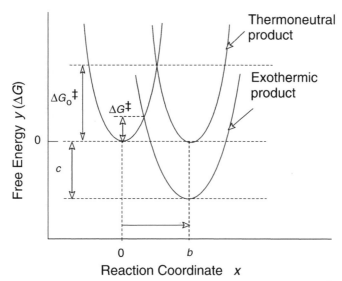

Figure 4 *Reaction coordinate considered as system of intersecting parabolas*[d]

$$y = \Delta G^{\ddagger} = a(ab^2 - c)^2/4b^2a^2$$

$$= (ab^2 - c)^2/4b^2a \quad (8)$$

When the reaction is thermoneutral $c = 0$ and $\Delta G^{\ddagger} = \Delta G_o^{\ddagger}$, the corresponding transition state energy is given by Equation (9).

$$\Delta G_o^{\ddagger} = (ab^2)^2/4b^2a = ab^2/4 = \lambda/4 \quad (9)$$

Since $c = -\Delta G_o$, substituting the simplifying $\lambda = ab^2$ into Equation (8) gives Equation (10) which is the Marcus equation[6] (for reaction within the encounter complex).

$$\Delta G^{\ddagger} = (\lambda + \Delta G_o)^2/4\lambda = \lambda(1 + \Delta G_o/\lambda)^2/4 \quad (10)$$

The Leffler α may be obtained by differentiation of Equation (10) (Equation 11).

$$\alpha = \partial\Delta G^{\ddagger}/\partial\Delta G_o = 2[\lambda(1 + \Delta G_o/\lambda)/4](1/\lambda)$$

$$= 0.5(1 + \Delta G_o/\lambda) = 0.5(1 + \Delta G_o/4\Delta G_o^{\ddagger}) \quad (11)$$

---

[d] The obvious disadvantage of this model is that the cross-over point is sharp and cannot have a smooth surface as required at the transition structure. Nevertheless the model is successful in predicting α values. Other equations model the shape of the energy surface at the transition structure (Chapter 5).

Equation (11) indicates that the slope of the free energy relationship will vary as a natural consequence of the intersecting free energy curves and varying ΔG_o.[e]

The component diffusion processes in the simple reaction (Equation 12) (k_1 and k_3 in Equation 1) can be described by *work terms* (ω) which represent the energies required to bring reactants together and to separate the products (Equation 13).[f]

$$A \ + \ B{-}C \ \rightleftharpoons \ A{-}B \ + \ C \qquad (12)$$

$$A + B{-}C \ \underset{\omega_R}{\rightleftharpoons} \ A.B{-}C \ \xrightarrow{\Delta G^{\ddagger}} \ \overset{\text{TS}}{\underset{}{\rightleftharpoons}} \ C.A{-}B \ \underset{\omega_P}{\rightleftharpoons} \ A{-}B + C \qquad (13)$$

$$\xrightarrow{\hspace{4cm}}$$
$$\Delta G^{\ddagger}_{\text{observed}}$$

Expanding equation (10) gives Equation (14) and inclusion of work terms (ω) yields Equation (15) from which the parameters may be obtained experimentally.

$$\Delta G^{\ddagger} = \Delta G_o^{\ddagger} + \Delta G_o/2 + (\Delta G_o)^2/16\Delta G_o^{\ddagger} \qquad (14)$$

$$\Delta G^{\ddagger}_{\text{observed}} = (\Delta G_o^{\ddagger} + \omega_R) + (\Delta G_o + \omega_R - \omega_P)/2$$
$$+ (\Delta G_o + \omega_R - \omega_P)^2/16\Delta G_o^{\ddagger} \qquad (15)$$

Differentiating Equation (11) yields Equation (16) which indicates that the transition structure changes as a function of the free energy of the reaction (ΔG_o).

$$\partial\alpha/\partial\Delta G^0 = (8\Delta G_0^{\ddagger})^{-1} \qquad (16)$$

The intersecting parabolas model is consistent with the Hammond postulate[7] that *two states (such as reactant and transition structures) occurring consecutively in a reaction and having nearly the same energy content will involve interconversions with only a small reorganisation of the molecular structure.* Consideration of Equation (11) reveals that for a

[e] Qualitatively the slope, α, changes as ΔG_o varies, even though the shapes of the parabolas are assumed to be invariant, because the cross-over point has slopes which alter (see Appendix 1, Section A1.1.3). The Marcus equation, although simple in concept, is remarkably successful, probably because parabolas are good approximations to the shapes of energy surfaces over small regions of space.

[f] This equation is analogous to Equation (1).

thermoneutral reaction ($\Delta G_o = 0$) the Leffler slope, α, equals 0.5. The Equation predicts a Leffler α less than 0.5 for an exothermic reaction ($\Delta G_o < 0$) and greater than 0.5 for an endothermic reaction. Figure 4 shows that shifting the location of the product parabola up or down moves the intersection point (the transition structure) between reactant and product states (see also Appendix 1, Section A1.1.3 and Figure A3c).

The adoption of a model for the free energy surface derived from two intersecting parabolas gives rise to exact equations which have predictive power as indicated above. The model is a good approximation but the obvious disadvantages are that the intersection point does not correspond exactly to the energy of the transition structure. Bell suggested that the energy is overestimated by a small but essentially constant factor, owing to resonance between the states represented by the two potential energy curves.[8] This is not a significant problem as most studies are concerned with *changes* in energy. A more serious assumption is that the "*a*" values (Equations 4 and 5) are the same for both parabolas. Any mathematical model possessing curvature with a positive differential ($\partial^2 y/\partial x^2$) will predict a variable Leffler parameter as shown in the case of intersecting Morse curves described in Chapter 1. The mathematics of these models (some of which are discussed in more detail in Chapter 5) are much more complicated than those of the model with intersecting parabolas.

6.2 THE REACTIVITY–SELECTIVITY POSTULATE

The Hammond postulate is often accepted as a general principle; an increase in reactivity is accompanied by a decrease in selectivity because the transition structure becomes closer to that of the reactant state as the energy barrier decreases. This idea has some truth for a hypothetical A to B reaction model where it is implicit that only a single bond change occurs; moreover the Hammond postulate is predicted by the Marcus theory (above). The postulate often breaks down for reactions where more than one major bond change results in product formation. It should be emphasised that any discussion of the reactivity–selectivity relationship has to be confined to those reactions where there is no change in rate-limiting step or mechanism.

The effect of variation in structure for example changing, the leaving group in the series fluoride, chloride, bromide, iodide and azide ion, has been used extensively in mechanistic studies but results should be viewed only in a confirmatory sense[9] because the essentially gross structural change could cause equally gross mechanistic changes. The selectivity of the reaction between 2-substituted pyridines and methyl iodide appears to follow reactivity but the Brønsted β_{nuc} for reaction of 3- and 4-substituted pyridines does not change over a range of 10^8-fold in reaction rates.[4] Changing the structure at the reaction site by small increments such as by variation of substituent in aryl esters or in pyridine nucleophiles should not cause significant changes in mechanism and the effects should therefore be applicable to diagnosis.

The manifest linearity of the majority of free energy plots is paradoxical because the constant slope would indicate that the transition structure is substituent independent.[10] Variation in structure might be expected to occur due to the change in substituent and thus yield curved relationships as indicated in Section 6.1. The relationship between reactivity and selectivity[11] is based on a very simple model, to which most reactions do not conform because they involve not only at least two major bonding changes but solvation changes as well.

The rate-limiting steps of many reactions are composed of several major, fundamental, bonding changes and hence it is not surprising that the transition structure is dependent on several variables. The elimination reaction (Equation 17) provides a useful example.

$$B \;+\; H{-}\underset{(2)}{C}{-}\underset{(1)}{C}{-}Lg \;\longrightarrow\; BH^+ \;+\; Lg^- \;+\; \bigg\rangle\!\!=\!\!\bigg\langle \qquad (17)$$

The 1,2-elimination involves C–H bond fission and proton transfer to a base, C–Lg bond fission (where Lg^- is the leaving group) and C=C bond formation. These steps can occur separately involving the formation of carbenium ion intermediate (E1), carbanion intermediate (E1cb) or a concerted process (E2).[12] These processes may be illustrated by a More O'Ferrall–Jencks diagram (Figure 5).[13] Appendix 1, Section A1.2, contains a description of More O'Ferrall–Jencks diagrams. The two fundamental, major bond fission steps are represented by axes drawn orthogonal to each other so that the reactant structure is placed in the bottom left-hand corner (0,0), the product in the top right-hand corner (1,1) and the putative reactive intermediates are placed in the remaining opposite corners (0,1 and 1,0 in Figure 5).[g]

The degree of proton transfer to the base in the transition structure is recorded as the Brønsted β-value obtained by varying the structure of

---

[g] In this illustration the 0,1 and 1,0 structures correspond to, albeit unstable, molecules. In some cases the structures are so unstable that they are not able to exist as discrete entities (see Figure 7 and pp. 54 and 130 of A. Williams, *Concerted Organic and Bio-Organic Mechanisms*, CRC Press, Boca Raton, FL, 2000.

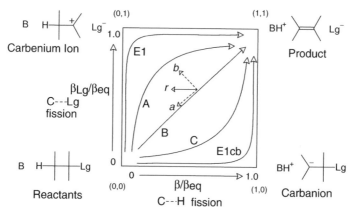

Figure 5 *More O'Ferrall–Jencks diagram for the elimination reaction (Equation 17). A, E1-like E2 mechanism; B, synchronous E2 mechanism; C, E1cb-like E2 mechanism. The vector (a) corresponds to a* Hammond *(parallel) effect and (b) to a* Thornton *(perpendicular) effect*

the base. The degree of bond fission to the leaving group in the transition structure is registered by a Leffler coefficient α_{Lg} (from the Brønsted β_{Lg} obtained by changing the nature of the nucleofuge, Lg). Provided that the selectivity coefficients are normalised, changes in transition structure are often predictable by the Hammond postulate for the two step-wise processes E1 and E1cb, along the edges of the diagram. However, for the *concerted* E2 mechanism changes in transition structure and selectivity parameters are more difficult to predict.

The Hammond postulate predicts that stabilisation of the reactants or products moves the transition structure *away* from the corner which is becoming more stable – an effect which is *parallel to the reaction coordinate* (see Appendix 1, Figure A3c).[14,15] However, stabilisation of the 0,1 or 1,0 corner structures moves the transition structure *towards* the corner becoming more stable. The direction of the movement is perpendicular to the reaction coordinate and is called a *perpendicular* effect (sometimes called a *Thornton*[14] or an *anti-Hammond* effect) (see Appendix 1, Figure A3b).

Prediction of the transition structure variation are embodied in Thornton's "rules"[14,15] which resolve the change into components along and perpendicular to the reaction coordinate. These respond to changes in energy of structures at the corners of the reaction surface.

When structural changes are predicted to cause both *parallel* (vector "*a*") and *perpendicular* (vector "*b*") effects, the net change in transition structure is the resultant (*r*) of the two vectors[16] (Figure 5). Increasing the stability of the leaving group, Lg⁻, would tend to make the transition structure for the almost synchronous E2 process have less C–Lg bond fission by the parallel effect (vector "*a*") but more by the perpendicular effect (vector "*b*"). The resultant (*r*) could indicate no change in the observed β_{Lg}.

There are some classical systems where the reactivity–selectivity postulate does not seem to apply.[3] The Hammett dependence (ρ_{Ar}) for alkaline hydrolysis of phenyl esters of substituted benzoic acids is independent of the stability of the phenolate ion leaving group (Ar'O⁻).[17] The Hammett ($\rho_{Ar'}$) for the alkaline hydrolysis of substituted phenyl esters of benzoic acids is independent of the substituent on the benzoyl group. These data are consistent with a transition structure located on the edge of a More O'Ferrall–Jencks diagram (Figure 6) where changing the stability of Ar'O⁻ will have little effect on the transition structure parallel to the y coordinate. Variation of the substituent in Ar will not substantially change the horizontal coordinate of the transition structure.

6.3 BORDWELL'S ANOMALY

The equilibrium constant for a proton transfer reaction from a series of proton donors (SH) to base (B) (Equation 18) has a β_{eq} value of unity for the Brønsted dependence on the pK_a of SH (pK_a^{SH}).

$$SH + B \underset{k_r}{\overset{k_f}{\rightleftharpoons}} S^- + BH^+ \qquad (18)$$

The equilibrium constant is given by Equation (19):

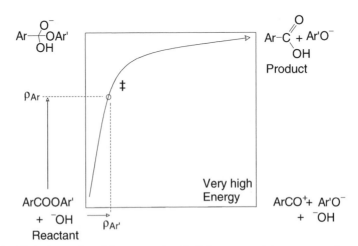

Figure 6   *Hydrolysis of aryl benzoates by alkali has a transition structure located at an edge of the reaction surface*

$$K_{eq} = [BH^+][S^-]/[SH][B] = K_a^{SH}/K_a^{BH} \tag{19}$$

where K_a^{SH} is the dissociation constant of the varying proton donor and K_a^{BH} is the dissociation constant of BH^+. Thus $\log K_{eq} = \log K_a^{SH} - \log K_a^{BH}$ and it therefore follows that $\beta_{eq} = \partial \log K_{eq}/\partial pK_a^{SH} = \partial \log K_a^{SH}/\partial pK_a^{SH} = -1$. Reference to Chapter 2 (Section 2.4) indicates that β_f and β_r are governed by Equation (20) which is general and applies to coefficients of other free energy relationships.

$$\beta_{eq} = \beta_f - \beta_r \tag{20}$$

Thus in the Hammett format $\rho_{eq} = \rho_f - \rho_r$ and in the normalised (Leffler) format: $\alpha_{eq} = \alpha_f - \alpha_r$. The parameter α_{eq} is unity[h] and α_f and α_r should both therefore be numerically less than 1 because of the negative sign of α_r.

Bordwell and his co-workers discovered that the formation of the pseudo-base from nitromethane in hydroxide ion solutions (Equation 21) has a ρ_f of 1.28 and a ρ_{eq} of 0.83 giving rise to an α_f^i of 1.54.[18] Similarly the α_f values for reaction of bases with other nitroalkanes are greater than unity also violating the Leffler assumption.

$$Ar{-}CH_2{-}NO_2 + HO^- \underset{\alpha_{eq} = 1.00 \text{ (by definition)}}{\overset{\alpha_f = 1.54}{\rightleftharpoons}} Ar{-}CH{=}NO_2^- + H_2O \tag{21}$$

The value of α_f indicates that the transition structure has a negative charge *greater* than that of the product. The pseudo-base formation reaction from nitroalkanes comprises C–H bond fission and changes in delocalisation (Figure 7). When these processes have made unequal progress at the transition structure the reaction is considered to be *imbalanced* and this is the cause of the apparent anomaly. Pseudo-base formation from nitroalkanes may be represented as in Figure 7 by consideration of carbanion formation (y axis) and the geometrical rearrangement from tetrahedral to trigonal carbon to allow delocalisation (x axis). If delocalisation were to lag behind proton transfer the carbon would bear greater negative charge in the transition structure than in the product giving rise to $\alpha_f > 1$.

There are two important points to be made about Figure 7: (1) The structure at (0,1) is a carbanion and if the value of β_{eq} for its formation (without rehybridisation) were known then a good assessment of the progress (at the transition structure) along the vertical axis would be possible. Values of β_{eq} can be determined from reference reactions not involving such a complication. (2) The structure at (1,0) has

---

[h] Because it is the slope of a plot of $\log K_{eq}$ versus $\log K_{eq}$.
[i] The value of α_f is given by ρ_f/ρ_{eq}.

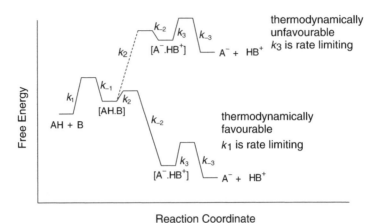

Figure 7 *Proton transfer from nitroalkanes to base. See text for details of the structure at (1,0).*

pentavalent carbon and should therefore have a very high energy distorting the reaction surface, forcing the reaction pathway to go near (0,1). If justification is needed for the inclusion of such a maverick, high-energy structure, then the reader could consult Olah's work on onium ion structures such as CH_5^+.[19]

The study of substituent effects on bond formation and bond fission in non-proton transfer reactions requires knowledge of the value of β_{eq}, which is unity only for proton transfer equilibria. The Brønsted β_{Lg} exceeds β_{eq} in some non-proton transfer reactions and it is often assumed that the effect is due to bond fission's being in advance of solvent reorganisation. If such were the case in, for example, departure of aryl oxide ions it is conceivable that the lack of solvation would cause the charge on the oxygen to be greater than in its solvated product state; the solvation process would form one of the coordinates of the More O'Ferrall–Jencks diagram with bond fission comprising the other.

6.4 MICROSCOPIC MEDIUM EFFECTS AND DEVIANT POINTS

Free energy correlations often exhibit scatter plots where the deviations from a linear regression fall outside experimental error. These deviations may be attributable to differences in microscopic environment between the standard equilibrium and the reaction being studied and are called *microscopic medium* effects.

The reaction of 3- and 4-substituted pyridines with an *N*-phosphoryl-isoquinolinium ion (isq–PO_3^{-2}, Equation 22) is a good example of the influence of such *microscopic medium* effects.

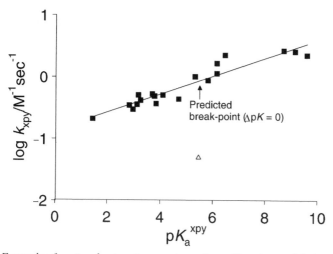

$$\text{(22)}$$

The reactivity is compared with the equilibrium constant for dissociation of pyridinium molecules under the same conditions[20,21] and the Brønsted-type plot (Figure 8) illustrates the extent of scatter usually obtained in such relationships.

The cause of the scatter is the non-systematic influence of the substituent on the microscopic environment of the transition structure. The linear free energy relationship between product state XpyH$^+$ (Equation 22) and the transition structure (Xpy ... PO$_3^{2-}$ ... isq) will be modulated by second-order non-systematic variation because the microscopic environment of the reaction centre in the standard (XpyH$^+$) differs slightly from that (Xpy–PO$_3^{2-}$) in the reaction under investigation giving rise to specific substituent effects. These effects are mostly small. An unusually dramatic intervention of the microscopic medium effect may be found in Myron Bender's extremely scattered Hammett dependence of the reaction of cyclodextrins with substituted phenyl acetates.[22] The cyclodextrin reagent complexes the substrate and interacts

Figure 8 *Example of scatter due to microscopic medium effects in a well-behaved system – the nucleophilic reaction of pyridines with N-phosphoryl-isoquinolinium ion; see text for the significance of Δ*

with both substituent and reaction centre, unlike the reference reaction where the interaction with the substituent is negligible. Large deviations, such as those found in the cyclodextrin case, are often easy to interpret by inspection of molecular models of the transition state.

Deviations of data from linear correlations can be very informative of mechanism as will be seen in Chapter 7. One of the most general causes of deviant points is where the bulk of the substituent has a different effect on the standard equilibrium compared with that of the reaction under investigation. Proton transfer equilibria, the most generally used reference reactions, are usually not very susceptible to steric effects. The observation of the substantial negative deviation shown in Figure 8 for the sterically hindered 2,6-dimethylpyridine attack (point Δ) is consistent with a nucleophilic mechanism rather than one where proton transfer is significant. A further example can be found in Chapter 2 (Figure 8, Equation 41) which illustrates phenolate ion attack on a triazine nucleus; 2,6-dichloro and 2,6-dimethyl phenolate ions are nearly two orders of magnitude less reactive than expected from the linear correlation due to the bulky substituents at the *ortho* positions hindering approach to the electrophile.

Microscopic medium effects are usually regarded as small and the question of which data points to include in the linear correlation is best dealt with by inspection. The minimum deviation between observed and predicted rate constants is generally accepted to be between one and two orders of magnitude if a different mechanism is to be assigned.[23]

One of the most powerful mechanistic techniques is the design and synthesis of models which will demonstrate a particular mechanistic hypothesis making use of such deviations. As an example of this technique we shall consider a proposed mechanism of serine proteases. Bruice and his co-workers devised a chemical model[24] of the *catalytic triad* hypothesis[25] (Scheme 1) which was advanced to explain the catalytic advantage of serine proteases.

The model exhibits only a three-fold acceleration over its counterparts having no catalytic triad (Figure 9). It is concluded that the mechanism

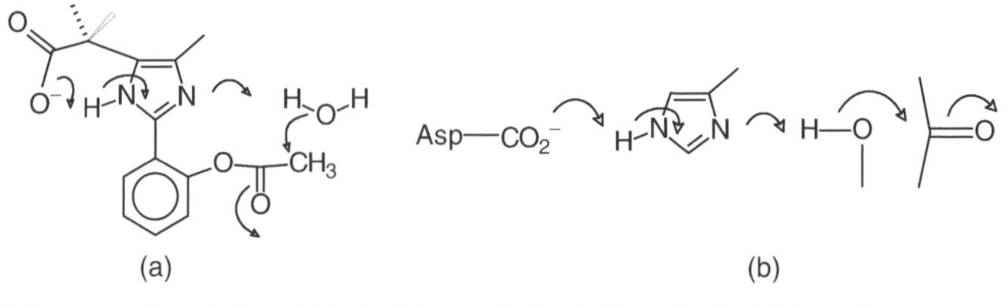

(a)   (b)

Scheme 1   *Chemical model (a) of the catalytic triad hypothesis (b) for serine proteases*

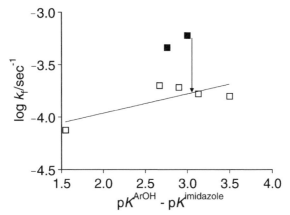

Figure 9 *Test of the catalytic triad hypothesis. □, Systems with imidazolyl but without the carboxylate component; ■, systems with both carboxylate and imidazolyl components. The arrow corresponds to an approximately three-fold advantage*

of the triad hypothesis contributes little to the catalytic advantage of serine proteases.[24] Free energy relationships with small numbers of points should be treated with suitable caution if useful conclusions are to be made, although the example quoted (Figure 9) has more data points than most. Small deviations have to be subjected to critical experimentation if they are to be used as evidence for different mechanisms. Confidence in the parameters such as slope and curvature is a function of the number of points in the correlation; such consideration becomes crucial if small variations in β or ρ or curvature in plots are under investigation. A large number of data points will make it easier to decide, on a statistical basis, if a point deviates significantly. On the contrary if there is evidence that a system is "well behaved" the use of small numbers of data points is valid to obtain selectivity parameters.

Free energy relationships for proton transfer reactions often exhibit little scatter because the proton being transferred at the transition structure has a microscopic environment similar to that of the completely transferred proton used in the standard dissociation equilibrium (Scheme 2).

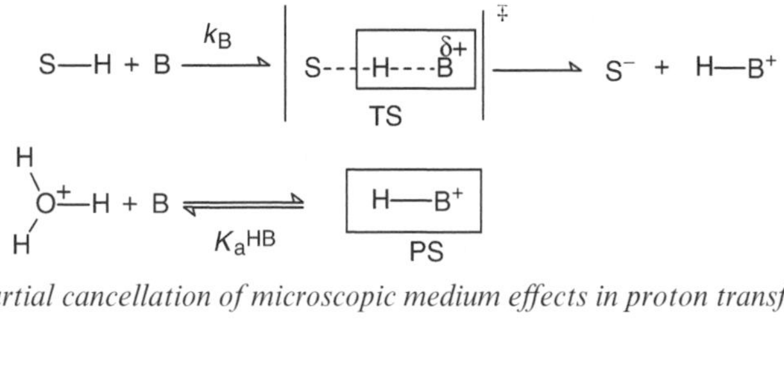

Scheme 2 *Partial cancellation of microscopic medium effects in proton transfer reactions*

The microscopic environment of the variant components (boxed in Scheme 2) would be approximately the same for both transition structure (TS) and protonated state (PS). As a corollary the microscopic medium effect in a general free energy relationship should be reduced if the model reaction chosen as standard closely resembles the reaction in question. In some cases it is useful to compare similar reactions, and an example of such a correlation is given in Problem 3.

6.5 STATISTICAL TREATMENT OF BRØNSTED PLOTS

Scatter in free energy correlations may also be caused by a statistical effect when more than one reaction centre can be involved.[26,27] The reactivity of a base or nucleophile possessing q identical basic or nucleophilic sites compared with a dissociation equilibrium where the conjugate acid possesses p identical acidic sites requires the following statistical correction to the simple Brønsted type relationship (Equations 23 and 24).

$$\log(k_{base}/q) = -\beta\log(K_a q/p) + C' = \beta pK_a + \beta\log(p/q) + C' \qquad (23)$$

$$k_{base}/q = G_B(K_a q/p)^{-\beta} \qquad (24)$$

A similar correction applies to reactivity of acid-catalysed reactions (Equations 25 and 26)

$$\log(k_{acid}/p) = \alpha\log(K_a q/p) + C = -\alpha pK_a + \alpha\log(q/p) + C \qquad (25)$$

$$k_{acid}/p = G_A(K_a q/p)^{\alpha} \qquad (26)$$

These equations factor all the acid–base types in a series to that where $p = q = 1$. For example, if triethylamine and diaminoethane monocation are to be interpreted in the same correlation they need corrections because the former has $p = 1$ and $q = 1$ whereas the latter has $p = 2$ and $q = 1$. The assignment of p and q is not satisfactory for acid–base pairs with uncertain structures (such as H_3O^+ and HO^-) or species with multiple centres of differing reactivity; hydronium and hydroxide ions almost always show anomalous reactivities in corrected Brønsted correlations, but these are partly due to solvation effects. The protons attached to a single atom such as ammonium ($R-NH_3^+$) or oxonium ion ($R-OH_2^+$) are regarded as having $p = 1$ rather than the number of identical protons. A similar convention selects the q value for hydroxide ion as unity even though there are three lone pairs free to accept a proton. The statistical correlation should also be applied to nucleophilic attack generally but in these cases the microscopic medium effect is often more dominant than the statistical effect. The most reliable Brønsted-type correlations are those for nucleophilic reactions in which the reagents have a common structure (such as all pyridines or all phenoxide ions) so that p and q are constant throughout.

The statistical correction is important for correlations with small numbers of points over a small range; when the range and number of points increases, the statistical correction becomes less important and the similarity coefficient shows very little difference between corrected and uncorrected treatments (see Problem 4).

6.6 ARE FREE ENERGY RELATIONSHIPS STATISTICAL ARTIFACTS?

The equilibrium constant for a reaction is related to the rate constants of forward and reverse reactions by the equation $K_{eq} = k_f/k_r$. The correlation of $\log k_f$ with $\log K_{eq}$, a Class I type free energy relationship, therefore has two variables which are not completely independent.[28] If the change in $\log k_r$ were small compared with $\log k_f$ the plot would be mainly $\log k_f$ versus *itself* and would therefore show a reasonable linear correlation. Even when the ranges of a set of computer generated random values of k_f and k_r are similar, a plot of $\log k_f$ correlates quite well with $\log k_f/k_r$ and the correlation improves as the range of k_r decreases (see worked answer to Problem 5).

These arguments do not invalidate the immense array of linear free energy relationships gathered over the past 70 years or so.[29] Class II systems such as Hammett or Taft relationships can be excluded from these considerations because the two variables are independent. The following arguments can be employed to demonstrate experimentally that a Class I correlation such as a Brønsted or a Leffler Equation does not arise from a statistical artifact in a system under investigation. Rearranging the Leffler Equation (Equation 27) yields Equation (29).

$$\log k_f = \alpha \log K_{eq} + C = \alpha \log(k_f/k_r) + C \tag{27}$$

$$= \alpha \log k_f - \alpha \log k_r + C \tag{28}$$

Thus by transposing Equation (28)

$$\log k_f = C/(1 - \alpha) - [\alpha/(1 - \alpha)] \log k_r \tag{29}$$

Since the coordinates $\log k_f$ and $\log k_r$ are independent of each other, a correlation between $\log k_f$ and $\log k_r$ as in Equation (29) would demonstrate the existence of a linear free energy relationship.[29] The demonstration of a linear free energy relationship by this means is not necessary for chemical systems because these usually involve substituent change directly adjacent to the reaction centre; the correlation is thus chemically reasonable. Application of Equation (29) is significant when the substituent changes may not be so obviously connected with the reaction centre as in the case of the effect of point mutation on conformational changes in enzymes.[29,30]

An interesting corollary is that the value of α obtained from Equation (27) is likely to be more accurate than that from Equation (29), because in the former case errors in the two coordinates tend to cancel each other.

6.7 THE *ORTHO* EFFECT [31]

It is generally recommended that Hammett correlations omit data for *ortho* substituents if information on polar effects is needed. This is because steric interaction between reaction centre and substituent and specific interactions such as hydrogen-bonding are likely to be enhanced for *ortho* substituents. Systems with *ortho* substituents should be treated with care because the steric interaction between substituent and reaction centre is not a substantial component of the Hammett σ in the standard dissociation reaction. However, there are cases where the *ortho* substituent can be included in a correlation and one of these is where a series of reactants has a constant *ortho* group, such as all chloro or all nitro groups; in those cases a Hammett relationship usually holds for substituent change in the other positions. It should be remembered that hydrogen is no different in principle from any other substituent, and that the Hammett relationship holds for the large number of reactions where hydrogen is an *ortho* substituent. For this reason it is possible to generate sets of σ values for the *ortho* position which enable pK_a values to be predicted for benzoic acids, phenols and anilinium ions (see Problems in Chapter 4).

Despite their associated problems, attempts have been made to incorporate results from *ortho* substituents in free energy correlations. It might be necessary, for example, to predict physico-chemical properties of a compound with an *ortho* substituent which acts as a drug or is a putative intermediate in a proposed mechanism. Quantitation of the *ortho* substituent effect requires that both steric and polar effects be considered. The *ortho* effect can be fitted to a modified version of the Pavelich–Taft equation (Chapter 2, Equation 15) where the values $\sigma_o{}^*$

and E_S are derived from the acid- and base-catalysed hydrolysis of alkyl esters of substituted benzoic acids. There is an increased possibility that some through-space interaction occurs, such as hydrogen-bonding, which is not possible in *meta* or *para* substituent effects and the incorporation of *ortho* substituents in free energy correlations is never wholly satisfactory.

Apart from the use of a set of benzene derivatives with a constant *ortho* substituent, another case where the *ortho* substituent can be employed is where the steric effect is likely to be small or where the *ortho* group can take up a conformation during reaction where it does not interact with the reaction centre. A good example of this is the nucleophilic reaction of phenolate ions with 4-nitrophenyl acetate (Figure 10) where phenolate ions with single *ortho* substituents fit the Brønsted line defined by the *meta* and *para* substituent points.

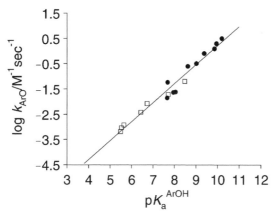

Figure 10 *Reaction of phenolate ions with 4-nitrophenyl acetate. Phenolate ions possessing* ortho *substituents are marked with open squares.*

6.8 TEMPERATURE EFFECTS[32]

The similarity coefficient, a, can be temperature dependent although reference dissociation constants are determined at $25\,°C$ under standard conditions which usually involve water solvent and zero ionic strength. It is therefore the aim to carry out all measurements of equilibrium constants and rate constants under these conditions or to extrapolate from other temperatures. The temperature effect on the similarity coefficient, a, is only meaningful if the standard dissociation equilibria are for the standard temperature. Measuring a values for different temperatures against standard equilibria at these same temperatures introduces the uncertainty due to the temperature variation of the standard a.

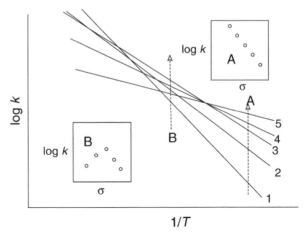

Figure 11   *Arrhenius plot without a single cross-over point; each line 1–5 corresponds
to a reaction bearing a single substituent. The insets are the Hammett plots
corresponding to the temperatures marked A and B*

The value of the similarity coefficient often bears an inverse relation-
ship to temperature (Equation 30) and moreover $\Delta H^{\ddagger}$ and $\Delta S^{\ddagger}$ are often
related by the isokinetic relationship (Equation 31).

$$\rho_1/\rho_2 = T_2/T_1 \qquad (30)$$

$$\Delta H^{\ddagger} = T_i \,\Delta S^{\ddagger} + b \qquad (31)$$

Reactions conforming to Equation (31) have Arrhenius correlations
($\log k$ versus $1/T$) for rate constants of the reaction of each substituent
which pass through a single point at the isokinetic temperature (T_i). Thus
at $T = T_i$ the value of ρ would be zero and the sign of ρ would reverse as
the isokinetic temperature is traversed.

If the Arrhenius plots for the reaction for each substituent do not cross
at a single point (Figure 11), resulting from breakdown of Equation (31),
then a curved free energy relationship could result at certain temperatures.

The temperature dependence of the similarity coefficient has been
studied at length but no significant conclusions have been deduced except
to delineate the range of validity of the various free energy relationships.

6.9 FURTHER READING

W.J. Albery *et al.*, Marcus-Grunwald Theory and the Nitroalkane
Anomaly, *J. Phys. Org. Chem.*, 1997, **1**, 29.
C.F. Bernasconi, The Principle of Non-Perfect Synchronisation, *Adv.
Phys. Org. Chem.*, 1992, **27**, 119.

C.F. Bernasconi, The Principle of Non-Perfect Synchronisation: More Than a Qualitative Concept? *Acc. Chem. Res.*, 1992, **25**, 9.

F.G. Bordwell and W.J. Boyle, Brønsted Coefficients and ρ Values as Guides to Transition-state Structures in Deprotonation Reactions, *J. Am. Chem. Soc.*, 1971, **93**, 511.

F.G. Bordwell and W.J. Boyle, Acidities, Brønsted Coefficients and Transition-state Structures for 1-Arylnitroalkanes, *J. Am. Chem. Soc.*, 1972, **94**, 3907.

B. Giese, Basis and Limitations of the Reactivity–Selectivity Principle, *Angew. Chem. Int. Et.*, 1977, **16**, 125.

C.D. Johnson, The Reactivity-Selectivity Principle: Fact or Fiction, *Tetrahedron Report*, 1980, **36**, 3461.

M.J. Kamlet and R.W. Taft, Linear Solvation Energy Relationships. Local Empirical Rules – or Fundamental laws of Chemistry? A Reply to the Chemometricians, *Acta Chem. Scand.*, 1985, **B 39**, 611.

A.J. Kresge, The Brønsted Relation – Recent Developments, *Chem. Soc. Rev.*, 1973, **2**, 475.

W. Linert, Mechanistic and Structural Investigations based on the Isokinetic Relationship, *Chem. Soc. Rev.*, 1994, 429.

L. Liu and Q-X. Guo, Isokinetic Relationship, Isoequilibrium Relationship, and Enthalpy–Entropy Compensation, *Chem. Rev.*, 2001, **101**, 673.

R.A. Marcus, Skiing the Reaction Rate Slopes, *Science*, 1992, **256**, 1523.

D.J. McLennan, Hammett ρ Values. Are they an Index of Transition State Character? *Tetrahedron Report*, 1978, **34**, 2331.

J. Shorter, *Correlation Analysis of Organic Reactivity*, Research Studies Press, New York, 1982, pp. 112–120.

J. Shorter, The Separation of Polar, Steric and Resonance Effects by the Use of Linear Free Energy Relationships, in *Advances in Linear Free Energy Relationships*, N.B. Chapman and J. Shorter (eds.), Plenum Press, New York, 1972, pp. 103–110.

M. Sjöström and S. Wold, Linear Solvation Energy Relationships. Local Empirical Rules – or Fundamental laws of Chemistry? *Acta Chem. Scand.*, 1981, **B 35**, 537.

M. Sjöström and S. Wold, Linear Solvation Energy Relationships. Local Empirical Rules – or Fundamental laws of Chemistry? A Reply to Kamlet and Taft, *Acta Chem. Scand.*, 1986, **B 40**, 270.

6.10 PROBLEMS

1 Differentiate the Marcus Equation (10) with respect to ΔG_o to obtain Leffler's α. Describe the conditions of ΔG_o where the Leffler α approximates to 0, 0.5 and 1.0 respectively. Under what conditions of ΔG_o

relative to $\Delta G_o^{\ddagger}$ does the Marcus equation, derived from intersecting parabolas, break down?

2 Estimate the value of ρ_{eq} for formation of the carbanion at position (0,1) in Figure 7 for reaction of base with substituted phenylnitromethanes. Use this value to calculate effective charges on carbon-1 in the transition structure and in the product pseudo-base. The values of ρ_{eq} and ρ_f for formation of pseudo-base are respectively 0.83 and 1.28.

3 The rate constants for reaction of phenolate ion and 4-cyanophenolate ion with substituted phenyl acetate esters are given in Table 1.[33]

(a) Graph the Brønsted dependence of k_{PhO^-} against pK_a^{ArOH}.

(b) Plot $\log k_{PhO^-}$ against $\log k_{4\text{-cyanophenolate}}$.

Demonstrate that the reaction of 4-cyanophenolate ion is a better reference standard than is the dissociation of phenols, commenting on the relative scatter in the two plots.

Table 1

Substituent	pK_a^{ArOH}	$\log k_{PhO^-}/M^{-1}sec^{-1}$	$\log k_{4\text{-cyanophenolate}}/M^{-1}sec^{-1}$
4-acetyl	8.05	−0.553	−2.34
4-formyl	7.66	−0.284	−2.07
2-nitro	7.23	−0.187	−1.77
4-nitro	7.14	0.093	−1.638
3-chloro-4-nitro	6.80	0.479	−1.131
3,5-dinitro	6.68	0.805	−0.854
4-chloro-2-nitro	6.46	0.185	−1.114
2-chloro-4-nitro	5.45	0.444	−1.143
3,4-dinitro	5.42	1.057	−0.398
2,5-dinitro	5.22	1.053	−0.301
2,4-dinitro	4.11	1.230	−0.194

4 Is a statistically corrected Brønsted treatment necessary when all the variant reagents have similar structures (*e.g.* all phenols or all pyridines)?

Table 2 shows rate constants for the base-catalysed decomposition of nitramide at 25°C.[34] Graph the data for $\log k_B$ versus pK_a^{HB} with and without the statistical correction and comment briefly on your result. What would happen to the correlation if a shorter range of pK_a^{HB} values (say between 8 and 11) were employed?

Table 2

pK_a^{HB}	$pK_a^{HB} +$ $log(p/q)$	$logk_B/$ $M^{-1}sec^{-1}$	$log(k_B/q)/$ $M^{-1}sec^{-1}$	Base	p	q
4.93	5.41	−1.06	−1.07	$CCl_3PO_3^{2-}$	1	3
5.6	6.08	−0.587	−0.587	$CHCl_2PO_3^{2-}$	1	3
6.59	7.07	0.265	0.265	$CH_2ClPO_3^{2-}$	1	3
7.2	7.38	0.978	0.677	$HOPO_3^{2-}$	2	3
7.36	7.84	0.852	0.852	$HOCH_2PO_3^{2-}$	1	3
8.0	8.48	1.38	1.38	$CH_3PO_3^{2-}$	1	3
8.71	9.19	2.19	2.19	$(CH_3)_3CPO_3^{2-}$	1	3
10.33	10.81	3.49	3.49	CO_3^{2-}	1	3
4.76	5.06	−1.19	−1.192	$CH_3CO_2^-$	1	2
6.23	6.23	−0.142	−0.142	$C_6H_2Cl_3O^-$	1	1
4.6	4.60	−1.34	−1.336	CH_3ONH_2	1	1
5.59	5.59	−0.426	−0.426	$CF_3CH_2NH_2$	1	1
7.8	7.80	1.13	1.13	$CN(CH_2)_2NH_2$	1	1
8.07	8.07	0.959	0.959	$(HOCH_2)_3CNH_2$	1	1
8.54	8.54	1.42	1.42	$(CH_3O)_2CHCH_2NH_2$	1	1
9.25	9.25	1.48	1.48	NH_3	1	1
9.78	9.78	1.95	1.95	$^-O_2CCH_2NH_2$	1	1
9.87	9.87	1.81	1.81	$^-O_2CCH(CH_3)NH_2$	1	1
8.35	8.35	1.71	1.71	$(CH_3)_3N^+(CH_2)_3NH_2$	1	1
9.39	9.39	2.05	2.05	$(CF_3)_2CHO^-$	1	1

5 Generate three sets of 20 random numbers for $logk_f$ and $logk_r$ with $logk_f$ ranging from 1.00 to 3.00 and $logk_r$ 1.00 to 1.1; 1.00 to 2.00 and 1.00 to 3.00. The QBASIC programs ($x = logk_r$ and $y = logk_f$) which send these ranges of random numbers to file n$ are:

```
CLS
INPUT "Enter Filename:"; n$
OPEN n$ FOR OUTPUT AS #1
RANDOMIZE TIMER
FOR i = 1 TO 20
x = RND*2000 (or 1000 or 100)
x = x + 1000
y = RND*2000
y = y + 1000
x = x/1000
y = y/1000
PRINT #1, x, y
NEXT i
CLOSE
```

The designated file n$ can be imported directly into graphing software or can be utilised by manual plotting. Using the random data plot $logk_f$

versus $\log k_r$ for the case where $\Delta \log k_r = 0.01$; plot $\log k_f$ against $\log(k_f/k_r)$ for the three cases $\Delta \log k_r = 2.00$, 1.00 and 0.01.

6 Using the σ values for *ortho* substituents given in Appendix 3, Table 10 calculate the pK_a values of the following (experimental values given in brackets): 2-chloro-4-nitrobenzoic acid (1.96), 2,6-dimethylpyridine (6.77), 2,4-dichloroanilinium ion (2.05) and 2,3,5-trichlorophenol (6.43). Hammett equations for the dissociation of substituted phenols, pyridinium ions, anilinium ions and benzoic acids are given in Appendix 4, Table 1.

7 Plot $\log k_f$ against ΔpK_a ($pK_a^{SH} - pK_a^{AH}$) for the reaction of acids (AH) with the enolate ion form of acetylacetone (SH) using data from Table 3.[35] Comment on the origin of the curvature.

Table 3

ΔpK_a	$\log k_f / M^{-1} sec^{-1}$	Acid
10.60	−3.52	H_3O^+
6.01	0.0414	$ClCH_2CO_2H$
5.12	0.362	HCO_2H
4.59	0.511	$PhCH_2CO_2H$
4.12	0.556	CH_3CO_2H
4.00	1.09	$CH_3CH_2CO_2H$
3.19	2.15	malonic acid
2.68	2.04	cacodylic acid
0.92	2.75	4-cyanophenol
0.39	3.20	2-chlorophenol
−0.15	3.11	3-chlorophenol
−1.13	3.78	phenol
−3.43	4.30	glucose
−5.13	4.60	H_2O

8 Use Bodenstein's steady-state assumption for the concentrations of the reactive intermediates AH.B and A⁻.BH⁺ to derive the rate law (Equation 2) for k_f and k_r and for its special case when $k_2 > k_{-1}$.

9 The data in Table 4 were obtained by Bednar and Jencks[36] for the protonation of CN⁻ by a series of general acids (AH⁺, Equation 32). Graph $\log k_a$ (k_a = overall second-order rate constant) versus pK^{AH}; fit the data to the Eigen Equation $\log k_a = \log k_1 - \log(1 + 10^{pK_{AH} - pK_o})$ (assuming $k_{-1} = k_3$) and comment on the results. There is no need to correct the data statistically.

$$AH^+ + CN^- \underset{k_{-1}}{\overset{k_1}{\rightleftharpoons}} [AH^+.CN^-] \underset{k_{-2}}{\overset{k_2}{\rightleftharpoons}} [A.HCN] \underset{k_{-3}}{\overset{k_3}{\rightleftharpoons}} A + HCN \qquad (32)$$

10 Does the energy of the transition structure lie exactly at the intersection point of the parabolas in Figure 4? (See, for example, reference 8.)

Table 4

General acid (AH^+)	pK^{AH}	$\log k_a$ ($M^{-1}sec^{-1}$)
H_3O^+	−1.74	10.6
H_2O	15.74	3.20
Cyanoacetic acid	2.23	9.11
Chloroacetic acid	2.65	9.02
Methoxyacetic acid	3.33	8.99
Acetic acid	4.65	8.67
$NH_2C(O)NHNH_3^+$	3.86	9.40
$MeONH_3^+$	4.76	9.28
$CF_3CH_2NH_3^+$	5.81	8.90
$Me_3N^+CH_2CH_2NH_3^+$	7.32	8.81
$H_3N^+CH_2CH_2NH_3^+$	7.50	8.67
$H_2NC(O)CH_2NH_3^+$	8.25	8.05
$ClCH_2CH_2NH_3^+$	8.81	7.70
$MeOCH_2CH_2NH_3^+$	9.72	6.86
$HOCH_2CH_2NH_3^+$	9.87	6.94
$C_2H_5NH_3^+$	10.97	5.73
(piperazinium, ^+HN...NH^+)	3.47	9.75
(N-methyl piperazinium, $^+N(Me)H$...NH_2^+)	4.64	9.51
(piperazinium, ^+H_2N...NH_2^+)	6.01	9.35
$Me_2NH^+CH_2CH_2NHMe_2^+$	6.47	9.40
Imidazolium ion	7.24	8.54
$(HOCH_2CH_2)_3NH^+$	7.99	7.97
$Me_2NH^+CH_2CH(OH)CH_2NHMe_2^+$	8.01	8.71
$Me_2NH^+(CH_2)_3NHMe_2^+$	8.39	8.09
$Me_2NH^+(CH_2)_4NHMe_2^+$	9.27	7.22
Me_3NH^+	10.16	6.22
4-Nitrophenol	7.14	8.61
3-Nitrophenol	8.35	8.43
$(CF_3)_2CHOH$	9.30	7.67
Phenol	9.99	7.82
$CF_3C(OH)_2CF_3$	10.5	7.07
CF_3CH_2OH	12.4	5.88

11 A 10^4-fold *increase* in the rate constant for the E2 elimination of HBr from alkyl bromides, brought about by a more electron withdrawing group (X) at C(2), is accompanied by an *increase* in the sensitivity to the strength of the catalysing base – the Brønsted β value increases from 0.39 to 0.67.[37] Explain this contradiction of the reactivity–selectivity postulate by a qualitative consideration of a More O'Ferrall–Jencks diagram of the reaction.

$$H \overset{X}{\underset{(2)}{\rule{0pt}{0pt}}} \overset{\rule{0pt}{0pt}}{\underset{(1)}{\rule{0pt}{0pt}}} Br \quad \xrightarrow{\quad B \quad} \quad HB^+ \;+\; Br^- \;+\; \overset{X}{\diagdown}\!\!=\!\!\diagup \qquad (33)$$

12 Verify Equation (20).

13 The alkaline hydrolysis of methyl esters of substituted benzoic acids has a Hammett ρ of 2.23. What would be the Brønsted β for the reaction when measured against the pK_a of the corresponding benzoic acid? Comment on the magnitude of the effective charge at the reaction centre in the transition structure compared with that of the carboxylate anion product in the reference dissociation reaction?

14 Use Equation (31) to demonstrate that ρ could become zero when T = isokinetic temperature, T_I.

6.11 REFERENCES

1. J.N. Brønsted and K. Pedersen, The Catalytic Decomposition of Nitramide and its Physico-Chemical Applications, *Z. Physik. Chem.*, 1923, **108**, 185; see also A.J. Kresge, The Brønsted Relation – Recent Developments, *Chem. Soc. Rev.*, 1973, **2**, 475.

2. M. Eigen, Proton Transfer, Acid-Base Catalysis, and Enzymatic Hydrolysis, *Angew. Chem. Int. Ed.*, 1964, **3**, 1.

3. D.S. Kemp and M.L. Casey, Physical Organic Chemistry of Benzisoxazoles. II. Linearity of the Brønsted Free Energy Relationship for the Base-Catalysed Decomposition of Benzisoxazoles, *J. Am. Chem. Soc.*, 1973, **95**, 6670.

4. E.M. Arnett and R. Reich, Electronic Effects in the Menschutkin Reaction. A Complete Kinetic and Thermodynamic Dissection of Alkyl Transfer to 3- and 4-Substituted Pyridines, *J. Am. Chem. Soc.*, 1980, **102**, 5892.

5. J.L. Kurz, The Relationship of Barrier Shape to "Linear" Free Energy Slopes and Curvatures, *Chem. Phys. Lett.*, 1978, **57**, 243; J.C. Harris & J.L. Kurz, A Direct Approach to the Prediction of Substituent Effects in Transition-state Structures, *J. Am. Chem. Soc.*, 1970, **92**, 349; R.P. Bell, *Proc. Roy. Soc. London*, 1936, **154A**, 414.

6. R.A. Marcus, Theoretical Relations among Rate Constants, Barriers, and Brønsted Slopes of Chemical Reactions, *J. Phys. Chem.*, 1968, **72**, 891; E.F. Caldin, *The Mechanisms of Fast Reactions in Solution*, IOS Press, Amsterdam, 2001, pp. 227–318.

7. G.S. Hammond, A Correlation of Reaction Rates, *J. Am. Chem. Soc.*, 1955, **77**, 334.

8. R.P. Bell, *The Proton in Chemistry*, Chapman & Hall, London, 2nd edition, 1973, p. 205.
9. T.H. Lowry and K.S. Richardson, *Mechanism and Theory in Organic Chemistry*, Harper and Row, New York, 3rd edition, 1987, p. 596.
10. C.D. Johnson and K. Schofield, A Criticism of the Use of the Hammett Equation in Structure-Reactivity Correlations, *J. Am. Chem. Soc.*, 1973, **95**, 270; T.J. Gilbert and C.D. Johnson, Acid-Catalysed Hydrogen Exchange of Acetophenones. Evidence for the Inapplicability of the Reactivity-Selectivity Principle, *J. Am. Chem. Soc.*, 1974, **96**, 5846.
11. A. Pross, The Reactivity-Selectivity Principle and its Mechanistic Applications, *Adv. Phys. Org. Chem.*, 1977, **14**, 69.
12. R.D. Guthrie and W.P. Jencks, IUPAC recommendations for the Representation of Reaction Mechanisms, *Acc. Chem. Res.*, 1989, **22**, 343.
13. W.P. Jencks, Acid-Base Catalysis of Complex Reactions in Water, *Chem. Rev.*, 1972, **72**, 705; see also Section 6.1, Chapter 6.
14. E.R. Thornton, A Simple Theory for Predicting the Effects of Substituent Changes on Transition-State Geometry, *J. Am. Chem. Soc.*, 1967, **89**, 2915.
15. W.P. Jencks, A primer for the Bema Hapothle. An Empirical Approach to the Characterisation of Changing Transition-State Structures, *Chem. Rev.*, 1985, **85**, 511.
16. D.A. Jencks and W.P. Jencks, On the Characterisation of Structure-Reactivity Coefficients, *J. Am. Chem. Soc.*, 1977, **99**, 7948.
17. J.F. Kirsch *et al.*, Multiple Structure-Reactivity Correlations. The Alkaline Hydrolyses of Acyl- and Aryl-Substituted Phenyl Benzoates, *J. Org. Chem.*, 1968, **33**, 127.
18. F.G. Bordwell *et al.*, Brønsted Coefficients Larger than 1 and less than 0 for Proton removal from Carbon Acids, *J. Am. Chem. Soc.*, 1969, **91**, 4002; A.J. Kresge, The Nitroalkane Anomaly, *Can. J. Chem.*, 1974, **52**, 1897.
19. G.A. Olah and G. Rasul, From Kekulé's tetravalent methane to five-, six-, and seven-coordinate protonated methanes, *Acc. Chem. Res.*, 1997, **30**, 245.
20. M.T. Skoog and W.P. Jencks, Reactions of Pyridines and Primary Amines with N-Phosphorylated Pyridines, *J. Am. Chem. Soc.*, 1984, **106**, 7597.
21. N. Bourne and A. Williams, Evidence for a Single Transition State in the Transfer of the Phosphoryl Group ($-PO_3^{2-}$) to Nitrogen Nucleophiles from Pyridine-N-phosphonates, *J. Am. Chem. Soc.*, 1984, **106**, 7591.

22. R.L. Van Etten *et al.*, Acceleration of Phenyl Ester Cleavage by Cycloamyloses. A Model for Enzymatic Specificity, *J. Am. Chem. Soc.*, 1967, **89**, 3242.

23. B. Capon, Neighbouring Group Participation, *Quart. Rev. Chem. Soc.*, 1964, **18**, 48.

24. G.A. Rogers and T.C. Bruice, Synthesis and Evaluation of a Model for the So-Called "Charge-relay" System of the Serine Proteases, *J. Am. Chem. Soc.*, 1974, **96**, 2473.

25. D.M. Blow, Structure and Mechanism of Chymotrypsin, *Acc. Chem. Res.*, 1976, **9**, 145.

26. R.P. Bell, *The Proton in Chemistry*, Chapman & Hall, London, 2nd edition, 1973, p. 197.

27. S.W. Benson, Statistical Factors in the Correlation of Rate Constants and Equilibrium Constants, *J. Am. Chem. Soc.*, 1958, **80**, 5151; V. Gold, Statistical Factors in the Brønsted Catalysis and Other Free Energy Correlations, *Trans. Faraday Soc.*, 1964, **60**, 738; E. Pollack and P. Pechukas, Symmetry Numbers, not Statistical Factors, should be used in Absolute Rate Theory and in Brønsted Relations, *J. Am. Chem. Soc.*, 1978, **100**, 2984.

28. D.A. Estell, Artifacts in the Application of Linear Free Energy Analysis, *Protein Engineering*, 1987, **1**, 441.

29. A.R. Fersht, Linear Free Energy Relationships *are* Valid, *Protein Engineering*, 1987, **1**, 442.

30. A.R. Fersht *et al.*, Structure–Activity Relationships in Engineered Proteins: Analysis of Use of Binding Energy by Linear Free Energy Relationships, *Biochemistry*, 1987, **26**, 6030.

31. M. Charton, The Quantitative Treatment of the Ortho Effect, *Progr. Phys. Org. Chem.*, 1971, **8**, 235.

32. O. Exner, *Correlation Analysis of Chemical Data*, Plenum Press, New York, 1988, pp. 99–111; O. Exner, The Enthalpy-Entropy Relationship, *Progr. Phys. Org. Chem.*, 1973, **10**, 411.

33. S. Ba-Saif *et al.*, Concerted Acetyl Group Transfer between Substituted Phenolate Ion Nucleophiles: Variation of Transition State Structure as a Function of Substituent, *J. Am. Chem. Soc.*, 1989, **111**, 2647.

34. C.H. Arrowsmith *et al.*, The Base-catalysed Decomposition of Nitramide: A New Look at an Old Reaction, *J. Am. Chem. Soc.*, 1991, **113**, 1172.

35. M.L. Ahrens *et al.*, Relaxation Studies of Base-Catalysed Keto-Enol Rearrangement of Acetylacetone, *Ber Bunsensges.*, 1970, **74**, 380.

36. R.A. Bednar and W.P. Jencks, Is HCN a Normal Acid? Proton

Transfer from HCN to Bases and Small Inhibition of Proton Exchange by Acid, *J. Am. Chem. Soc.*, 1985, **107**, 7117.

37. R.F. Hudson and G. Klopman, Nucleophilic Reactivity. Part IV. Competing Bimolecular Substitution and β-Elimination, *J. Chem. Soc.*, 1964, 5; B.D. England and D.J. McLennan, Elimination Promoted by Thiolate Ions. Part I. Kinetics and Mechanism of the Reactions of DDT with Sodium Benzenethiolate and other Nucleophiles, *J. Chem. Soc. B*, 1966, 696; R.F. Hudson, Nucleophilic Reactivity, in *Chemical Reactivity and Reaction Paths*, G. Klopman (ed.), John Wiley, New York, 1974, p. 167.

CHAPTER 7

Applications

7.1 DIAGNOSIS OF MECHANISM

7.1.1 Mechanism by Comparison

The similarity coefficient of Hammett and other Class II free energy correlations often bears no direct relationship to the transition structure because of dissimilarity between the model equilibrium and the reaction being studied. The closer the model is to the reaction under investigation the more reliable is any mechanistic conclusion from the value of the similarity coefficient. Some representative free energy relationships and their similarity coefficients are collected in Appendix 4.

The simplest use of the similarity coefficients of any of the polar substituent effect plots is to indicate qualitatively the sign of the change in charge at the reaction centre from reactant to transition structure. The sign of the similarity coefficient is in most cases only necessary as confirmation of chemical intuition and of the rate law governing the stoichiometry of the transition structure. Thus the reaction of methyl iodide with anilines has a rate law (rate = $k[CH_3I]$ [aniline]) indicating that the transition structure is composed of the atoms of methyl iodide and of a single aniline. The identity of the products and chemical intuition indicate that only two major bonding changes are occurring (N–CH$_3$ bond formation and CH$_3$–I bond fission) and that there is increased positive charge on the nitrogen in the transition structure due to its donating its pair of electrons to the methyl group to form the N-methylanilinium ion product. The ρ value for this reaction is predicted to be negative because the build-up of +ve charge would be stabilised by electron donating substituents (which have negative σ values). Knowledge of the sign of the similarity coefficient is helpful in those cases, such as cyclisations, where it is not possible to indicate *a priori* which is the nucleophile and which is the electrophile.

The magnitudes of the similarity coefficients may be used to indicate the size of the change in charge and this requires a knowledge of similarity coefficients for known reactions. For example, the alkaline hydrolysis of substituted phenyl phosphoramidates (Scheme 1) has a ρ value of 2.84[1] which is more positive than the value of 1.47 observed for the attack of hydroxide ion on substituted phenyl phosphate esters (which possesses an *associative* mechanism).[1] A *dissociative* pathway involving a metaphosphoramidate intermediate (**1**, Scheme 1) is compatible with generation of a much larger negative charge on the aryl oxygen leaving group in the transition structure than found for the associative process.

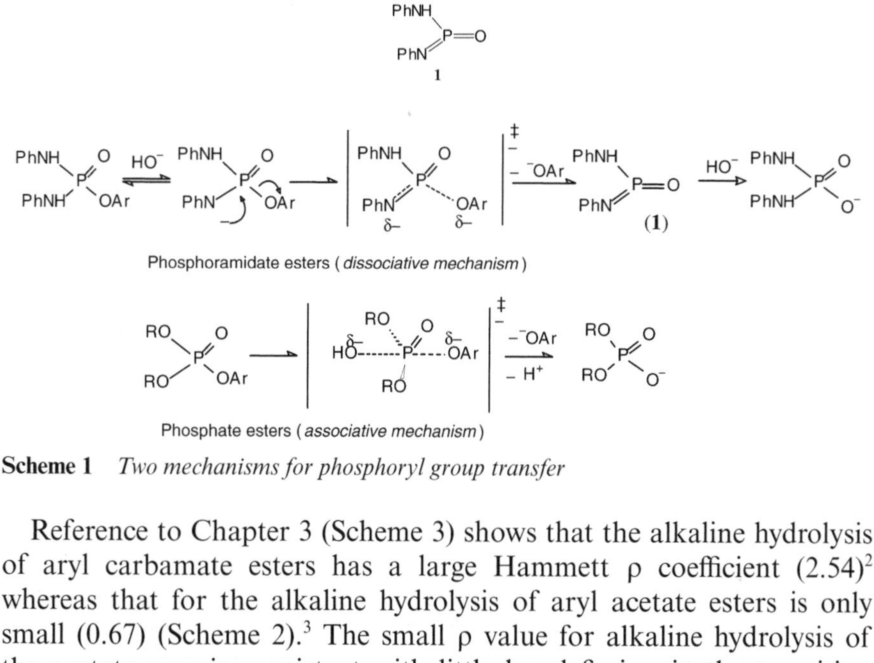

Scheme 1   *Two mechanisms for phosphoryl group transfer*

Reference to Chapter 3 (Scheme 3) shows that the alkaline hydrolysis of aryl carbamate esters has a large Hammett ρ coefficient (2.54)[2] whereas that for the alkaline hydrolysis of aryl acetate esters is only small (0.67) (Scheme 2).[3] The small ρ value for alkaline hydrolysis of the acetate case is consistent with little bond fission in the transition structure and the large ρ for the carbamate case (see Chapter 3 for mechanism) indicates substantial bond fission and a gross difference in mechanism between the two reactions. The simple Hammett relationship only correlates the alkaline hydrolysis rates of the carbamate when *meta* substituents are employed. *Para* substituents, which withdraw negative charge by resonance (4-NO$_2$, 4-CN, *etc.*) require σ^- values to obtain a good correlation. This is interpreted to mean that the transition structure has sufficient aryl oxide ion character for resonance to occur. In the case of the acetate hydrolyses the ArO–C bond fission is not sufficiently advanced in the transition structure for negative charge to develop enough for resonance to occur, and the rate constants follow a simple Hammett relationship for both *meta* and *para* substituents.

7.1.2 Effective Charge Distribution of the Transition Structure

The effective charge distributions for alkaline hydrolysis of carbamate esters and acetates are illustrated in Chapter 3 (Scheme 3) and in Scheme 2 respectively; the change in charge on the carbonyl oxygen from neutral ester to transition structure is +0.14 and –0.70. A substantial development of negative effective charge on the carbonyl oxygen would be reasonable for the acetate case if the transition structure were close in structure to that of a tetrahedral intermediate ($CH_3C(OH)O^-(OAr)$). The small development of effective charge in the carbamate case is expected for a transition structure where the oxygen has substantial carbonyl character as in the reactant.

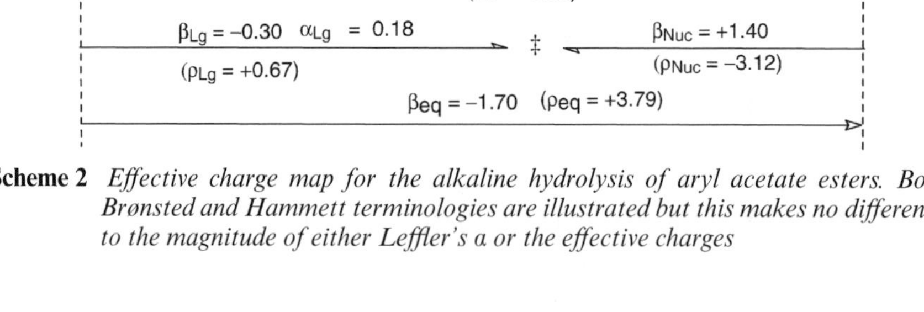

Scheme 2   *Effective charge map for the alkaline hydrolysis of aryl acetate esters. Both Brønsted and Hammett terminologies are illustrated but this makes no difference to the magnitude of either Leffler's α or the effective charges*

7.1.2.1 Effective Charge Maps and Conservation of Effective Charge. Maps such as that of Scheme 2 may be constructed where effective charge is assigned for each atom or group as it passes through the various stages of the mechanism. Scheme 3 illustrates the effective charge map for the alkaline hydrolysis of fluorenoyl esters[4] and shows how β values may be estimated making use of the principle of conservation of effective charge.

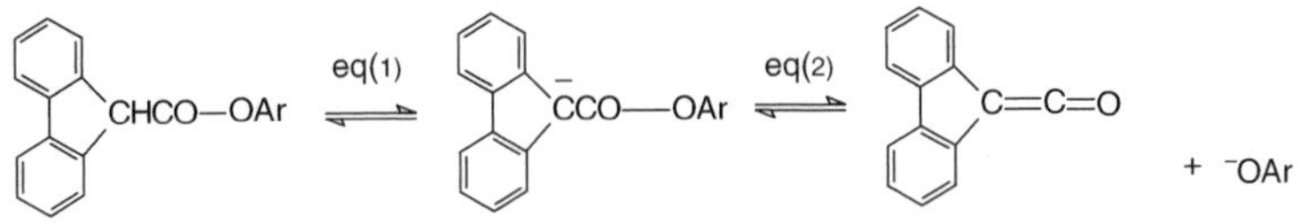

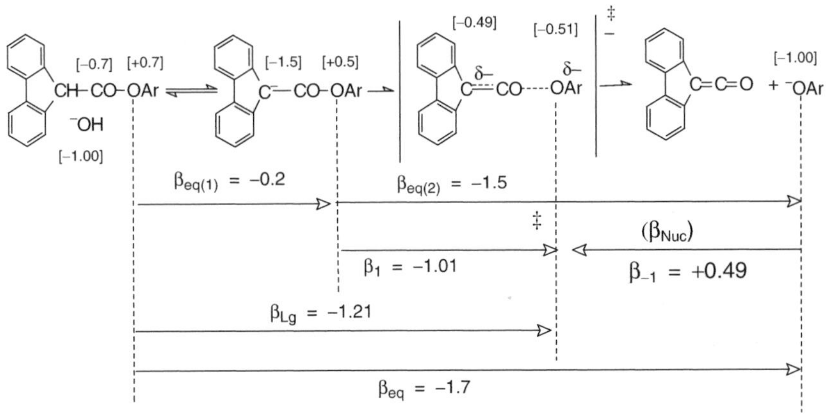

Scheme 3 *Effective charge map for aryl oxygen in the reactions of fluorenoyl esters. The ketene intermediate reacts rapidly with water and hydroxide ion to give the acid product*

The overall value of β_{eq} (−1.7) is derived from data in Chapter 3 (Scheme 1); it is reasonably assumed to be the same as that for acetate esters (see examples in Chapter 3).

Assuming that the changes in effective charge shown in Scheme 3 are *conserved*[5–7] the β values relating rate and equilibria are given in Equation (1). The principle of conservation of effective charge enables the calculation of charges for reactions not easily accessible to direct measurement.

$$\beta_{eq} = \beta_{eq1} + \beta_{eq2} = \beta_{eq1} + \beta_1 - \beta_{-1} \tag{1}$$

The magnitude of the charge difference on the acyl function is based on the standard change in charge in the calibrating equilibrium even though the atom structure in the carbonyl function does not resemble the proton in the dissociation reaction used for standardisation. The positive effective charge on the atom adjacent to the acyl function is balanced by an equal negative charge on the acyl function in order to conserve total charge. In the fluorenoyl case the value of β_{-1} (+0.49) is estimated from β_{eq} (−1.7) and β_{Lg}(−1.21) and is a reasonable value when compared with that (+0.66) directly measured for reaction of ArO⁻ with HN=C=O (Chapter 3, Scheme 3).

7.1.2.2 Charge and Bond Order Balance. Bond order *balance* is the relative extent of individual bond changes in a transition structure. In a reaction such as nucleophilic substitution involving two major bond changes, the transition structure could involve a *range* of structures from that where bond formation is largely complete and fission incomplete (transition structure B in Figure 1) to that where formation is incomplete and fission is complete (C in Figure 1).

The term *synchronous* is reserved to denote a mechanism where formation and fission occur to the *same extent* in the transition structure of a concerted mechanism; the transition structure is anywhere on the diagonal dashed line (Figure 1). Bond order balance can be deduced from the changes in effective charge in the transition structure provided the charge is defined by the same standard dissociation equilibrium. This is illustrated by the displacement reaction of pyridines on *N*-phosphopyridinium species (Scheme 4); the standard equilibrium for both bond formation and fission is the dissociation of substituted pyridinium ions ($XpyH^+$) and hence the effective charges can be directly compared for both processes.

Displacement of pyridines from *N*-phosphopyridinium ions by nucleophilic attack is a concerted process[8,9] with a substantial imbalance of effective charge between attacking and leaving pyridine (Scheme 4, $\Delta\varepsilon = -0.92$ for bond fission and $\Delta\varepsilon = +0.15$ for bond formation). The decrease in positive charge on the leaving nitrogen is supplied by donation of negative charge from the $-PO_3^{2-}$ group. Comparison of the effective charges with those expected for putative intermediates (Scheme 4) indicates that the transition structure is metaphosphate-like (see Scheme 4) and does not resemble the structure of a pentacoordinate intermediate.

The ratio $\alpha_{formation}/\alpha_{fission}$ may be taken as a formal indicator of the degree of balance, where deviation from unity indicates the amount of

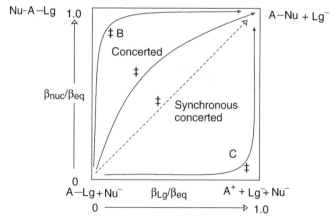

Figure 1   *Reaction map for the reaction $A–Lg + Nu^- \rightarrow A–Nu + Lg^-$*

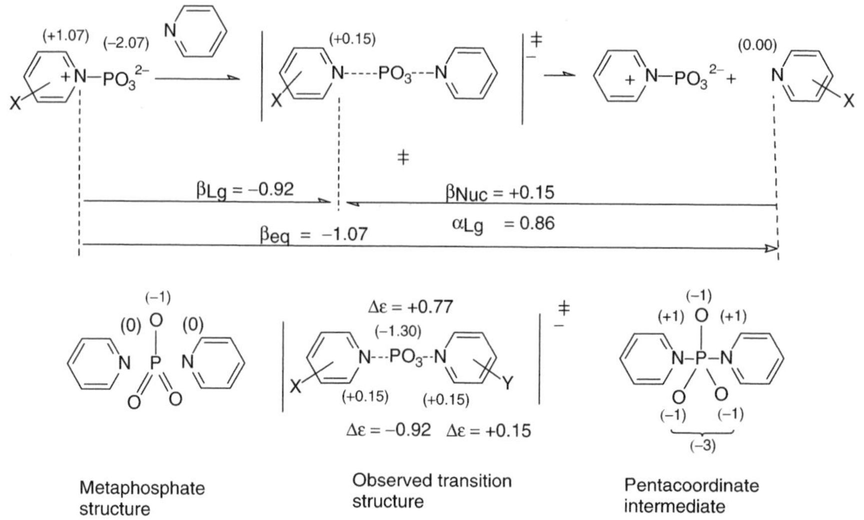

Scheme 4 *Transfer of the phosphoryl group between pyridine nucleophiles – an "open concerted" mechanism; shown also is the charge distribution in the transition structure and that expected for two putative intermediates*

imbalance in the transition structure; for example for reactions such as in Scheme (5)[10] which do not involve identical formation and fission processes the ratio equals $0.2/0.87 = 0.23$. Transition structures at positions *anywhere* on the synchronous concerted diagonal (Figure 1) possess a ratio of $\alpha_{formation}/\alpha_{fission} = 1.00$.

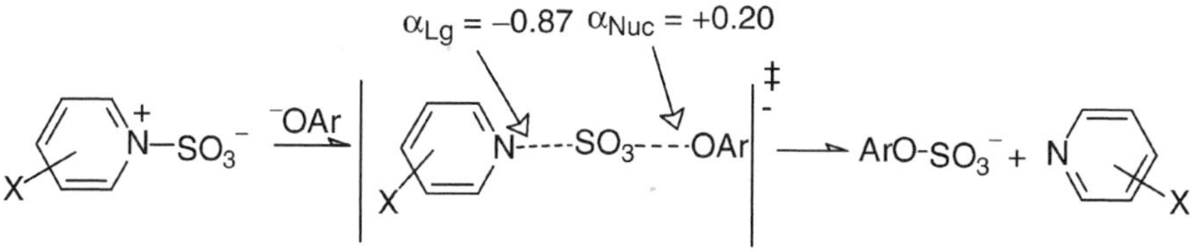

Scheme 5 *Balance of effective charge in a sulphyl group transfer reaction. The value $a_{Nuc} = a_{formation} = +0.2$; $a_{Lg} = a_{fission} = +0.87$*

7.2 DEMONSTRATION OF INTERMEDIATES

The build-up of measurable concentrations of an intermediate in a *stepwise reaction* does not occur in the majority of cases and the concentration can be *very* small compared with that of reactant and product. Under these conditions, the Bodenstein steady-state assumption is valid and Equation (2), the mechanism involving an intermediate, can be solved to give Equation (3).

$$A \xrightleftharpoons[k_{-1}]{k_1} I \xrightleftharpoons[]{k_2} P \tag{2}$$

$$d[I]/dt = [A]k_1 - [I](k_{-1} + k_2) = 0$$

Thus

$$[I] = [A]k_1/(k_{-1} + k_2)$$

and

$$\text{Rate} = k_2[I] = [A]\,k_1\,k_2/(k_{-1} + k_2) = k_{\text{overall}}\,[A]$$

$$k_{\text{overall}} = k_1 k_2/(k_{-1} + k_2) \tag{3}$$

Equation (3) is not valid when the concentration [I] builds up to values commensurate with the initial concentration of reactant [A]; the conditions for this build-up are $(k_{-1} + k_1) > k_2$ and $k_1 > k_{-1}$ and in this case the intermediate should be demonstrated by direct observation.[11] The problem of demonstrating an intermediate becomes acute when, as is often observed, its concentration does not build up and it remains at too low a level to be detected analytically.[a] Equation (3) can be employed to demonstrate the presence of an intermediate present in only trace amounts if variation in some factor such as substituent, concentration or other condition can change k_{-1} and k_2 independently; this allows the ratio k_{-1}/k_2 to vary between *less than* and *greater than* unity and thus to change the *rate-limiting step*.[12] Such behaviour is not, in general, predictable but if a change in rate-limiting step occurs there will be a *break* in the free energy plot (at $k_{-1}/k_2 = 1$) which demonstrates the presence of an intermediate. The individual steps (k_n) in Equation (2) should have linear Hammett dependencies where ρ_n and C_n are the parameters of the Equation for k_n.[b] The overall Hammett dependence is given by Equation (4).[c]

$$k_{\text{overall}} = (10^{\{\rho_1\,\sigma\,+\,C_1\}})/(10^{\{(\rho_{-1}\,-\,\rho_2)\sigma\,+\,C_{-1}\,-\,C_2\}} + 1) \tag{4}$$

[a] Other major techniques for demonstrating high-energy intermediates are "trapping" and positional isotopic exchange and are described elsewhere together with other techniques (A. Williams, *Concerted Mechanisms of Organic and Bio-organic Chemistry*, CRC Press, Boca Raton, FL, 2000, Chapter 2).

[b] In this case we employ Hammett terminology but any free energy relationship such as a Brønsted or a Swain–Scott equation could be used.

[c] Similar equations govern a plot where the individual steps of Equation (2) obey Brønsted equations (see Equations 11 and 12). $\log k_n = \beta_n \sigma + C_n$ ($n = 1$, -1 or 2 corresponding to k_1, k_{-1} and k_2).

A reaction studied by this classical method is the formation of aromatic semicarbazones (Scheme 6). The data (Figure 2) fit Equation (4) and $\Delta\rho$ is the difference in slope of the two linear portions of the correlation ($\rho_{-1} - \rho_2$). A similar Equation may be written for the Brønsted correlation or any other free energy relationship including those for concentration changes which might affect k_2 or k_{-1}.

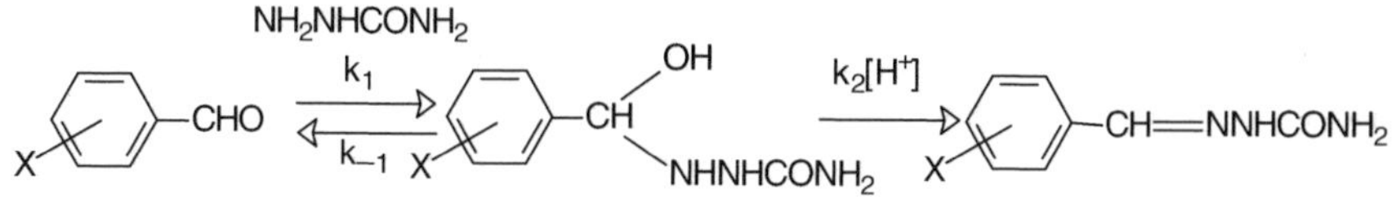

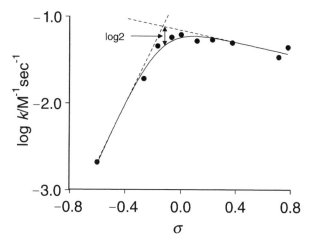

Scheme 6 *The formation of aromatic semicarbazones*

Figure 2 *Demonstration of a stepwise mechanism for aromatic semicarbazone formation; the theoretical line is obtained from Equation (4); data are from Noyce et al.*[13] *The intersection of the two linear Hammett plots falls log2 units above the smooth line as shown*

There are *three* types of non-linear free energy relationships arising from two intersecting linear plots in stepwise mechanisms. These plots can easily be derived as shown in Problem 1b (Figure A24 in Appendix 2). These plots conform to a general requirement for a changeover in rate-limiting step: an intermediate is diagnosed by a non-linear free energy correlation if the observed rate constant at the break-point is *less* than that representing the intersection point of the two linear sections.[d]

---

[d] Alternative definitions are: an intermediate is diagnosed if the observed rate constant on one side of the break-point in a non-linear free energy relationship is *less* than that calculated from the correlation on the other side. Another definition is that a stepwise mechanism gives a non-linear correlation with a *convex upwards* curvature to the free energy line; in this case the sense of convexness must be defined.

7.2.1 Curvature in pH-profiles

Changes in reaction parameters such as pH, solvent, substituent, concentration, *etc.* are essentially changes in free energy and, if the logarithm of a rate constant is plotted against these, a free energy correlation ensues. A change in pH can cause a change in rate-limiting step if the magnitude of either k_{-1} or k_2 in the standard Equation (3) is dependent on hydrogen ion concentration. A break in the pH-rate profile which gives a smaller rate constant than that predicted from extrapolation of the rate from lower pH values is consistent with an intermediate. The hydrolysis of 2-methyl-4,6-diphenylpyrylium perchlorate (Equation 5) provides a good example of a pH-induced change in rate-limiting step.[14]

The break occurs when $k_H[H_3O^+] = k_2$ in Equation (3) ($k_H[H_3O^+] = k_{-1}$). At high pH values the first step, the pH-independent addition of water, is rate limiting because $k_2 > k_H[H_3O^+]$ (see Figure 3). It is essential in such experiments to prove that the inflexion is not due to the dissociation of the conjugate acid of a buffer component which undergoes reaction with the substrate. Moreover, the break deriving from a dissociation step prior to the rate-limiting step could be demonstrated by separate pH-titration studies.

$$\text{(5)}$$

At even higher pH values a mechanism involving attack of hydroxide ion starts to predominate (see Section 7.3).

7.3 DEMONSTRATION OF PARALLEL REACTIONS

A mechanism composed of parallel reactions (Scheme 7) can be diagnosed by non-linear free energy relationships. The kinetic rate law is Equation (6) and this can be written in the form of Equation (7). If the individual free energy relationships for k_A and k_B have different similarity coefficients then a non-linear free energy relationship could ensue for the measured rate constant if the condition $\log k_A = \log k_B$ occurs within the range. The observed value of $\log k_{obs}$ at the break-point is log2 units above that at the intersection of the two free energy relationships $\log k_A$ and $\log k_B$.

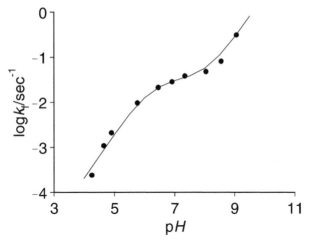

Figure 3 *Hydrolysis of 2-methyl-4,6-diphenylpyrylium salt over a pH-range exhibiting a change in rate-limiting step and mechanism. The line is calculated from* $k_f = k_1[H_2O]/(1 + [H_3O^+]k_{-1}/k_2) + k_{OH}·K_w/[H_3O^+]$

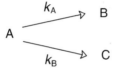

Scheme 7 *A mechanism with parallel reactions*

$$\text{Rate} = (k_A + k_B)[A]$$

$$k_{obs} = k_A + k_B \qquad (6)$$

$$\log k_{obs} = \log(k_A + k_B) \qquad (7)$$

Three types of non-linear free energy relationships due to change in the predominant mechanism in parallel reactions are illustrated in Appendix 2 (Figure A23). The plots conform to the general requirement for a change in mechanism: parallel reactions are diagnosed if the observed rate constants at the breakpoint *exceed* that representing the intersection of the two intersecting limbs of the free energy correlation.[e]

The hydroxide ion-catalysed hydrolysis of aryl 4-hydroxybenzoates (Scheme 8) exhibits a change in mechanism caused by a change in substituent.[15]

---

[e] Alternative definitions: parallel mechanisms can be diagnosed when the observed rate on one side of the break-point in a free energy plot is *greater* than that calculated from the correlation on the other side. Parallel mechanisms are diagnosed if the non-linear free energy correlation exhibits a *concave upwards* curvature.

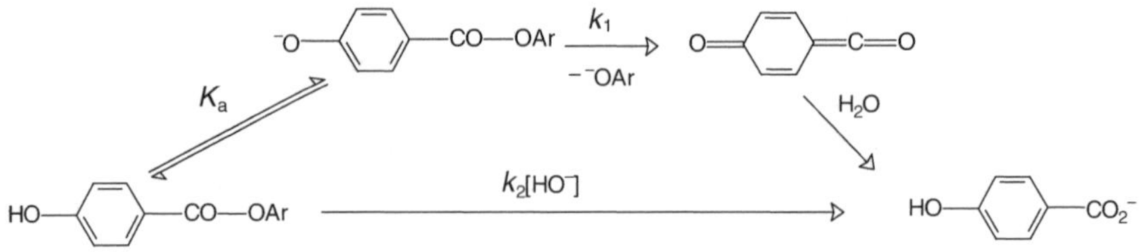

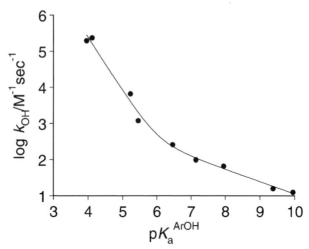

Scheme 8 *Parallel mechanisms for the alkaline hydrolysis of aryl 4-hydroxybenzoate esters*

The rate of hydrolysis is faster for esters with good leaving groups (phenols of low pK_a) than is predicted from the Brønsted plot of esters with weak leaving groups (phenols with high pK_a) (Figure 4). This is indicative of a change in mechanism from a simple bimolecular one for esters of weakly acidic phenols to an elimination–addition mechanism through an unsaturated intermediate (for esters of strongly acidic phenols) where the good leaving group is expelled in a unimolecular step from the conjugate base.

Figure 4 *Brønsted-type plot for the alkaline hydrolysis of substituted phenyl 4-hydroxy-benzoate esters – results diagnostic of parallel mechanisms*

7.3.1 Concave Curvature in pH-profiles

The existence of parallel mechanisms can be diagnosed by the observation of breaks in pH-profiles at which points the mechanisms have equal rate constants. The *concave* (*upwards*) break-point in Figure 3 is the pH at which the water and hydroxide ion pathways compete equally in the hydrolysis of the 2-methyl-4,6-diphenylpyrylium salt.

7.4 DEMONSTRATION OF CONCERTED MECHANISMS

Although the observation of a *convex* break in a free energy correlation indicates a stepwise process the observation of a linear correlation does *not exclude* an intermediate unless it can be shown that the condition $k_{-1} = k_2$ for the putative step-wise process occurs within the range of experimental substrates studied; usually there is no way of predicting the value of the σ or pK_a parameters for this condition. A linear plot results when *either* the break-point condition does not occur within the range of substrates studied *or* the polar substituent effects for the k_{-1} and k_2 rate constants are similar leading to small difference in slopes ($\Delta\rho$ or $\Delta\beta$) commensurate with the error in measuring the slope.

7.4.1 Technique of Quasi-symmetrical Reactions[16]

The substituent parameters for the condition $k_{-1} = k_2$ *can* be predicted for *quasi-symmetrical* reactions. The decomposition steps of a quasi-symmetrical intermediate to reactants (k_{-1}) and to products (k_2) therefore obey identical linear free energy relationships. A change in rate-limiting step will occur when the intermediate is perfectly symmetrical (and $k_{-1} = k_2$). In the addition-elimination mechanisms of Scheme 9 the ratio k_{-1}/k_2 becomes unity when entering and leaving groups are identical.

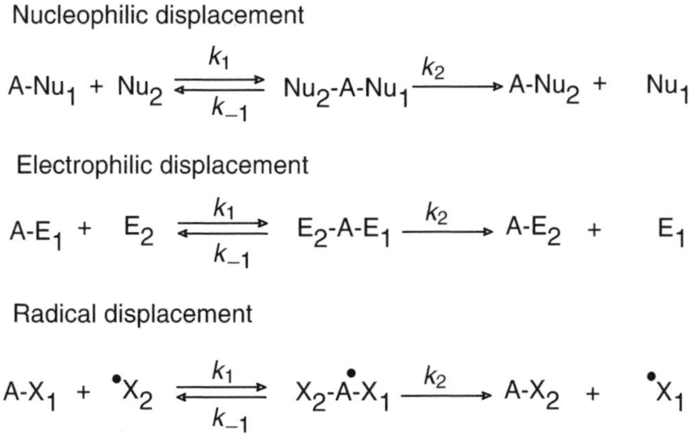

Nucleophilic displacement

$$A\text{-}Nu_1 + Nu_2 \underset{k_{-1}}{\overset{k_1}{\rightleftharpoons}} Nu_2\text{-}A\text{-}Nu_1 \xrightarrow{k_2} A\text{-}Nu_2 + Nu_1$$

Electrophilic displacement

$$A\text{-}E_1 + E_2 \underset{k_{-1}}{\overset{k_1}{\rightleftharpoons}} E_2\text{-}A\text{-}E_1 \xrightarrow{k_2} A\text{-}E_2 + E_1$$

Radical displacement

$$A\text{-}X_1 + {}^{\bullet}X_2 \underset{k_{-1}}{\overset{k_1}{\rightleftharpoons}} X_2\text{-}A\overset{\bullet}{\text{-}}X_1 \xrightarrow{k_2} A\text{-}X_2 + {}^{\bullet}X_1$$

Scheme 9 *Addition-elimination reactions with quasi-symmetrical intermediates*

Thus, in a reaction of a series of nucleophiles (Nu_2) with $A\text{-}Nu_1$ a plot of $\log k_{\mathrm{overall}}$ versus a polarity parameter of the nucleophile (for example pK_a or σ) will yield a break when the value of the parameter for Nu_2 is that for the constant nucleofuge Nu_1.[f] A similar result would occur for

---

[f] We identify the leaving group as Nu_1 to emphasise the symmetry of the reaction; the leaving group in a nucleophilic displacement reaction, although it is a nucleophile, is usually identified as Lg.

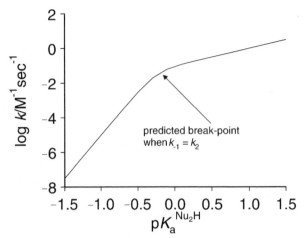

Figure 5 *Brønsted-type dependence for a nucleophilic displacement reaction involving a quasi-symmetrical intermediate (Scheme 9). Predicted break-point is at the pK_a of the leaving group (Nu_1H) for the symmetrical reaction*

electrophilic or homolytic displacement reactions. A free energy correlation as shown in Figure 5 illustrates the expected results for a quasi-symmetrical nucleophilic displacement reaction with putative intermediates.

Quasi-symmetrical bimolecular reactions involving dissociative mechanisms are illustrated in Scheme 10. These mechanisms involve the intervention of pre-association complexes (such as $Nu_2^-.A-Nu_1$).

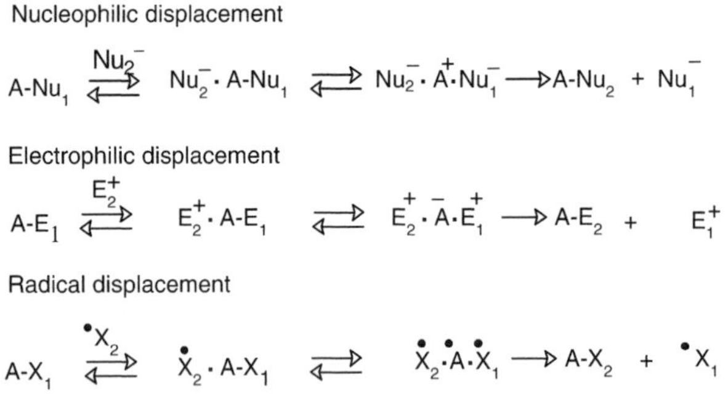

Nucleophilic displacement

$$A\text{-}Nu_1 \xrightleftharpoons{Nu_2^-} Nu_2^-\cdot A\text{-}Nu_1 \rightleftharpoons Nu_2^-\cdot \overset{+}{A}\cdot Nu_1^- \longrightarrow A\text{-}Nu_2 + Nu_1^-$$

Electrophilic displacement

$$A\text{-}E_1 \xrightleftharpoons{E_2^+} E_2^+\cdot A\text{-}E_1 \rightleftharpoons \overset{+}{E_2}\cdot \overset{-}{A}\cdot \overset{+}{E_1} \longrightarrow A\text{-}E_2 + E_1^+$$

Radical displacement

$$A\text{-}X_1 \xrightleftharpoons{{}^{\bullet}X_2} \overset{\bullet}{X_2}\cdot A\text{-}X_1 \rightleftharpoons \overset{\bullet}{X_2}\cdot \overset{\bullet}{A}\cdot \overset{\bullet}{X_1} \longrightarrow A\text{-}X_2 + {}^{\bullet}X_1$$

Scheme 10 *Quasi-symmetrical bimolecular reactions with dissociative mechanisms and termolecular intermediates*

The absence of a break-point at the predicted value of the polarity parameter can in principle exclude an association intermediate ($Nu_2^-.A^+ Nu_1^-$) and thus indicate that the mechanism involves only a single transition structure.

The Brønsted equations for the individual steps in the two-step mechanisms of Scheme 9 are written generally as Equation (8) where $k^o_n =$ the value of k_n at the location of the break-point ($pK_a - pK_o = \Delta pK_a = 0$).

$$\log k_n/k^o_n = \beta_n(pK_a - pK_o) = \beta_n \Delta pK_a \qquad (8)$$

Thus $\log k_1/k^o_1 = \beta_1 \Delta pK_a$, $\log(k_2/k^o_2) = \beta_2 \Delta pK_a$ and $\log(k_{-1}/k^o_{-1}) = \beta_{-1} \Delta pK$ and Equations (9) and (10) for $k_{overall}$ follow by substituting into Equation (3).

$$k_{forward} = (10^{\beta_1 \Delta pKa}\, k^o_1 10^{\beta_2 \Delta pKa}\, k^o_2)/(10^{\beta_{-1} \Delta pKa}\, k^o_{-1} + 10^{\beta_2 \Delta pKa}\, k^o_2) \qquad (9)$$

$$= 10^{\beta_1 \Delta pKa}\, k^o_1/[(k^o_{-1}/k^o_2)10^{(\beta_{-1} - \beta_2)\Delta pKa} + 1] \qquad (10)$$

Since $k_{-1} = k_2$ at $\Delta pK = 0$ (*i.e.* $k^o_{-1} = k^o_2$) Equation (10) reduces to Equation (11); by analogy $k_{reverse}$ is given by Equation (12).

$$k_{forward} = 10^{\beta_1 \Delta pKa}\, k^o_1/(10^{-\Delta\beta \Delta pKa} + 1) \qquad (11)$$

$$k_{reverse} = 10^{\beta_{-2} \Delta pKa}\, k^o_{-2}/(10^{\Delta\beta \Delta pKa} + 1) \qquad (12)$$

The parameter $\Delta\beta$ is defined as $\beta_2 - \beta_{-1}$. A similar equation can be written for Hammett dependencies with ρ replacing β and σ replacing pK_a. If the investigator is fortunate enough to be studying a system where the reverse reaction can be followed, Equation (12) can also be fitted to the data. Only one such system is at present known to have been investigated in this way.[17]

The technique of *quasi-symmetrical reactions* can be illustrated by application to the reaction of substituted phenolate ions with 4-nitro-phenyl acetate (Figure 6). This reaction possesses a linear free energy relationship over the range of pK_a of the nucleophiles encompassing the predicted break-point. This result was very surprising at the time as it was thought that all esters reacted *via* tetrahedral intermediates and that a break-point should therefore occur at $pK_a \sim 7$, the pK_a of the conjugate acid of the 4-nitrophenolate leaving group. The results indicate that there is no discrete intermediate and that these displacement reactions have concerted mechanisms.

The quasi-symmetrical technique relies on a relatively smooth variation of rate with substituent parameter. In practice, due attention must be paid to uncertainties in data fitting which arise from micro-scopic medium effects (Chapter 6). A reliable conclusion demands a large number of data which moreover cover a substantial range above

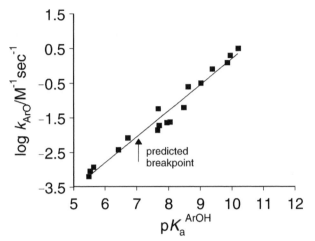

Figure 6 *Displacement reaction of 4-nitrophenyl acetate by substituted phenolate ion nucleophiles*

and below the value of the parameter for the break-point where ΔpK_a or $\Delta\sigma = 0$.

The technique can be applied to dissociative processes (Scheme 10) and distinguishes between a concerted mechanism and a stepwise process. The break-point in the free energy correlation would occur at the value of the parameter giving a symmetrical termolecular intermediate for example ($Nu_2.PO_3^-.Nu_1$) in the pyridinolysis of *N*-phospho-isoquinolinium ion (Figure 7). The example shown is the first system to be studied by the *quasi-symmetrical* technique.

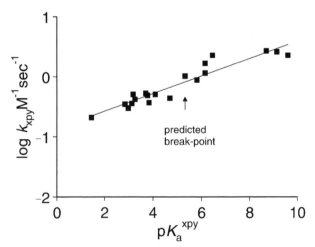

Figure 7 *Displacement of isoquinoline from N-phospho-isoquinolinium ion by substituted pyridines*

7.4.2 Criteria for Observing the Break

The sharpness of the break-point in Figure 5 (the angle between the intersecting energy lines) depends on the value $\Delta\beta$ (or $\Delta\rho$) which is the relative sensitivity of k_{-1} and k_2 to change in the attacking group or leaving group. Changing the leaving group will have a larger effect on k_2 than on k_{-1}, whereas change of attacking group has a larger effect on k_{-1} than on k_2; $\Delta\beta$ (or $\Delta\rho$) is expected to be relatively large in a stepwise substitution reaction.

  If the fit has a value $\Delta\rho$ or $\Delta\beta$ no greater than the confidence limit the putative intermediate may not have a significant barrier to decomposition. The value of $\Delta\beta$ or $\Delta\rho$ then gives an *upper limit* to the charge difference between intermediate and either forward or return transition structures. This value can be compared with the *overall* change in charge for the reaction. The upper limit of $\Delta\beta$ in phenoxide attack on phenyl acetates (0.05) may be compared with the maximal change of 1.7 units and is considered to be too small to accommodate significant barriers for a stable intermediate.[18]

7.4.2.1 Effective Charge Map for a Putative Stepwise Process.   Effective charge maps can also be employed to discount a stepwise process if estimates of the effective charge change from reactant and product to putative intermediates are not consistent with expectation. Consider the Brønsted dependence for reaction of substituted phenolate ions with 4-nitrophenyl acetate (Figure 6).[19] The value of β_{Nuc} is approximately 0.80 for attack of substituted phenolate ions on 4-nitrophenyl acetate when the second step of the putative two-step mechanism (decomposition of the putative tetrahedral intermediate) would be rate limiting ($pK_a^{Nuc} <$ pK_a^{Lg}) (Scheme 11).

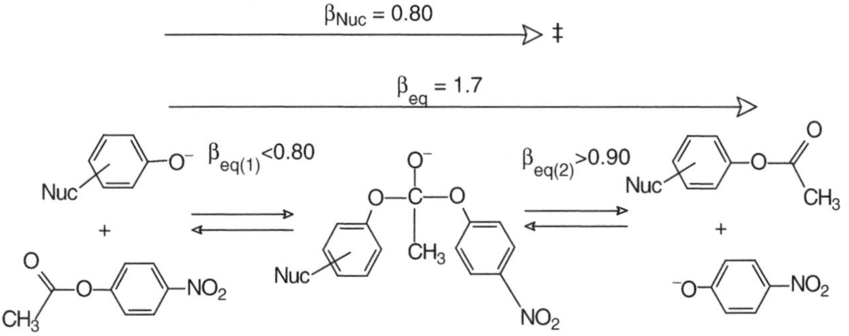

Scheme 11   *Effective charge map for putative stepwise mechanism for phenoxide ion attack on 4-nitrophenyl acetate*

Since k^{ArO} is $k_1 k_2 / k_{-1}$ under these conditions, the value of β_{eq1} for formation of the intermediate (k_1/k_{-1}) *must be less than* 0.80. Since the overall β_{eq} is 1.70 (see Scheme 11) the equilibrium constant for formation of product from intermediate must have $\beta_{eq2} > 0.9$ (because $\beta_{eq} = \beta_{eq1} + \beta_{eq2}$). The second step in the putative two-step mechanism would thus be *more* sensitive to substituent on the attacking nucleophile than is the first step where the substituent is directly linked to the bond formation. These conclusions exclude the stepwise process since they are contrary to the expectation that β_{eq2} should be less than β_{eq1} because the substituent effect of the nucleophile on the leaving bond should be less than that on bond formation.

7.5 CALCULATION OF PHYSICO-CHEMICAL CONSTANTS

7.5.1 Dissociation and Equilibrium Constants

Linear free energy relationships are particularly useful if it is necessary to have accurate knowledge of the pK_a of an intermediate which is unstable, or for which other experimental difficulties such as solubility prevent its experimental determination. Information about the pK_a values of putative intermediates is essential in mechanistic studies of reactions where proton transfers are involved.[g]

The simplest calculations utilise a free energy equation and the appropriate substituent constant both obtained from tables. Thus the Hammett, Taft and Charton relationships (see Appendix 4) may be combined with σ, σ^* and σ_I values (see Appendix 1) to arrive at a calculated pK_a for a substrate.

Some calculations, implicitly using free energy relationships, involve summing substituent constants in substrates possessing more than one substituent. This additivity concept can be employed in a different way where the calculation is anchored on the pK_a of a standard structure which may be drawn from tables of experimental values[h] or calculated by use of free energy correlations. The technique has been discussed in Chapter 4 (Section 4.2) for substituent constants where the effects of substituents on the pK_a of an aromatic acid are *additive*.

Examples are shown in Scheme 12 for calculating the pK_a values of some acids. The unstable hydrate of formaldehyde is a model of many putative intermediates in biosynthetic pathways, and an indirect experimental method gives its pK_a as 13.7. The pK_a of primary alcohols (RCH_2OH) is governed by the Equation $pK_a = 15.9 - 1.42\sigma^*$.

---

[g] A good example is found in J.P. Fox and W.P. Jencks, *J. Am. Chem. Soc.*, 1974, **96**, 1436.
[h] See footnotes to tables in Appendices 3 and 4 for details of data sources.

Allowing for a statistical factor of 0.3 (there are two identical hydroxyl functions in the hydrate) substitution of σ^* for R = OH gives a calculated pK_a of 13.7.

Hydroxymethylmethoxyammonium ion (by Charton method)

$$HO-CH_2-NH_2\overset{+}{-}OCH_3$$

Route 1

Standard ionisation: $H-CH_2-NH_2\overset{+}{-}OCH_3$ $pK_{a}(obs) = 4.75$

ρ_I for $RCH_2NH_3^+ = -8.0$ and σ_I for $HO^- = 0.24$

$pK_a (calc) = 4.75 - 8.0 \times 0.24 = 2.83$

Route 2

Standard ionisation: $H-NH_2\overset{+}{-}OCH_3$ $pK_{a}(obs) = 4.6$

ρ_I for $RNH_3^+ = -20$

Substitute $HOCH_2$ ($\sigma_I = 0.11$) for H ($\sigma_I = 0$)

$pK_a (calc) = 4.6 - 20 \times 0.11 = 2.4$

Scheme 12 *Application of the Charton equation for the pK_a of the ammonium group of hydroxymethyl-methoxyammonium ion ($HOCH_2NH_2^+OCH_3$). More than one calculation is carried out to obtain a consensus where an experimentally obtained value is not possible. See text for details*

Despite the reasonable result for the formaldehyde hydrate the nature of the method means that the calculated pK_a carries a relatively large degree of uncertainty which should be assessed by carrying out several calculations starting with pK_a values for different standard dissociations. Scheme 12 illustrates calculations for the dissociation of the ammonium group in the hydroxymethylmethoxyammonium cation ($HOCH_2N-H_2^+OCH_3$), an example of a common intermediate in studies of acyl group transfer. Errors can also be reduced by using as standard the dissociation of an acid with structure as close as possible to that of the acid in question. In *Route 1*, the ammonium ion used as standard ($MeN^+H_2(OMe)$) has a pK_a of 4.75; the pK_a values of ammonium ions have a Hammett ρ value of -3.2 (equivalent to $\rho_I = -20$). Allowing a factor of 0.4 for the attenuation coefficient through a single methylene group gives a value $\rho_I = -8$ for $XCRRN^+HR(OMe)$ and substitution for the σ_I of the substituents gives the calculated pK_a of 2.83 for HOCH-$_2N^+H_2OMe$. In *Route 2*, a calculated pK_a of 2.4 is obtained starting with $MeONH_3^+$ ($pK_a = 4.6$) and substituting CH_2OH for H (20×0.11). The use to which such results can be put depends very much on the spread of calculated pK_a values from the different calculation routes. Spreads of up to 0.4 units are to be expected from these methods and even though these seem to be relatively large the results are still useful.

Calculations employing the additivity of contributions from fragments are illustrated in Scheme 13 for the dissociation of the $C_{(1)}$ hydroxyl of glucose and a secondary amine. Tables of ΔpK_a for fragments are given in Appendix 3 (Tables 9a and 9b); they may be used to calculate pK_a values by a simple additive protocol (the ΔpK_a method).[20]

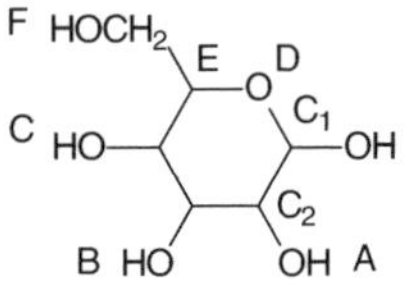

Glucose

σ^* for fragments of the glucose moiety by path C_2:

A	0.54	(CH_2OH)
B	0.54 x 0.4 = 0.22	(CH_2CH_2OH)
C	0.54 x $(0.4)^2$ = 0.086	($CH_2CH_2CH_2OH$)
E	1.81 x $(0.4)^4$ = 0.046	($CH_2CH_2CH_2CH_2OCH_3$)
F	0.54 x $(0.4)^4$ = 0.014	($CH_2CH_2CH_2CH_2CH_2OH$)

σ^* for fragments of the glucose moiety by path D:

A	0.54 x $(0.4)^3$ x 0.64 = 0.022	($-OCH_2CH_2CH_2OH$)
B	0.54 x $(0.4)^2$ x 0.64 = 0.055	($-OCH_2CH_2OH$)
C	0.54 x 0.4 x 0.64 = 0.14	($-OCH_2OH$)
D	(model is EtCH(Me)O—) 1.62	($-OCH(CHOH)-CH_2OH$)
F	0.54 x 0.4 X 0.64 = 0.14	($-OCH_2CH_2OH$)

$$\Sigma\sigma^* = 2.88$$

By Taft Equation for C_1 hydroxyl pK_a = 15.7 − 1.32 x 2.88 = 11.90

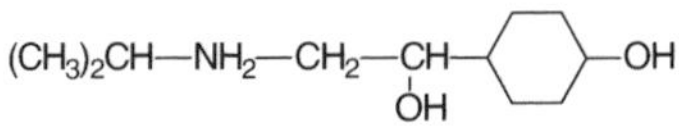

Secondary amine

pK_a of aliphatic secondary amine	11.15
ΔpK_a for β −OH	−1.1
ΔpKa for −OH on cyclohexyl 4 carbons from the β carbon 1.1x$(0.4)^4$ x 2 (two routes to nitrogen)	−0.06
pK_a^{calc}	9.99
pK_a^{exp}	10.23

Scheme 13 *Calculations of the pK_a of glucose and the conjugate acid of a secondary amine assuming additivity of substituent effects. The attenuation factor is taken as 0.4 per carbon and 0.64 per −O− (Table 1, Chapter 4) and the substituent effects operate along both arms of the six-membered rings*

Cross-checking of calculated pK_a values by using other calculating routes would give confidence in the result; however, the technique is time consuming and requires considerable background skill in obtaining and choosing standards. If the pK_a values of a large number of structures are needed for a *rough* guide then recourse to software is possible and a computer program, CAMEO®, has been developed[21] for pK_a values of acids in dimethylsulfoxide solvent[22] which starts with a structure building routine assuming that the pK_a is an additive property of the effects of functional groups. Further software, pKalc,® may be employed to calculate pK_a values.[23]

Data bases which possess routines for obtaining pK_a values[i] are deposited in the internet. The origin of the data in the software/website sources is not always divulged so that the resultant pK_a values should be treated with caution. It is not certain, for example, if the sources function as databases or as calculation routines. There are major literature data collections of pK_a values (see Further Reading) which list conditions of the measurements and which are therefore preferred sources.

Linear free energy relationships also enable the calculation of equilibrium constants (as opposed to dissociation constants). There have been no extensive compilations of these free energy correlations because ranges of equilibrium constants are difficult to measure. For the limited data available the reader is referred to the article by Jaffé[24] and the books of Hine[25] and Leffler and Grunwald;[26] techniques similar to those elaborated above enable equilibrium constants to be calculated with reasonable accuracy. Selected equations are listed in Appendix 4, Table 4.

The change in effective charge from reactants to products is reflected by the sensitivity of the equilibrium constant to change in the group which bears substituent change; it can be employed to calculate equilibrium constants for quasi-symmetrical group transfer reactions. The equilibrium constant for the transfer of the phosphoryl group ($-PO_3^{2-}$) between donor and acceptor (Scheme 14) may be deduced from Equation (13)[j] where $\Delta\varepsilon$ is the difference in effective charge between ArO^- acceptor and product ester (+1.36 in this case)[k] and $pK_a^{Acceptor-H}$ and $pK_a^{Donor-H}$ are the pK_a values of the conjugate acids of the phenolate ion of the donor molecule and acceptor phenolate ion respectively.

$$\log K_{eq} = \Delta\varepsilon(pK_a^{Acceptor-H} - pK_a^{Donor-H}) \qquad (13)$$

---

[i] See for example: http://www.chemweb.com/databases
[j] For a derivation see Problem 20.
[k] Refer to Scheme 1, Chapter 3, for effective charge data.

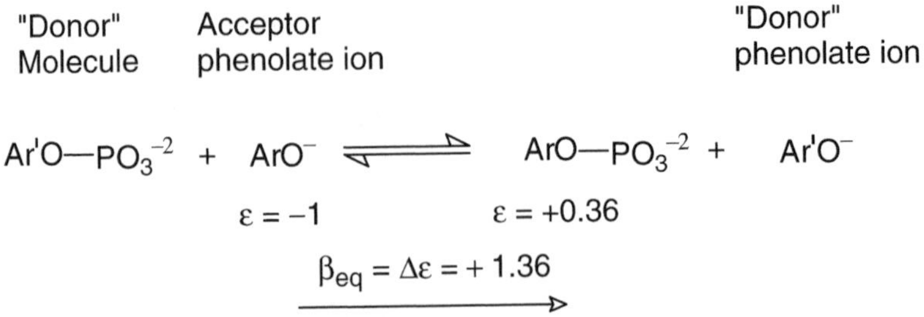

"Donor"   Acceptor   "Donor"
Molecule   phenolate ion   phenolate ion

$$Ar'O{-}PO_3{}^{-2} \; + \; ArO^- \; \rightleftharpoons \; ArO{-}PO_3{}^{-2} \; + \; Ar'O^-$$

$$\varepsilon = -1 \qquad\qquad\qquad \varepsilon = +0.36$$

$$\beta_{eq} = \Delta\varepsilon = +1.36$$

Scheme 14   *The Brønsted* β_{eq} *is related to the change in effective charge*

The equilibrium constant for a group transfer reaction where the effective charge has not been experimentally determined could be obtained by estimating the effective charges on the donors and acceptors as described in Chapter 3 and then applying Equation (13).

In the case of unsymmetrical reactions the equilibrium constant can be calculated for any donor provided it is known for any one donor. The equilibrium constant $(K_{eq}{}^{4NPOH} = 0.37)^{27}$ for the reaction of imidazole with 4-nitrophenyl acetate (Equation 14, Ar = 4-nitrophenyl = 4NP) can easily be measured by use of forward and reverse rate constants $(K_{eq} = k_f/k_r)$. The equilibrium constant for the reaction is governed by a Brønsted Equation $\log K_{eq} = \beta p K_a{}^{ArOH} + C$ where β is the value $\Delta\varepsilon$, the difference in effective charge between phenolate ion product and aryl acetate ester reactant ($\Delta\varepsilon = -1.7$). The constant, C, is computed to be $C = \log K_{eq}{}^{4NPOH} - \Delta\varepsilon p K_a{}^{4NPOH}$. The value of K_{eq} for any other phenol can then be calculated by taking $pK_a{}^{4NPOH} = 7.14$ (Equation 15).

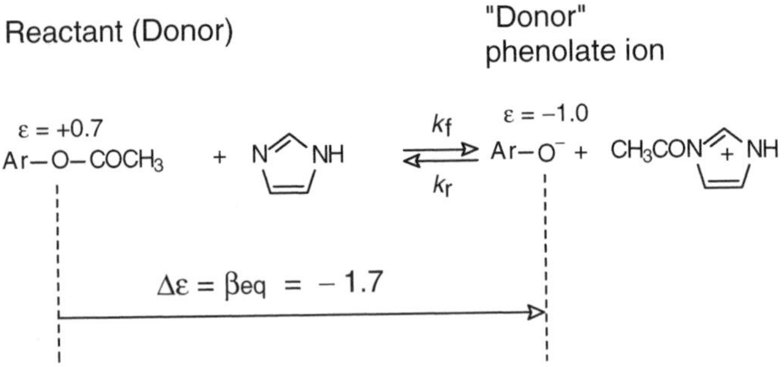

Reactant (Donor)   "Donor"
                                              phenolate ion

$$\varepsilon = +0.7 \qquad\qquad\qquad\qquad\quad k_f \quad\; \varepsilon = -1.0$$
$$Ar{-}O{-}COCH_3 \;\; + \;\; N{\overset{\frown}{\underset{\smile}{}}}NH \;\; \underset{k_r}{\overset{}{\rightleftharpoons}} \;\; Ar{-}O^- + \; CH_3CON{\overset{\frown}{\underset{\smile}{}}}{+}\,NH \qquad (14)$$

$$\Delta\varepsilon = \beta_{eq} \; = \; -1.7$$

$$\log K_{eq} = \Delta\varepsilon(pK_a{}^{ArOH} - pK_a{}^{4NPOH}) + \log K_{eq}{}^{4NPOH}$$

$$= -1.7 pK_a{}^{ArOH} + \log K_{eq}{}^{4NPOH} + 1.7 pK_a{}^{4NPOH}$$

$$= -1.7 pK_a{}^{ArOH} - 0.43 + 1.7 \times 7.14$$

$$= -1.7 pK_a{}^{ArOH} + 11.71 \qquad\qquad\qquad (15)$$

7.5.2 Rate Constants

Methods for calculating rate constants are of undoubted use in estimating times to completion for reactions in large-scale industrial processes. Estimated rate constants for parallel reactions are of importance in calculating expected yields when there are likely to be by-products. Calculated rate constants may be used in mechanistic studies to estimate the lifetimes of reactive, putative, intermediates to gauge the significance of postulated mechanisms. Free energy relationships are of substantial use in all such calculations.

An interesting mechanistic application of free energy relationships in calculating rate constants concerns the reactivity of the putative intermediate in the reaction of imidazole with 4-nitrophenyl acetate. The putative stepwise process (Scheme 15) involves a zwitterionic tetrahedral intermediate which would decompose to the product *N*-acetylimidazole, subsequently hydrolysing to acetate ion and imidazole.

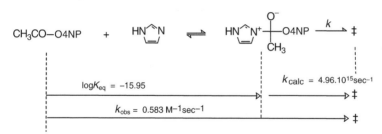

Scheme 15 *Equilibria in the reaction of imidazole with 4-nitrophenyl acetate; the equilibrium data are from ref. 28*

The equilibrium constant for zwitterion formation ($\log K_{eq} = -15.95$) may be obtained from the equilibrium constant for formation of the neutral adduct ($\log K_{eq} = -13.39$)[28] and the pK_a values of hydroxyl and imidazolyl of the adduct (8.12 and 5.56 respectively), which are determined by calculation. The decomposition rate constant k_{calc} can be determined for the putative zwitterion intermediate by combining the equilibrium constant and the observed rate constant (0.583 M^{-1} sec^{-1}) to give 4.96×10^{15} sec^{-1}. This rate constant exceeds the value expected for a vibration ($10^{13} sec^{-1}$) and indicates that the intermediate cannot have a discrete existence and that the mechanism must be a concerted process *enforced* by the short lifetime of the intermediate. In other words the putative intermediate cannot pass enough reaction flux to support the observed reactivity.

7.5.3 Partition Coefficients and Other Constitutive Molecular Properties

Partition coefficients are of extreme value in medicinal chemistry as they may be related to drug transfer across membranes and binding at active centres. The parameter "$\log P$" models various kinds of transport between aqueous and lipid phases[29] and P is defined as the partition coefficient of a substance between water and n-octanol. The free energy of the transport process is often linearly related to standard "hydrophobic" parameters (π) for substituents (X) determined from P_X values for substituted phenoxyacetic acids and that for the parent acid P_H (Equation 16).

$$\pi = \log P_X - \log P_H \tag{16}$$

Linear free energy equations for a variety of transport processes are collected in Appendix 4, Table 5; these include both polar and hydrophobic parameters. $\log P$ values may be calculated by a "fragmental hydrophobicity" system.[29] Partition coefficients (P) expressed on a logarithmic scale are an additive constitutive molecular property (like, for example, molecular volume) and a selection of additivity values (f) which relate to fragments such as groups, functions and bonds making up the structure of a molecule are recorded in Appendix 3, Table 8. The overall value of $\log P$ is given by Equation (17) where a_i indicates the population of a given fragment in a structure and f_i is its absolute contribution to the total lipophilicity.

$$\log P = \Sigma_{i = 0 \to n} a_i f_i \tag{17}$$

Simply adding contributions from the constituent fragments neglects the effect of the extra bonds associated with these fragments, and the terms f_b and f_{cbr} are employed to allow for a decrease in $\log P$ associated with a bond and branching in a carbon chain respectively.[29,30] Another correction (f_{gbr}) is needed where the molecule possesses branching at a group (see Problem 15(a)). Every single bond *after the first one* makes a negative contribution to $\log P$. The calculation of the partition coefficient from f values is exemplified in Equation (18) for PhCH$_2$-CH$_2$CH$_2$Cl:

$$\log P = f_{Ph} + 3f_{CH2} + 3f_b + f_{aliph.Cl} \tag{18}$$

$$= 1.90 + 1.98 - 0.36 + 0.06$$

$$= 3.58 \text{ (experimental } \log P = 3.55)$$

The calculations can also be carried out with software, which is useful for the large numbers of substances often requiring processing in the pharmaceutical industry. A standard software package (MEDCHEM)® is available in which the CLOGP® algorithm is employed to calculate the log$P$ values.[31] The confidence in a particular calculated value will derive from comparison of a number of estimates (which can be carried out manually) starting with experimental log$P$ values of different standard compounds.

7.6 RESOLUTION OF KINETICALLY EQUIVALENT MECHANISMS

An *experimentally* determined rate law describes the *composition* of the rate-limiting transition structure but not its structure or atom connectivity. Even though the composition of the transition structure of a rate-limiting step is uniquely described by the rate law, kinetic equations for a given reaction can often be written in different but equivalent ways; these cannot be distinguished by studying the reaction kinetics under varying conditions. Such ambiguity most frequently occurs in reactions where acid–base reagents are involved. The rate Equation can be predicted uniquely from the mechanism of a reaction but the mechanism is not unique for the rate equation. Free energy relationships can sometimes be used to resolve kinetic ambiguities by techniques shown in the following examples.

7.6.1 General Base Versus Nucleophilic Reaction[32,33]

Both nucleophile and base are essentially similar reagents as they each have a lone pair of electrons to bond to either proton or electrophile. A lone pair donor acting as a nucleophile or as a general base (Scheme 16) in an overall hydrolysis reaction leads to identical rate laws.

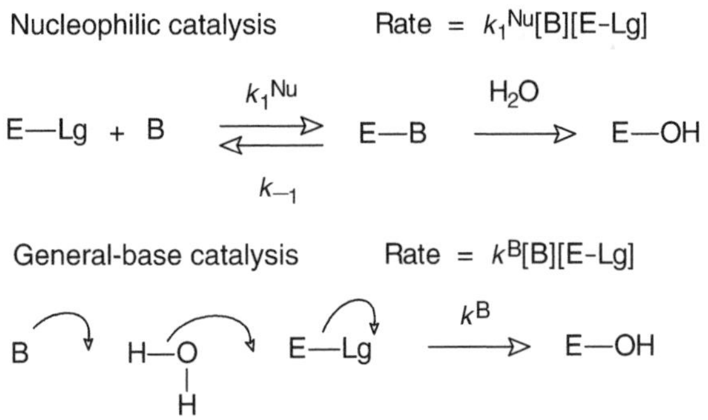

Scheme 16 *Kinetic ambiguity in hydrolysis reactions catalysed by a base or nucleophile (B)*

During nucleophilic attack the reagent must approach the electrophile more closely than when acting as a base. A free energy relationship for the general base mechanism is therefore expected to be insensitive to steric effects of the base whereas that for the nucleophilic mechanism is expected to be very sensitive to steric bulk. The points for the sterically hindered nucleophiles will fall below the line determined for sterically unhindered nucleophiles. Figure 8 in Chapter 2 illustrates the large deviation for attack of 2,6-disubstituted phenolate ions at a 4-nitro-phenoxytriazine compared with the line defined for 3- and 4-substituted phenolate ion nucleophiles. These results confirm that the reaction is nucleophilic substitution and not general base catalysed hydrolysis.[1] Similar arguments can be made for reaction of substituted phenolate ions with 4-nitrophenyl acetate. Substitution at a *single ortho* position often does not cause a substantial deviation. The attacking phenolate ion can swing its single *ortho* substituent away from the reaction centre during reaction so that it incurs little steric penalty; this is not possible with the 2,6-disubstituted species. The 2,6-difluoro substituted phenolate ions behave as if there were no steric effect in accord with the known small size of the fluoro group. In Scheme 16 the kinetic ambiguity could be resolved if the intermediate E–B were observed but this is often not possible.

7.6.2 General Acid Catalysis Versus Specific Acid–General Base Catalysis[32]

The rate of a bimolecular reaction between acid (HA) and a nucleophilic substrate (Nu) could obey Equation (19).

$$\text{Rate} = k_2[\text{HA}][\text{Nu}] \tag{19}$$

If the proton transfer steps are not rate limiting HA and Nu are respectively in equilibrium with their respective conjugate base and acid (Equations 20 and 21) and Equation (19) may be rewritten as Equation (22):

$$K_a^{HA} = [\text{H}^+][\text{A}^-]/[\text{HA}] \tag{20}$$

$$K_a^{HNu} = [\text{H}^+][\text{Nu}]/[\text{HNu}^+] \tag{21}$$

---

[1]   In this case the evidence is confirmatory because product analysis demonstrates ether formation rather than hydrolysis.

$$\text{Rate} = k_2[\text{A}^-][\text{HNu}^+]K_a^{\text{HNu}}/K_a^{\text{HA}} \equiv k_2' [\text{A}^-][\text{HNu}^+] \qquad (22)$$

Equations (19) and (22) refer to mechanisms involving reaction of HA with Nu or A$^-$ with HNu$^+$ respectively. The respective rate constants k_2' and k_2 are related by the Equation $k_2' = k_2 K_a^{\text{HNu}}/K_a^{\text{HA}}$ and the experimental rate laws do not distinguish between A$^-$ reacting with HNu$^+$ or HA reacting with Nu.

Attempting to distinguish the two equivalent rate laws by changing the reaction conditions to favour the concentration of the reactive species leads to failure. At high pH values, the concentration of A$^-$ relative to that of HA increases; it would be incorrect to assume that an observed increase in rate constant with pH indicates Equation (22) as the rate law because increasing pH has an *opposite* effect on [HNu$^+$]. Conversely, increasing the pH decreases [HA] relative to [A$^-$] according to Equation (21) but increases that of [Nu] relative to [HNu$^+$]. It is therefore not possible to distinguish between equations (21) and (22) and they are deemed to be *kinetically indistinguishable*.

The kinetic ambiguity of general acid/specific acid–general base catalysis is a consequence of the mobile proton transfer equilibrium prior to the rate-limiting step; it arises most frequently in reactions involving proton transfer because proton transfer is often a labile process and it could be regarded as a special case of the ambiguity seen in systems to which the Curtin–Hammett[12] principle is applied. Since many bioorganic reactions require proton transfer, kinetic ambiguity is a very important topic and is one which is often forgotten by over-enthusiastic students of mechanism.

The kinetic ambiguity registered between general acid and specific acid–general base mechanisms in carbonyl addition reactions with nucleophiles (Scheme 17) is a classic mechanistic problem.[m]

The rate laws for mechanisms A and B of Scheme (17) are Equations (23) and (24):

$$\text{Rate} = k[\text{HNu}][{>}\text{C}{=}\text{O}][\text{HB}] \qquad (23)$$

$$= k'[\text{HNu}][{>}\text{C}{=}\text{O}][\text{H}_3\text{O}^+][\text{B}] \qquad (24)$$

$$\equiv k'[\text{HNu}][{>}\text{C}{=}\text{O}][\text{HB}]K_a^{\text{HB}}$$

---

[m] Although classic, this continues to be a contemporary problem in mechanistic studies of fundamental organic as well as enzymatic catalyses.

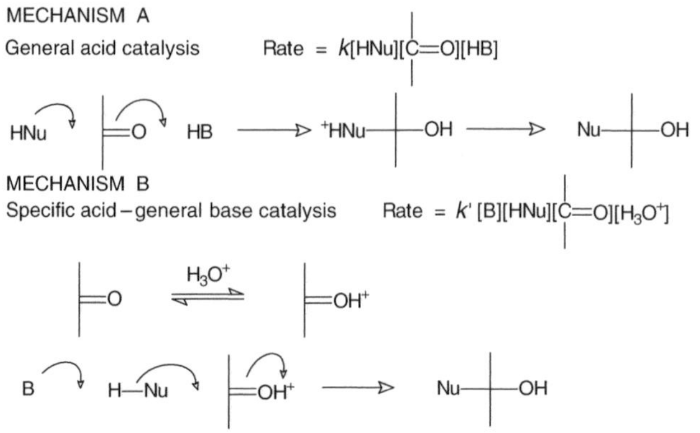

MECHANISM A
General acid catalysis   Rate = $k[HNu][C{=}O][HB]$

MECHANISM B
Specific acid–general base catalysis   Rate = $k'[B][HNu][C{=}O][H_3O^+]$

Scheme 17   *The general acid/specific acid–general base ambiguity*

Thus

$$k \equiv k'K_a^{HB} \qquad (25)$$

The Brønsted slope (α) for the general acid-catalysed reaction (A) and (β) for the general base-catalysed reaction (B) are related according to Equation (25) as $\beta = 1 - \alpha$.[n]

The sensitivity of a substrate to the nucleophilicity of the attacking reagent should decrease with increasing basicity of a general base catalyst or with increasing acidity of the general acid catalyst. Thus mechanism (A) (Scheme 17) should possess a numerical value of α which decreases as the nucleophilic strength of HNu increases. Mechanism (B) should possess a β decreasing with increasing strength of HNu.

The numerical values of α for addition of nucleophiles to acetaldehyde and benzaldehydes are observed to decrease as the nucleophilic strength of the attacking reagent increases. The value β (determined from $\beta = 1 - \alpha$) therefore increases with increasing nucleophilic strength of the adding nucleophile ($\beta\downarrow = 1 - \alpha\uparrow$) ruling out mechanism (B).[34,35]

The change in α and β can be understood from the More O'Ferrall–Jencks diagrams (Figures 8 and 9) where the vector resulting from increasing nucleophilicity of HNu causes a shift to smaller α and smaller β for mechanisms A and B respectively.

7.7 SOME BIOLOGICAL APPLICATIONS

Application of linear free energy relationships to enzyme mechanisms has, naturally, been attempted but the influence of the substituent may

---

[n] The term α in these arguments is the Brønsted α and not the Leffler α.

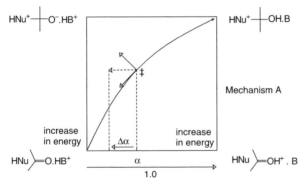

Figure 8 *Increase in nucleophilicity of HNu (SE and SW corners increase in energy) causes a Hammond movement to the SW and a Thornton (perpendicular) movement to the NW. The resultant vector (Δα) decreases α*

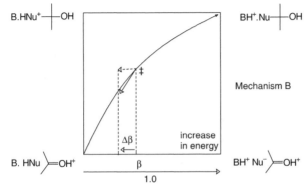

Figure 9 *Increase in nucleophilicity of HNu (SW corner increases) causes a movement to the West giving a resultant vector (Δβ) decreasing β*

not simply be through the transmission path of the regular Hammett or Brønsted standard; this is because of possible interaction between substrate and enzyme through space and through steric exclusion. The enzyme active site is generally a region which encourages multiple interactions between enzyme and substrate. Such interactions are liable to increase the probability and significance of special effects similar to those involving change in the microscopic medium. Such problems are seen in the extreme in a non-enzymic example namely the acylation of cyclo-dextrin by substituted phenyl acetates[36] where a combination of Hansch constants, Hammett σ and molar volumes are unable to correlate the rate constant data for a system manifestly less complex than an enzyme one.

In the case of chymotrypsin, the acylation rate constant k_{cat}/K_m for substituted phenyl acetates possesses a tolerably good Hammett correlation, but the plot is relatively scattered and there is probably substantial binding of the phenyl moiety within the "tosyl pocket", a lipophilic binding cavity in the enzyme. The binding of the side chain of substituted phenyl esters of *N*-acylamino acids is within the "tosyl pocket".[37] This binding will effectively force the phenoxy leaving group to reside in the bulk solvent where the interactions are with solvent alone and are thus similar to those of the standard dissociation reaction used to define the Hammett σ values. The low value of the Hammett ρ value for these substrates points to substantial electrophilic assistance in the transition structure for the acylation step. The obvious interpretation of this is that the electrophilic assistance comes from some sort of interaction at the carbonyl of the ester[38] because phenoxide ion leaving groups are unlikely to require assistance to enable them to depart from a tetrahedral intermediate: the interaction can be identified by X-ray crystallography of enzyme complexes with substrate-like inhibitors as deriving from hydrogen-bonding from peptide NH groups to the ester carbonyl oxygen.[37]

In the above example the development of negative charge in the transition structure is substantially less than the negative charge developed on the leaving oxygen in the product. It is not often possible to disentangle the effect of the binding step on Hammett or Brønsted exponents from that of the catalytic steps in the enzyme reaction, unlike the situation in non-enzymatic reactions. The second-order rate constant k_{cat}/K_m is probably the best enzymatic parameter to study; it reflects the free energy difference between the transition structure for acylation and the *free* enzyme and *free* substrate and registers enzyme substrate binding interactions only in the transition structure. Hence, the reactant state "structure" of the substrate in terms of effective charge distribution may be obtained from non-enzymatic data such as given in Scheme 1 of Chapter 3. When k_{cat} is used as the diagnostic tool it reflects the free energy difference between an unknown transition structure and the often unknown substrate structure in its bound state.

The Leffler parameter α has been applied to the effect of point mutations of amino acid residues on the binding step of tyrosyl-^{t}RNA synthetase, the enzyme responsible for catalysing formation of tyrosyl-^{t}RNA during protein biosynthesis. The logarithm of the rate constant for the reaction is linearly dependent on the logarithm of the equilibrium constant (Equation 26) for a range of directed point mutations of amino acid residues close to the proposed active site.[39]

$$\log k = \alpha \log K + C \qquad (26)$$

The Leffler α value (0.79) indicates that the change in binding is some 79% complete in the transition structure compared with that in the final bound system. The constraints which should be applied in the use of this type of approach are the same as those discussed in Chapter 3, namely that the effect of the changes in the structure should be relatively small; in addition they should not destroy the integrity of the enzyme structure if the results of such changes are to be easily interpreted.

Aspartate aminotransferase catalyses the transfer of the NH_2 group between aspartic acid and 2-oxoacids *via* the intervention of a pyridoxal coenzyme. The kinetically significant step, a 1,3-proton transfer, is thought to involve lysine-258 in a proton relay system (Scheme 18); replacement of lysine-258 with an alanine residue by site-directed muta-tion[40] yields a mutant enzyme which is only weakly active but which can have activity restored by added amines (Scheme 18).

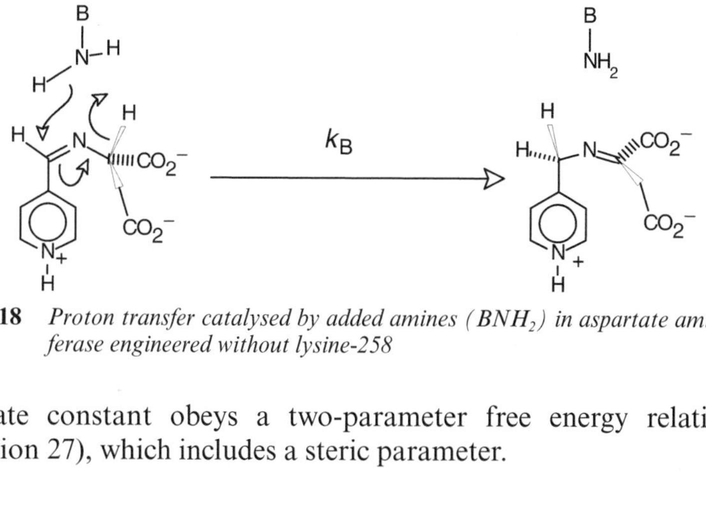

Scheme 18 *Proton transfer catalysed by added amines (BNH₂) in aspartate aminotrans-ferase engineered without lysine-258*

The rate constant obeys a two-parameter free energy relationship (Equation 27), which includes a steric parameter.

$$\log k_B = 0.39 \ pK_a - 0.055 \ \text{Molecular Volume} - 0.7 \qquad (27)$$

The molecular volume effect is relatively large and consistent with the requirement that the amine must be accommodated in a restricted cavity vacated when lysine-258 is exchanged for the alanine residue. The β value indicates that some 40% of the full charge is developed on the nitrogen base.

7.8 FURTHER READING

Reference Tables of pK_a Values

E.P. Serjeant, *Ionisation Constants of Organic Acids in Aqueous Solution*, Pergamon Press, London, 2nd edition, 1979.

W.P. Jencks and J. Regenstein, Ionisation Constants of Acids and Bases, in *The Handbook of Biochemistry*, H.A. Sober, (ed.), Chemical Rubber Publishing Co., Cleveland, OH, 1970, 2nd edition, Section i-187.

G. Kortüm *et al.*, *Dissociation Constants of Organic Acids in Aqueous Solution*, Butterworth, London, 1982.

See also http://www.compudrug.com/pkalc.html for software and information for calculating pK_a values and http://www.chemweb.com/ databases.

C.A. Carreira *et al.*, Estimation of the Ionisation pK_a of Pharmaceutical Substances using the Computer Program SPARC, *Talanta*, 1996, **43**, 607 (a related web-site is http://ibmlc2.chem.uga.edu/sparc).

Reference Tables of log*P* Values

C. Hansch and A. Leo, *Substituent Constants for Correlation Analysis in Chemistry and Biology*, Wiley-Interscience, New York, 1979.

See also http://www.Biobyte.com for software and information for calculating log*P* values.

General References

J.P. Guthrie, Correlation and Prediction of Rate Constants for Organic Reactions, *Can. J. Chem.*, 1996, **74**, 1283.

C. Hansch, A Quantitative Approach to Biochemical Structure-Activity Relationships, *Accs. Chem. Res.*, 1969, **2**, 232.

C. Hansch, Quantitative Structure–Activity Relationships and the Unnamed Science, *Accs. Chem. Res.*, 1993, **26**, 147.

C. Hansch *et al.*, Chem-Bioinformatics and QSAR: A Review of QSAR Lacking Positive Hydrophobic Terms, *Chem. Rev.*, 2001, **101**, 619.

W.P. Jencks, When is an Intermediate not an Intermediate? Enforced Mechanisms of General Acid Base Catalysed, Carbocation, Carbanion and Ligand Exchange Reactions, *Accs. Chem. Res.*, 1980, **13**, 161.

W.P. Jencks, Are Structure–Reactivity Correlations Useful? *Bull. Soc. Chim. France*, 1988, 218.

A.J. Leo, Calculating logP_{oct} from Structures, *Chem. Rev.*, 1993, **93**, 1281.

D.D. Perrin *et al.*, pK_a *Prediction for Organic Acids and Bases*, Chapman & Hall, London, 1981.

P.J. Taylor, On the Calculation of Tetrahedral Intermediate pK_a Values, *J. Chem. Soc., Perkin Trans. 2*, 1993, 1423.

A. Williams, The Diagnosis of Concerted Organic Mechanisms, *Chem. Soc. Rev.*, 1994, 93.

A. Williams, *Concerted Organic and Bio-Organic Mechanisms*, CRC Press, Boca Raton, FL, 2000.

7.9 PROBLEMS

1(a) Derive the Equation governing the overall rate constant for parallel mechanisms ($k_{overall} = k_1 + k_2$) given that $\log k_n = \beta_n pK_a + C_n$ ($n = 1$ or 2). Sketch the three main types of non-linear free energy relationships to illustrate a change in mechanism according to the equation, assuming that linear correlations for k_1 and k_2 have either positive or negative slopes or combinations of both.

1(b) Derive Equation (4) from Equation (3). Sketch the three main types of non-linear free energy relationships arising from a change in rate-limiting step in a step-wise mechanism (see Equations 4, 11 and 12). Each rate constant has its own linear free energy relationship and the correlations for k_1 and $k_1 k_2 / k_{-1}$ can have positive or negative slopes or combinations of both.

2 The hydrolysis of alkyl and aryl acetoacetates (HS) possesses pH-independent rate constants (Table 1) corresponding to the decomposition of the conjugate base of the ester ($CH_3CO-CH^--CO_2Ar$). Comment on the hydrolysis mechanism considering that the value of β_{Lg} for the alkaline hydrolysis of aryl acetates is -0.26.

Table 1

Acetoacetate ester	pK_a^{LgH}	$\log k_{plateau}/sec^{-1}$
phenyl	9.95	−1.25
4-chlorophenyl	9.38	−0.602
3-chlorophenyl	9.02	−0.0458
4-methoxycarbonylphenyl	8.50	0.914
4-acetylphenyl	8.05	1.26
3-nitrophenyl	8.35	0.819
4-nitrophenyl	7.14	2.29
2′2′2-trifluoroethyl	12.43	−3.04
propargyl	13.55	−3.12
2-methoxyethyl	14.7	−3.15
ethyl	16	−3.20

Compare the rate constant for the putative reaction of hydroxide ion with 4-nitrophenyl acetoacetate ($k_{OH} = k_{plateau} K_a^{HS} / K_w = 1.32\ 10^5\ M^{-1}\ sec^{-1}$)[41] with the value calculated from that for 4-nitrophenyl acetate ($9.5\ M^{-1}\ sec^{-1}$)[42] using an estimated ρ^* for the reaction of 4-nitrophenyl substituted acetates of 2.5, $\sigma^* = 0.62$ for CH_3COCH_2- and $\sigma^* = 0$ for CH_3-.

3 Construct an effective charge map for the alkaline hydrolysis of aryl acetate esters using data for β_{eq} from Chapter 3 and $\beta_{Lg} = -0.3$.

4 Assuming the conservation of effective charge, construct a full effective charge map for an identity reaction of phenolate ions with phenyl acetates. Assume that the reaction has a β_{Nuc} of +0.75 and β_{Lg} of −0.95 and that the corresponding β_{eq} is −1.70. Comment on these results regarding charge balance.

5　Construct a Brønsted-type graph for the hydroxide ion-catalysed elimination reaction (Equation 28) using the data[43] for $\log k_{OH}$ and pK_a^{Xpy} given in Table 2.

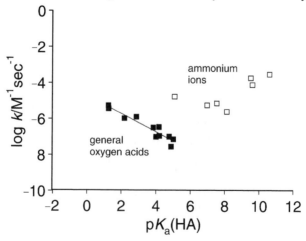

$$\overset{+}{N}-CH_2-CH_2-CN \quad \xrightarrow[^-OH]{k_{OH}} \quad N + CH_2{=}CH-CN \qquad (28)$$

Table 2

Substituent (X)	pK_a^{Xpy}	$\log k_{OH}/M^{-1}sec^{-1}$
3-Cl	2.84	−3.09
3-CH$_2$CN	4.08	−3.36
3-C$_6$H$_5$	4.88	−3.47
H	5.16	−3.46
3-Me	5.82	−3.56
2-Me	6.04	−3.65
4-Me	6.15	−3.65
3,4-Me$_2$	6.45	−3.74
3-Br-4-NH$_2$	7.15	−4.00
4-morpholino	8.80	−4.58
4-NH$_2$	9.21	−4.81
4-Me-2-NO$_2$	9.68	−5.00

The data should be fitted to Equation (4) suitably modified for Brønsted correlations. Comment on the non-linearity of the plot and suggest a mechanism fitting the observations.

6　The Brønsted plot for the general acid-catalysed halogenation of acetone is illustrated in Figure 10. Carboxylic acid catalysts exhibit

Figure 10　*Halogenation of acetone catalysed by oxygen and ammonium ion acids*

a linear regression with ammonium ion catalysts from primary and secondary amines deviate markedly from the line.[44] Suggest a possible reason for this deviation.

7 Refer to Chapter 2 (Figure 8), where the reaction of substituted phenolate anions with a 4-nitrophenoxytriazine is graphed against the pK_a of the attacking nucleophile. Explain the significance of this Brønsted-type relationship which exhibits no break-point when $pK_a^{NucH} = pK_a^{LgH}$. In this, and Problems 8 and 9, the two deviant points should not be considered as part of any correlations; however the significance of these points should be briefly discussed.

8 The reaction of substituted phenoxide ions with 2(4'-nitrophenoxy)-4,6-dimethoxy-1,3,5-triazine (Equation 29) exhibits a linear Brønsted correlation over a range of nucleophile pK_a from 5.5 to 10 (see Problem 7).[45]

$$\text{(29)}$$

The Brønsted β_{Nuc} in the range of $pK_a^{Nuc} > pK_a^{Lg}$ (7–10) is 0.95. The overall β_{eq} for the reaction is 1.42. Draw an effective charge map for the reaction and apply the conservation of effective charge to show that the β_{eq} for the second step ($K_{eq(2)}$) is too large to support the stepwise process shown.

9 Even though the data of Problems 7 and 8 fit a linear Equation (except for the deviant points) they can be forced to fit the Equation for the quasi-symmetric method (Equations 11 or 12) to yield a value for $\Delta\beta$ of 0.1 units (with an uncertainty of ±0.1). Indicate how this value affects the conclusions that can be drawn regarding the concertedness or otherwise of the displacement reaction. Compare the $\Delta\beta$ with that for the pyridinolysis of N-triazinyl-pyridinium ions.[46]

10 Calculate pK_a values of the following (experimental values given in brackets): 2,4-dichlorophenol (7.84), 3,4-dinitrobenzoic acid (2.82), 3,4-dibromo-anilinium ion (2.34) and 3,4-dimethylpyridinium ion (6.52). The pK_a values of the standard acids are: phenol, 9.95; 2-chlorophenol, 8.48; 4-chlorophenol, 9.38; benzoic acid, 4.20; anilinium ion, 4.62; 3-bromoanilinium ion, 3.51; 4-bromoanilinium ion, 3.91; pyridinium ion, 5.17; 4-methylpyridinium ion, 6.02; 3-methylpyridinium ion, 5.68. The Hammett σ values, representing differences in pK_a values of substituted benzoic acids, are to be found in Appendix 3, Table 1.

11 The equilibrium constant for formation of 4-methylpyridine and 4-nitrophenyl acetate from acetyl 4-methylpyridinium ion and

4-nitrophenolate ion is 3.37×10^6. Use the effective charge data in Chapter 3, Scheme 1, to calculate β_{eq} for the equilibria of Scheme 19 and hence determine $\log K_{eq}$ for the reaction of 4-chlorophenolate ion with acetyl 4-dimethylamino-pyridinium ion (pK_a values of 4-chlorophenol, 4-nitrophenol, 4-dimethyl-aminopyridinium and 4-methylpyridinium are respectively 9.38, 7.14, 9.68 and 6.14).

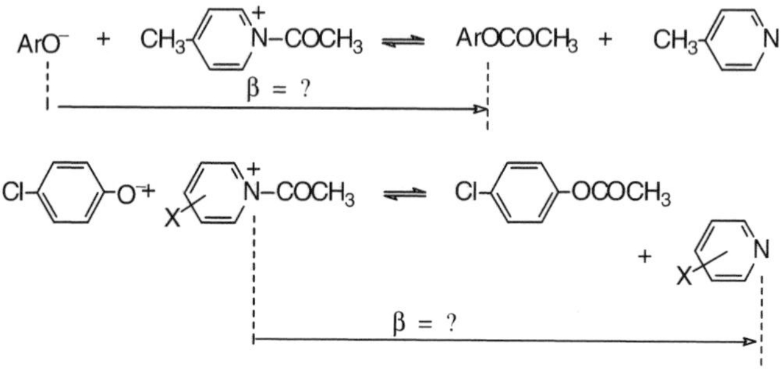

Scheme 19

12 The rate constants (k_2, Table 3) for decomposition of $PhCH^-SO_2$-OAr were determined from the second-order rate constants for reaction of hydroxide ion with the neutral ester and the dissociation constants for the acidity of the α-proton.[47]

Table 3

Substituent	$\log k_2 / sec^{-1}$	pK_a^{ArOH}
Parent	4.46	9.92
4-chloro	5.43	9.38
3-nitro	7.52	8.39
4-acetyl	7.62	8.05
4-nitro	9.7	7.14
2-nitro-4-chloro	11.15	6.46
2-nitro	10.40	7.23
3-chloro	6.20	9.02
2-chloro	7.08	8.48
4-cyano	8.20	7.95
4-ethoxycarbonyl	7.63	8.5

Graph the Brønsted-type relationship of $\log k_2$ against the pK_a^{ArOH} of the leaving phenol and calculate the pK_a^{ArOH} at which k_2 becomes equal to 10^{13} sec^{-1}. Comment on your results.

13 Calculate for the two cases $\alpha = -0.1$ and -1 the proportion of reaction flux passing through catalysis by general acid (pK_a 5), water (pK_a 15.7) and oxonium ion (pK_a −1.7) in an acid catalysed reaction which obeys

$\log k_{HA} = \alpha pK_a + C$. The general acid is at 0.1 molar and the solution is at pH 5 and [water] = 55.5 M. The arguments are from Bell's text.[48]

Show how these calculations bear on the accuracy of experimentally determined α values in the region of $\alpha = 0$ and $\alpha = -1$.

14 Estimate β values for reaction of acetylketene with ArO⁻ (Scheme 20). Make use of $\beta_{Lg} = -1.29$ for the displacement of the phenolate ion from the conjugate base of the ester. You will need to estimate β_{eq} for the overall reaction and for the dissociation of proton on the C(2)-carbon. The β_{eq} for dissociation of the ester can be calculated by inserting CO–C between H and O of ArO–H and assuming attenuation by two atoms $(1.00 \times 0.4 \times 0.4)$. Calculate β_{eq} for the overall reaction from data in Scheme 1, Chapter 3. Consult Scheme 3 (Chapter 7) and Scheme 3 (Chapter 3) for models of the procedure.

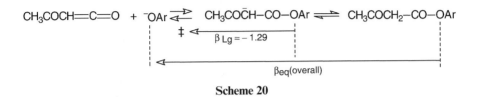

Scheme 20

15 Use fragmental constants in Appendix 3, Table 8 to calculate the log$P$ values of the following solutes (observed values in brackets):

(a) $(CH_3)_2CHNH_2$ (0.26)
(b) $C_6H_5CH_2Br$ (2.92)
(c) $(CH_3)_3COH$ (0.37)
(d) cyclohexane (3.44)

16 Calculate the pK_a values of the following acids given that the pK_a values of a simple carboxylic acid and alkylammonium ion are respectively 4.80 and 10.77 (observed values in brackets):

(a) acetoacetic acid $CH_3COCH_2CO_2H^*$ (3.58)
(b) citric acid $*HO_2CC(OH)(CH_2CO_2H)_2$ (3.58)
(c) 2-ammoniocycloheptanol (RNH_3^+ acid) (9.25)
(d) 2-ammonioethylamine (RNH_3^+ acid) (9.98)

(Employ the acid strengthening effects (ΔpK_a values) given in Appendix 3, Tables 9 and 10.)

17 The rate constants for transfer of hydride ion from 1-alkyl-3-cyano-quinolinium and 1-allyl-3-aminocarbonylquinolinium ions to 10-methyl-acridinium ion (Equation 30) are given in Table 4.[49] The overall equilibrium constants are given by $K_{ij} = k_{ij}/k_{ji}$.

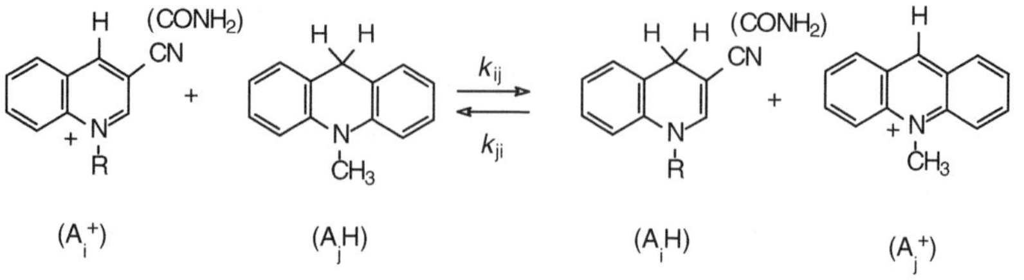

(30)

Table 4

log K_{ij}	*log* $k_{ij}/M^{-1}sec^{-1}$	*Substituent (R)*
1-Alkyl-3-cyanoquinolinium ions		
1.196	−1.400	4-methylbenzyl
1.35	−1.346	benzyl
1.42	−1.328	4-fluorobenzyl
1.61	−1.246	4-bromobenzyl
1.66	−1.177	3-fluorobenzyl
1.92	−1.161	4-cyanobenzyl
2.058	−1.066	3-trifluoromethylbenzyl
1.42	−1.153	4-trifluoromethylbenzyl
−0.495	−2.092	methyl
1-Alkyl-3-aminocarbonylquinolinium ions		
−3.347	−3.509	methyl
−2.699	−2.745	benzyl

This is a quasi-symmetrical reaction; graph log k_{ij} versus log K_{ij} and comment on the results.

18 Calculate the pH-dependence for logk_{obs} for the hydration of a pyrylium salt (Equation 5) over the pH range 1 to 10 assuming $k_H/k_2 = 10^6$ M^{-1} and $k_1[H_2O] = 0.1$ sec^{-1}.

19 Construct an Eigen plot (see Figure 2, Chapter 6) using Equation (31) and the data of Table 5 for the formation of carbinolamine from 4-methoxybenzaldehyde and methoxyamine catalysed by general acids (AH).[50]

$$k_{HA} = k_{max}/(1+10^{(pK_a^{HA} - pK_o')})$$

(31)

Calculate the pK_a of the alcohol of the conjugate acid of the zwitterion intermediate $CH_3OC_6H_4CH(N^+H_2OCH_3)O^-$ and compare it with the value (pK_o) obtained from the "Eigen" fit. Assume that the pK_a of $CH_3N^+H_2CH_2OH$ is 9.98[51] and that replacing H– by Ph– at –CH_2– has a ΔpK_a effect of –0.20.[52] The Hammett ρ for the pK_a of substituted benzyl alcohols is 1.0, the ρ_I for alcohol dissociation is 8.4[53] and an attenuation factor for transmission of σ_I through N is 0.5.[54]

Table 5

General acid AH	pK_a^{HA}	$logk_{HA}/M^{-2}sec^{-1}$
Cyanoacetic acid	2.33	2.53
Chloroacetic acid	2.7	2.43
Methoxyacetic acid	3.4	2.30
Formic acid	3.56	2.30
β-Chloropropionic acid	3.93	2.30
Acetic acid	4.65	2.33
Cacodylic acid	6.15	2.20
Triazolium ion	2.58	2.65
Methoxyammonium ion	4.73	2.50
Imidazolium ion	7.21	2.52
2-Cyanoethylammonium ion	8.20	2.18
2-Methoxyethylammonium ion	9.72	1.26
3-Methoxypropylammonium ion	10.46	0.699
Ethylammonium ion	10.97	0.114

20 Sum the equilibria in Scheme 21 and express K_{eq} in terms of K_1 and K_2.

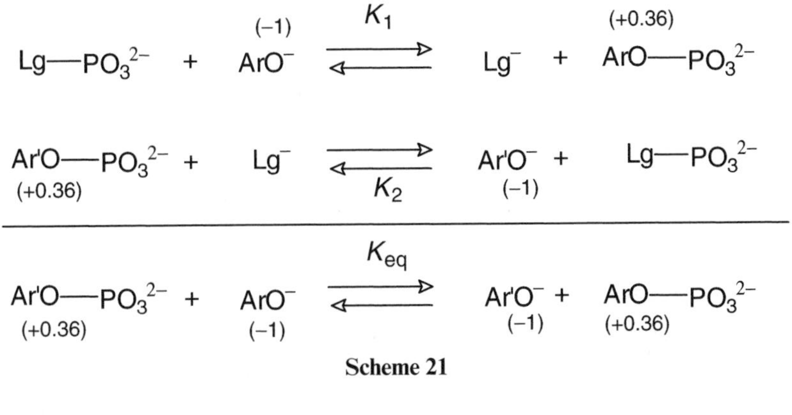

Scheme 21

Now use the Brønsted equations given below to derive Equation (13) for the transfer of the phosphoryl group between phenolate donors and acceptors. Note that K_1 and K_2 refer to *opposite* directions.

$$logK_1 = 1.36pK_a^{ArOH} + C \quad \text{and} \quad logK_2 = 1.36pK_a^{Ar'OH} + C$$

21 The reaction of aryloxide ions with 4-nitrophenyl diphenylphosphinate ester (Equation 32) has a β_{Nuc} of 0.46. Given that β_{eq} of the reaction is 1.25 construct an effective charge map for the concerted displacement reaction. Comment on the structure of the transition state for the identity reaction of the 4-nitrophenyl ester.[55]

$$X\text{—}C_6H_4\text{—}O^- + NO_2\text{—}C_6H_4\text{—}O\text{-PO(Ph)}_2 \rightleftharpoons X\text{—}C_6H_4\text{—}O\text{-PO(Ph)}_2 + NO_2\text{—}C_6H_4\text{—}O^- \quad (32)$$

22 The β_{Nuc} value for the displacement of ammonia from phosphoramidate ($^+NH_3PO_3^{2-}$) by substituted pyridines to yield the corresponding $XpyPO_3^{2-}$ (Equation 33) is 0.22.[56]

Construct an effective charge map and comment on

$$X\text{—py—}N + \ ^+H_3N\text{–}PO_3^{2-} \rightleftharpoons X\text{—py—}N\text{–}PO_3^{2-} + NH_3^+ \quad (33)$$

(Xpy)   (XpyPO$_3$H$^-$)

the charge distribution of the transition structure assuming a concerted mechanism.

7.10 REFERENCES

1. A. Williams and K.T. Douglas, E1$_{cb}$ Mechanism Part II. Base Hydrolysis of Substituted Phenyl Phosphorodiamidates, *J. Chem. Soc. Perkin Trans. 2*, 1972, 1454.
2. H. Al-Rawi and A. Williams, Elimination–Addition Mechanisms of Acyl Group Transfer: the Hydrolysis and Synthesis of Carbamate Esters, *J. Am. Chem. Soc.*, 1977, **99**, 2671.
3. J.F. Kirsch and W.P. Jencks, Nonlinear Structure-Reactivity Correlations. The Imidazole-catalysed Hydrolysis of Esters, *J. Am. Chem. Soc.*, 1964, **86**, 837.
4. M. Alborz and K.T. Douglas, Effect of an Aromatic Ester Conjugate Base on E1cB Ester Hydrolysis. Alkaline Hydrolysis of Fluorene-9-Carboxylate Esters, *J. Chem. Soc. Perkin Trans. 2*, 1982, 331.
5. A. Williams, Effective Charge and Transition State Structure, *Adv. Phys. Org. Chem.*, 1991, **27**, 1.
6. D.J. Hupe and W.P. Jencks, Nonlinear Structure–Reactivity Correlations. Acyl Transfer Between Sulphur and Oxygen Nucleophiles, *J. Am. Chem. Soc.*, 1977, **99**, 451.
7. J.M. Sayer and W.P. Jencks, Imine-Forming Elimination Reactions. 2.

Imbalance of Charge Distribution in the Transition State for Carbinolamine Dehydration, *J. Am. Chem. Soc.*, 1977, **99**, 464.

8. M.T. Skoog and W.P. Jencks, Reactions of Pyridines and Primary Amines with N-Phosphorylated Pyridines, *J. Am. Chem. Soc.*, 1984, **106**, 7597.

9. N. Bourne and A. Williams, Evidence for a Single Transition State in the Transfer of the Phosphoryl Group ($-PO_3^{2-}$) to Nitrogen Nucleophiles from Pyridine-N-Phosphonates, *J. Am. Chem. Soc.*, 1984, **106**, 7591.

10. A.R. Hopkins *et al.*, Sulphate Group Transfer between Nitrogen and Oxygen: Evidence Consistent with an Open "Exploded" Transition State, *J. Am. Chem. Soc.*, 1983, **105**, 6062.

11. C.F. Bernasconi *et al.*, Kinetics of the Reaction of β-Methoxy-α-stilbene with Thiolate Ions – 1st Direct Observation of the Intermediate in a Nucleophilic Vinylic Substitution, *J. Am. Chem. Soc.*, 1984, **111**, 6862.

12. M.I. Page and A. Williams, *Organic and Bio-Organic Mechanisms*, Addison Wesley Longman, Harlow, 1997; H. Maskill, *The Physical Basis of Organic Chemistry*, Oxford University Press, Oxford, 1985.

13. D.S. Noyce *et al.*, Carbonyl Mechanisms. III. The Formation of Aromatic Semicarbazones. A Non-Linear Rho-Sigma Correlation, *J. Org. Chem.*, 1958, **23**, 752; B.M. Anderson and W.P. Jencks, The Effect of Structure on Reactivity in Semicarbazone Formation, *J. Am. Chem. Soc.*, 1960, **82**, 1773.

14. G. Salvadori and A. Williams, Demonstration of Intermediates in the Hydrolysis of 4-Ethoxypyrylium Salts, *J. Am. Chem. Soc.*, 1971, **93**, 2727.

15. G. Cevasco *et al.*, A Novel Dissociative Mechanism in Acyl Transfer from Aryl 4-Hydroxybenzoate esters in Aqueous Solution, *J. Org. Chem.*, 1984, **50**, 479.

16. A. Williams, The Diagnosis of Concerted Organic Mechanisms, *Chem. Soc. Rev.*, 1994, 93.

17. N.R. Cullum *et al.*, Effective Charge on the Nucleophile and Leaving Group during the Stepwise Transfer of the Triazinyl Group between Pyridines in Aqueous Solution, *J. Am. Chem. Soc.*, 1995, **117**, 9200.

18. M.T. Skoog and W.P. Jencks, Reactions of Pyridines and Primary Amines with N-Phosphorylated Pyridines, *J. Am. Chem. Soc.*, 1984, **106**, 7597; P.E. Dietze and W.P. Jencks, Swain-Scott Correlations for Reactions of Nucleophilic Reagents and Solvents with Secondary Substrates, *J. Am. Chem. Soc.*, 1986, **108**, 4549.

19. S.A. Ba-Saif *et al.*, Concertedness in Acyl Group Transfer: A Single

Transition State in Acetyl Transfer Between Phenolate Ion Nucleophiles, *J. Am. Chem. Soc.*, 1987, **109**, 6362.

20. D.D. Perrin *et al.*, *pK$_a$ Prediction for Organic Acids and Bases*, Chapman & Hall, London, 1981.

21. W.L. Jorgensen *et al.*, CAMEO: A Program for the Logical Prediction of the Products of Organic Reactions, *Pure Appl. Chem.*, 1990, **10**, 1921.

22. A.J. Gushurst and W.L. Jorgensen, Computer-Assisted Mechanistic Evaluation of Organic Reactions. 12. pK$_a$ Predictions for Organic Compounds in Me$_2$SO, *J. Org. Chem.*, 1986, **51**, 3513.

23. F. Csizmadia *et al.*, Expert System Approach for Predicting pK$_a$, in "Trends in QSAR and Modelling", in *Trends in QSAR and Modelling*, C.G. Wermuth, (ed.), ESCOM, Leiden, 1993, p. 507.

24. H.H. Jaffé, A Re-examination of the Hammett Equation, *Chem. Rev.*, 1953, **53**, 191.

25. J. Hine, *Structural Effects on Equilibria in Organic Chemistry*, John Wiley, New York, 1975.

26. J.E. Leffler and E. Grunwald, *Rates and Equilibria of Organic Reactions*, John Wiley, New York, 1963.

27. J. Gerstein and W.P. Jencks, Equilibria and Rates for Acetyl Transfer among Substituted Phenyl Acetates, Acetylimidazole, O-Acylhydroxamic Acids, and Thiol Esters, *J. Am. Chem. Soc.*, 1964, **86**, 4655.

28. J.P. Guthrie and D.C. Pike, Hydration of Acylimidazoles: Tetrahedral Intermediates in Acylimidazole Hydrolysis and Nucleophilic Attack by Imidazole on Esters, *Can. J. Chem.*, 1987, **65**, 1951.

29. A.J. Leo *et al.*, Partition Coefficients and their Uses, *Chem. Rev.*, 1971, **71**, 525.

30. R. Rekker and R. Mannhold, *Calculations of Drug Lipophilicity*, VCH, Weinheim, 1992; H. Kubinyi, Lipophilicity and Drug Activity, *Prog. Drug Res.*, 1979, **23**, 97; A.J. Leo, *et al.*, Calculation of Hydrophobic constant (log$P$) from π and *f* constants, *J. Med. Chem.*, 1975, **18**, 865.

31. A.J. Leo, Calculating logP$_{oct}$ from Structures, *Chem. Rev.*, 1993, **93**, 1281; see also the Biobyte website given in Further Reading.

32. W.P. Jencks, *Catalysis in Chemistry and Enzymology*, McGraw-Hill, New York, 1969, pp. 221, 603–605.

33. M.L. Bender, *Mechanisms of Homogeneous Catalysis from protons to Proteins*, Wiley-Interscience, New York, 1971, Chapter 5.

34. C.J. Swain and J.C. Worosz, Mechanism of General Acid Catalysis of Additions of Amines to Carbonyl Groups, *Tetrahedron Lett.*, 1965, **36**, 3199.

35. R.L. Van Etten *et al.*, Acceleration of Phenyl Ester Cleavage by Cyclodextrins. A Model for Enzymatic Specificity, *J. Am. Chem. Soc.*, 1967, **89**, 3242.

36. D.M. Blow, Structure and Mechanism of Chymotrypsin, *Acc. Chem. Res.*, 1976, **9**, 145.

37. S. Thea and A. Williams, Measurement of Effective Charge in Organic Reactions in Solution, *Chem. Soc. Rev.*, 1986, **15**, 125; A. Williams, Effective Charge and Leffler's Index as Mechanistic Tools for Reactions in Solution, *Acc. Chem. Res.*, 1984, **17**, 465.

38. A.R. Fersht *et al.*, Quantitative Analysis of Structure–Activity Relationships in Engineered Proteins by Linear Free Energy Relationships, *Nature (London)*, 1986, **322**, 284; A.R. Fersht *et al.*, Structure–Activity Relationships in Engineered Proteins: Analysis of Use of Binding Energy by Linear Free Energy Relationships, *Biochem*, 1987, **26**, 6030.

39. M.D. Toney and J.F. Kirsch, Direct Brønsted Catalysis of the Restoration of Activity to a Mutant Enzyme by Exogenous Amines, *Science*, 1989, **243**, 1485.

40. R.F. Pratt and T.C. Bruice, The Carbanion Mechanism (E1$_{cb}$) of Ester Hydrolysis. III. Some Structure–Reactivity Studies and the Ketene Intermediate, *J. Am. Chem. Soc.*, 1970, *92*, 5956.

41. J.F. Kirsch and W.P. Jencks, Nonlinear Structure–Reactivity Correlations. The Imidazole-catalysed Hydrolysis of Esters, *J. Am. Chem. Soc.*, 1964, *86*, 837.

42. J.W. Bunting and J.P. Kanter, A Change in the Rate-Determining Step in the E1$_{cb}$ Reactions of N-(2-(4-Nitrophenyl)ethyl)pyridinium Cations, *J. Am. Chem. Soc.*, 1991, *113*, 6950.

43. M.L. Bender and A. Williams, Ketimine Intermediate in Amine-Catalysed Enolisation of Acetone, *J. Am. Chem. Soc.*, 1966, **88**, 2502.

44. A.H.M. Renfrew *et al.*, Stepwise Versus Concerted Mechanisms at Trigonal Carbon: Transfer of the 1,3,5-Triazinyl Group Between Aryl Oxide Ions in Aqueous Solution, *J. Am. Chem. Soc.*, 1995, **117**, 5484.

45. N.R. Cullum *et al.*, Effective Charge on the Nucleophile and Leaving Group during the Stepwise Transfer of the Triazinyl Group between Pyridines in Aqueous Solution, *J. Am. Chem. Soc.*, 1995, **117**, 9200.

46. M.B. Davy *et al.*, Elimination-Addition Mechanisms of Acyl Group Transfer: Hydrolysis and Aminolysis of Arylmethanesulphonates, *J. Am. Chem. Soc.*, 1977, **99**, 1196.

47. R.P. Bell, *Acid Base Catalysis*, Oxford University Press, Oxford, 1940, p. 94.

48. M.M. Kreevoy and I-S.H. Lee, Marcus Theory of a Perpendicular Effect on α for Hydride Transfer Between NAD⁺ Analogues, *J. Am. Chem. Soc.*, 1984, **106**, 2550.

49. S. Rosenberg *et al.*, Evidence for Two Concurrent Mechanisms and a Kinetically Significant Proton Transfer Process in Acid-catalysed O-Methoxime Formation, *J. Am. Chem. Soc.*, 1974, **96**, 7986.

50. J. Hine *et al.*, Kinetics and Mechanism of the Hydrolysis of N-iso-butylidene-methylamine in Aqueous Solution, *J. Am. Chem. Soc.*, 1970, **92**, 5194; J. Hine and F.C. Kokesh, Rates and Equilibrium Constants for Each Step in the Reaction of Trimethylammonium ions with Formaldehyde to give Formocholine Cations in Aqueous Solution, *J. Am. Chem. Soc.*, 1970, **92**, 4383.

51. J. Hine and G.F. Koser, The Mechanism of the Reaction of Phenylpropargyl Aldehyde with Aqueous Sodium Hydroxide to give Phenylacetylene and Sodium Formate, *J. Org. Chem.*, 1971, **36**, 1348.

52. J.P. Fox and W.P. Jencks, General Acid and General Base Catalysis of ther Methoxyaminolysis of 1-Acetyl-1,2,4-triazole, *J. Am. Chem. Soc.*, 1974, **96**, 1436.

53. See footnote 56 in J.M. Sayer and W.P. Jencks, Mechanism and Catalysis of 3-Methyl-3-thiosemicarbazone Formation. A Second Change in Rate-Determining Step and Evidence for a Stepwise Mechanism for Proton Transfer in a Simple Carbonyl Addition Reaction, *J. Am. Chem. Soc.*, 1973, **95**, 5637.

54. N. Bourne *et al.*, A single transition state in the reaction of aryl diphenylphosphinate esters with phenolate ions in aqueous solution, *J. Am. Chem. Soc.*, 1988, **111**, 1890.

55. W.P. Jencks and M. Gilchrist, Reactions of Nucleophilic Reagents with Phosphoramidate, *J. Am. Chem. Soc.*, 1965, **87**, 3199.

Equations and the More O'Ferrall–Jencks Diagram

A1.1 SUMMARY OF EQUATIONS

A1.1.1 The Morse Equation

The potential energy (E_x) of a bond dissociation process can be plotted as a function of the interatomic distance (x) making use of the empirical Morse equation $E_x = D_E\{1 - \exp[b(x_o - x)]\}^2 - \exp[b(x_o - x)]\}^2$ (see Figure A1). D_E is the bond dissociation energy and x_o is the mean bond length.

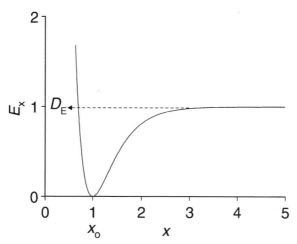

Figure 1 *Example of a Morse curve for a dissociation process* $(D_E = b = x_o = 1)$

A1.1.2 Derivation of Equation 7 (Chapter 1)

The inset of Figure 1 (Chapter 1) shows the region with cross-over points of Morse equations (Figure 1, Chapter 1) which have not been smoothed. Change in the cross-over point is reasonably assumed to be the same as that suffered by the smoothed line at the maximum (the transition structure). Smoothing does not invalidate the trigonometrical arguments given below. The tangents m_1 and m_2 are given by $m_1 = a/c$ and $m_2 = b/c$. Thus

$$\partial \Delta E_0 = (a + b) = m_1 c + m_2 c$$

and

$$\partial \Delta E_a = a = m_1 c$$

Therefore

$$\partial \Delta E_a / \partial \Delta E_0 = m_1 c/(m_1 c + m_2 c) = m_1/(m_1 + m_2)$$

A1.1.3 The Effect of ΔE_0 on the Leffler α Parameter

Consider the crossed Morse curves in Figure A2. In an *endothermic* reaction the similarity coefficient $a = m_1/(m_1 + m_2)$ will approach unity as ΔE_0 becomes positive because the tangent m_2 approaches zero. For the *exothermic* case ΔE_0 becomes negative and the tangent m_1 tends towards zero; this is due to the cross-over point approaching the minimum of the Morse curve for dissociation of the reactant.

A1.1.4 Fitting Data to Theoretical Equations

Most of the equations in the text can be fitted by use of software programs (such as ENZFITTER®, FigP® or MINITAB®). Alternative statistical packages might be available to readers through their institutional computing services. Some simple statistical routines have been included for the solution of the major equations employed in this text should the reader not have such access. A computer loaded with a computational language such as Qbasic® is all that is required to formulate a program from these equations.

   Data may be fitted to the equations using the least-squares condition that the sum of the squares of the residuals (E) from observed value y and the value of y calculated from the fitting equation ($y = f(x)$) is a minimum.

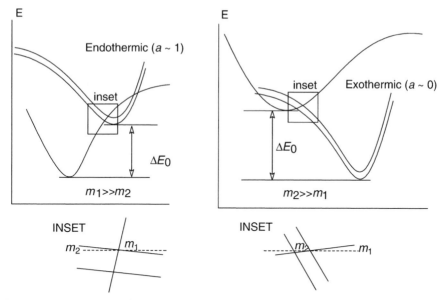

Figure 2 *Dependence of similarity coefficients on* ΔE_0 *for Class I relationships*

$$E = \Sigma(y - f(x))^2 \tag{1}$$

Details of the mathematics can be found in standard statistical texts.

A1.1.4.1 Linear Equation.

$$y = mx + c \tag{2}$$

For n pairs of data, x and y, the least squares fit to the equation yields:

$$m = (\Sigma xy - \Sigma x \Sigma y/n)/D \pm [E/(D(n-2))]^{0.5} \tag{3}$$

$$c = (\Sigma y - m\Sigma x)/n \pm [E\Sigma x^2/(D(n-2))]^{0.5} \tag{4}$$

where:

$$D = \Sigma x^2 - (\Sigma x)^2/n \quad \text{and} \quad E = \Sigma(y - mx - c)^2$$

The correlation coefficient, r, is given by equation (5):

$$r = (\Sigma xy - \Sigma x \Sigma y/n)/ \{D(\Sigma y^2 - (\Sigma y)^2/n)\}^{0.5} \tag{5}$$

A1.1.4.2 Bilinear Equation. This equation is used, for example, to fit the Jaffé equation – Chapter 4.

$$y = \log k = a + bx_1 + cx_2 \tag{6}$$

$$= \log k_H + \rho_A \sigma_A + \rho_B \sigma_B \tag{7}$$

For n triads of data y, x_1, and x_2 the coefficients a, b, and c may be obtained by substituting the following values into Equations (8) to (11):[a]

$$
\begin{array}{llll}
a1 = n & b1 = \Sigma x_1 & c1 = \Sigma x_2 & d1 = \Sigma y \\
a2 = \Sigma x_1 & b2 = \Sigma x_1^2 & c2 = \Sigma x_1 x_2 & d2 = \Sigma x_1 y \\
a3 = \Sigma x_2 & b3 = \Sigma x_1 x_2 & c3 = \Sigma x_2^2 & d3 = \Sigma x_2 y
\end{array}
$$

$$D = a1(b2c3 - b3c2) - b1(a2c3 - a3c2) + c1(a2b3 - a3b2) \tag{8}$$

$$a = (d1(b2c3 - b3c2) - b1(c3d2 - c2d3) + c1(b3d2 - b2d3))/D \tag{9}$$

$$b = (a1(c3d2 - c2d3) - d1(a2c3 - a3c2) + c1(a2d3 - a3d2))/D \tag{10}$$

$$c = (a1(b2d3 - b3d2) - b1(a2d3 - a3d2) + d1(a2b3 - a3b2))/D \tag{11}$$

The correlation coefficient, r, is given by the expression (Equation 12)

$$r = [\{ad1 + bd2 + cd3 - d1^2/n\}/\{\Sigma y^2 - d1^2/n\}]^{0.5} \tag{12}$$

The standard deviations on a, b, and c for Equation (6) are:

$$\delta a = \Delta^2(b2c3 - b3c2)/D \tag{13}$$

$$\delta b = \Delta^2(a1c3 - a3c1)/D \tag{14}$$

$$\delta c = \Delta^2(a1b2 - a2b1)/D \tag{15}$$

where:

$$\Delta^2 = \Sigma(y - a - bx_1 - cx_2)^2/(n - 3)$$

A1.1.4.3   Global Analysis of Two Linear Equations.   Plots of β_{Nuc} versus pK_{Lg} and β_{Lg} versus pK_{Nuc} obey linear equations (16) and (17) (see Chapter 5, Equations 31 and 32).

$$\beta_{Lg} = p_{xy} \, pK_{Nuc} + L \tag{16}$$

$$\beta_{Nuc} = p_{xy} \, pK_{Lg} + N \tag{17}$$

[a] See W.A. Pavelich and R.W. Taft, *J. Am. Chem. Soc.*, 1957, **79**, 4953.

The equations are simplified to (18) and (19)

$$y_1 = a\, x_1 + c \; (n \text{ points})\tag{18}$$

$$y_2 = a\, x_2 + b \; (m \text{ points})\tag{19}$$

and the following values computed which can be substituted

$$
\begin{aligned}
a1 &= \Sigma x_2^2 + \Sigma x_1^2 & b1 &= \Sigma x_1 & c1 &= \Sigma x_2 & d1 &= \Sigma y_1 x_1 + \Sigma y_2 x_2 \\
a2 &= \Sigma x_1 & b2 &= n & c2 &= 0 & d2 &= \Sigma y_1 \\
a3 &= \Sigma x_2 & b3 &= 0 & c3 &= m & d3 &= \Sigma y_2
\end{aligned}
$$

into Equations (8) to (11) to solve Equations (18) and (19) ($p_{xy} = a$, $N = b$ and $L = c$).

The correlation coefficient, r, is given by the expression (Equation 20)

$$r = \left\{ \frac{[ad1 + bd2 + cd3 - (d2 + d3)^2/(n+m)]}{[\Sigma y^2 - (d2 + d3)^2/(n+m)]} \right\}^{0.5}\tag{20}$$

The standard deviations on a, b, and c for Equations (18) and (19) are obtained by substituting

$$\Delta^2 = \sum \frac{(y - a(x_1 + x_2) - b - c)^2}{n + m - 3}$$

into Equations (13) to (15)

A1.1.4.4 Binomial Equation. This is employed, for example, in estimating p_x or p_y see Chapter 5.

$$\log k = a + bx + cx^2\tag{21}$$

Substituting the following into Equations (8–11) solves Equation (21) (n = number of data points).

$$
\begin{aligned}
a1 &= n & b1 &= \Sigma x & c1 &= \Sigma x^2 & d1 &= \Sigma y \\
a2 &= \Sigma x & b2 &= \Sigma x^2 & c2 &= \Sigma x^3 & d2 &= \Sigma yx \\
a3 &= \Sigma x^2 & b3 &= \Sigma x^3 & c3 &= \Sigma x^4 & d3 &= \Sigma yx^2
\end{aligned}
$$

The correlation coefficient, r, is given by the expression (Equation 22)

$$r = \left[\frac{(ad1 + bd2 + cd3 - d1^2/n)}{(\Sigma y^2 - d1^2/n)} \right]\tag{22}$$

Substituting $\Delta^2 = \Sigma(y - a - b\,x - c\,x^2)^2/(n - 3)$ into Equations (13–15) gives standard deviations on the parameters of Equation (21).

A1.1.4.5 Cross-correlation.   This is used, for example, in correlating $\log k^x/k^H = \rho_1\sigma_1 + \rho_2\sigma_2 + \rho_{12}\sigma_1\sigma_2$.

Substituting the following data into Equations (8–11) solves Equation (23).

$$y = ax_1 + bx_2 + cx_1x_2 \tag{23}$$

$$
\begin{array}{llll}
a1 = \Sigma x_1^2 & b1 = a_2 & c1 = a_3 & d1 = \Sigma yx_1 \\
a2 = \Sigma x_1 x_2 & b2 = \Sigma x_1^2 & c2 = b_3 & d2 = \Sigma yx_2 \\
a3 = \Sigma x_1^2 x_2 & b3 = \Sigma x_1 x_2^2 & c3 = \Sigma x_1^2 x_2^2 & d3 = \Sigma yx_1 x_2
\end{array}
$$

The correlation, r, is given by Equation (22) and substituting

$$\Delta^2 = \Sigma(y - ax_1 - bx_2 - cx_1x_2)^2/(n - 3)$$

into Equations (13–15) gives standard deviations on the parameters in Equation (23).

A1.1.5 Effective Charge from Hammett Slopes

The Hammett equation is often employed even though the reaction under investigation is not similar to the reference benzoic acid dissociation. Effective charge cannot be obtained directly from equation (7) (Chapter 3) but can be obtained from the Hammett ρ values as shown for the example given below.

$$CH_3CO\text{---}OAr \overset{\varepsilon_r}{\ } + \ Nu^- \ \rightleftharpoons \ CH_3CO\text{---}Nu \ + \ {}^{\varepsilon_p}_{\ }\text{-}OAr \qquad (d) \quad (\rho_1 = 3.8) \tag{24}$$

$$H\text{---}OAr \overset{\varepsilon_{rs}}{\ } + \ H_2O \ \rightleftharpoons \ H_3O^+ \ + \ {}^{\varepsilon_{ps}}_{\ }\text{-}OAr \qquad (d_S) \quad (\rho_2 = 2.23) \tag{25}$$

$$\overset{\varepsilon'_{rs}}{\ }\text{X-C}_6H_4\text{--CO}_2H + H_2O \ \rightleftharpoons \ H_3O^+ \ + \ \overset{\varepsilon'_{ps}}{\ }\text{X-C}_6H_4\text{--CO}_2H \quad (d'_S) \quad (\rho = 1.00) \tag{26}$$

$$\rho_1 = [(\varepsilon_p - \varepsilon_r)/(\varepsilon'_{ps} - \varepsilon'_{rs})]\,(d'_s/d) = 3.8$$

$$\rho_2 = [(\varepsilon_{ps} - \varepsilon_{rs})/(\varepsilon'_{ps} - \varepsilon'_{rs})]\,(d'_s/d_s) = 2.23$$

$$[(\varepsilon_p - \varepsilon_r)/(\varepsilon_{ps} - \varepsilon_{rs})]\,(d_s/d) = \rho_1/\rho_2 = 1.7$$

Since $d_s = d$

$$\varepsilon_p - \varepsilon_r = 1.7(\varepsilon_{ps} - \varepsilon_{rs}) = -1.7$$

A1.2 THE MORE O'FERRALL–JENCKS DIAGRAM

The More O'Ferrall–Jencks diagram[b] is a structure–reactivity surface defined as a three-dimensional energy contour diagram where the extent of reaction is described by structure–reactivity parameters (Figure A3a) rather than by the more fundamental (but less experimentally accessible) bond length or bond order. The diagram has the advantage that it directly relates the extent of reaction with experimental parameters. A disadvantage is that the relationship between structure–reactivity parameters and fundamental quantities is not straightforward. These diagrams were first introduced by More O'Ferrall but their experimental realisation was only possible when Jencks introduced similarity coefficients (such as Leffler's α_{Lg} and α_{Nuc}, β etc) as their x and y coordinates.[c]

For reactions in solution the coordinates at the corners of the diagram are usually taken to represent molecular structures within their reaction complexes. Throughout this text we use the convention for displacement reactions that the x coordinate registers the bond fission process while the y coordinate registers bond formation. The More O'Ferrall–Jencks diagram is excellent for visualising energy surfaces of reactions with two major bonding changes. Inclusion of a further dimension for reactions with three major bonding changes is much more difficult to visualise but this can be done if the dimensions can be combined to give the two coordinates x and y.[d] The example shown (Figure A3a) has x and y coordinates which correspond to bond length. The diagrams can also be applied to systems with more than two major bond changes such as base-catalysed elimination (see Chapter 6, equation 17) where B–H, H–C and C–Lg bonds undergo formation or fission and a C–C bond undergoes rehybridisation. The y coordinate (Figure 5, Chapter 6) is taken as the C–Lg bond change and the x coordinate combines the proton transfer from C to B and is easily measured by a normalised Brønsted exponent.

---

[b] R.A. More O'Ferrall, *J. Chem. Soc. (B)*, 1970, 274; W.P. Jencks, *Chem. Rev.*, 1972, **72**, 705; W.J. Albery and M.M. Kreevoy, *Adv. Phys. Org. Chem.*, 1978, **16**, 87.
[c] D.A. Jencks and W.P. Jencks, *J. Am. Chem. Soc.*, 1977, **99**, 7948.
[d] E. Grunwald, *J. Am. Chem. Soc.*, 1985, **107**, 4710; E. Grunwald, *J. Am. Chem. Soc.*, 1985, **107**, 4715; J.P. Guthrie, *Can. J. Chem.*, 1966, **74**, 1283; P. Kandanarachchi and M.L. Sinnott, *J. Am. Chem. Soc.*, 1994, **116**, 5601.

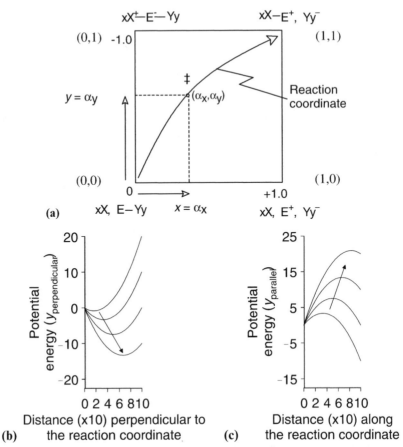

Figure 3 *(a) More O'Ferrall–Jencks diagram the structure–reactivity surface of the displacement reaction $(E - Yy + xX \rightarrow E - Xx + yY)$. Values of ρ or β are normalised by division by ρ_{eq} or β_{eq} for each fundamental bond-forming or -fission reaction to give α_x or α_y representing x and y coordinates respectively. (b) A quadratic equation (with positive curvature – see text) models the energies perpendicular to the reaction coordinate at the saddle point (dashed line in Figure 4a, Chapter 5). The arrow shows that changing the energies shifts the transition structure towards the decreasingly energetic end. (c) A quadratic equation (with negative curvature – see text) models the energies along the reaction coordinate at the saddle point (line r.c. in Figure 4(a), Chapter 5). The arrow shows that changing the energies shifts the transition structure towards the increasingly energetic end*

The saddle point in the More O'Ferrall–Jencks diagram has a shape which approximates a quadratic equation with a *minimum* in a direction perpendicular to the reaction coordinate (Figure A3b) and a quadratic equation with a *maximum* along the reaction coordinate (Figure A3c). In the illustrations the potential energies for movement in parallel and perpendicular directions are given by $y_{\text{parallel}} = -0.3x^2 + (b/10 + 3)x$

and $y_{\text{perpendicular}} = 0.3x^2 + (b/10 + 3)x$ respectively; b is the potential energy at $x = 10$. Varying the potential energy, b, changes the positions of the maxima and minima as shown in Figures A3b and A3c. The position of the minimum, corresponding to motion perpendicular to the reaction coordinate, moves towards the end coordinate with *decreasing* energy (Thornton effect, Figure A3b). Motion parallel to the reaction coordinate is towards the end coordinate of *increasing* energy (Hammond effect, Figure A3c). The Hammond and Thornton motions are thus seen as direct consequences of the shape of the reaction surface at the saddle-point. Although the illustrations are for quadratic equations the quartic equation (Chapter 5) also exhibits these properties.

Answers to Problems

The figures illustrating the answers can be drawn manually or by use of a graphics software package such as FigP. The regression analyses may be carried out by use of suitable software or programs written in Basic making use of the equations in Section A1.1.4 of Appendix 1. Where a graph is required and there is a similar plot included in the main text we have omitted a figure in the answer.

Chapter 2

2.1 Data from the tables in Appendix 3 indicate that the Hammett equations for the anilinium ions and arylmercaptoacetic acids are $pK_a = 4.58 - 2.21 \Sigma\sigma$ and $pK_a = 3.57 - 0.3 \Sigma\sigma$ respectively. Solving for equal pK_a gives the equation $4.58 - 2.21.\Sigma\sigma = 3.57 - 0.3.\Sigma\sigma$ and hence $\Sigma\sigma = (4.58 - 3.57)/(2.88 - 0.30) = 0.39$; this is close to the σ value for the *meta*-bromo substituent.

2.2 The triangles in Figure 1 fit a Hammett relationship where $\rho = 2.35$, intercept $= -2.53$ and refer to *meta* and *para* substituents. The squares represent *ortho* substituents which show various degrees of steric hindrance due to the increase in steric requirements from reactant to transition state.

2.3 The σ_m values for the substituents are in the series $CH_3- < CH_3O- < H- < Cl- \approx CH_3CO- \approx CH_3OCO- < NC- < O_2N-$. Note that these are *inductive* effects and do not include resonance interactions. A slightly different order would be obtained if σ_p were used, owing to the possible incursion of resonance transmission effects.

2.4 The Hammett equations for the reaction in the two solvents are:

$$CH_3CN: \log k/M^{-1} \sec^{-1} = -0.84\sigma - 1.27$$

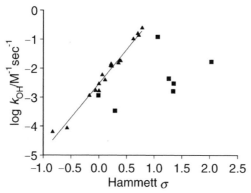

Figure 1 *Hammett graph of logk_{OH} versus σ for the alkaline hydrolysis of ethyl esters of substituted benzoic acids*

$$C_6H_5Cl: \log k/M^{-1} \sec^{-1} = -1.45\sigma - 2.50$$

The different selectivities are due to the different solvation effects. The more polar solvent stabilises the charge formed in the transition state of the reaction relative to that in the reactant and causes a smaller Hammett ρ value. The stabilisation of the developing charge also increases the rate of the reaction. Note that the standard dissociation reaction has water solvent in both cases.

2.5 The standard dissociation of benzoic acids involves the hydroxyl group where the base atom, oxygen, is one atom removed from the benzene ring. If the dissociation of benzohydroxamic acids involved the fission of the O–H group, an attenuated sensitivity would be expected relative to the benzoic acid dissociation because the OH group is two atoms removed from its benzene ring. Because the sensitivity to substituent change is similar to that of the dissociation of benzoic acids it is concluded that the NH group is the acidic function because the nitrogen atom is one atom removed from the benzene ring.

2.6 The substituent has to interact with the reaction centre through a longer distance in the case of the dissociation of the 2-phenylpropionic acids compared with that of benzoic acids and hence its effect on the dissociation equilibrium constant is less.

2.7 The pK_a and logk_{OH} for the dissociation of the conjugate acid and hydroxide ion catalysed elimination from substituted ethyl benzimidate esters obey the Hammett correlations:

$$pK_a = -1.71\sigma + 6.47 \qquad \log k_{OH} = 0.54\sigma - 2.00$$

The map of the Hammett ρ values for the E1cB reaction is shown in Scheme 1.

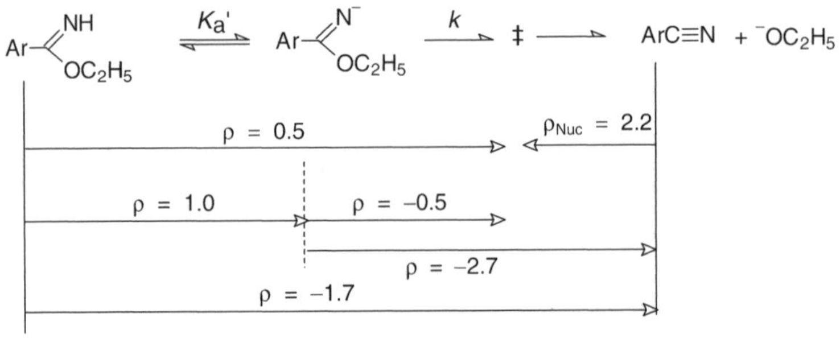

Scheme 1   *Map of ρ values for hydroxide ion catalysed elimination of substituted ethyl benzimidate esters. Note that K_a' is not the same as the dissociation constant for the conjugate acid $ArC(NH_2)^+OC_2H_5$)*

The ρ value for the pK_a of the neutral imidate is estimated to be unity as the nitrogen atom is one atom removed from the benzene ring. The ρ value of 1.7 found for the dissociation of the protonated imidate is possibly due to the delocalisation of the positive charge into the benzene ring by resonance. Resonance interaction is negligible for the dissociation of the neutral imidate (Scheme 2).

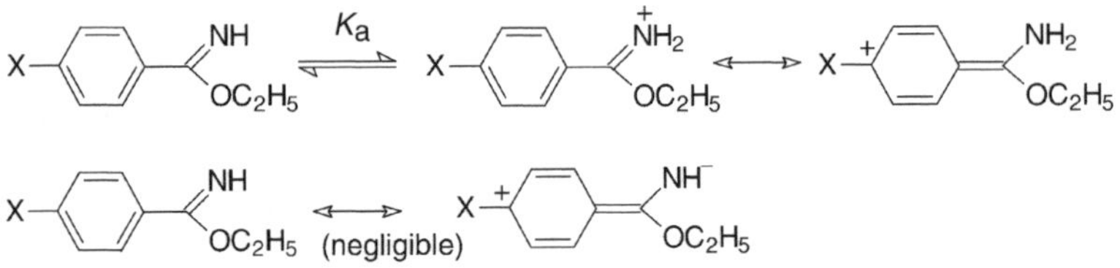

Scheme 2   *Dissociation of benzimidate esters*

2.8   The data fit the Taft–Pavelich equations where the polar effect (ρ*σ*) is constant.

2,4-dinitrochlorobenzene

$$\log k_{obs} = -1.75 E_s - 2.37$$

allyl bromide

$$\log k_{obs} = -0.813 E_s - 2.12$$

The correlations possess δ values consistent with the differing steric requirements; the numerical magnitude of δ for the more hindered aromatic species is more than that for the allyl bromide. The range of σ* for the amine substituents is not significant and the polar effect is therefore constant throughout.

2.9   The $σ_I$ and $σ_m$ values drawn from tables in Appendix 3 are related by the equation: $σ_I = 0.718 σ_m + 0.129$ (see Figure 2).

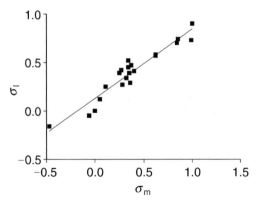

Figure 2   *Linear dependence of σ_I upon σ_m*

Provided there is no resonance transmission in the effect of substituents on the dissociation of acetic acids (as would be expected) then the linear equation indicates that the transmission of the effect of the *meta* substituent does not involve resonance either. The correlation shows that σ_m provides a useful secondary definition of σ_I. This relationship is a Class II free energy correlation between the dissociation of substituted acetic acids and *meta*-substituted benzoic acids.

2.10   The parameter σ_I is linear in σ^* (see Figure 3):

$$\sigma_I = 0.178\sigma^* - 0.054.$$

The correlation provides a further useful secondary definition of σ_I.

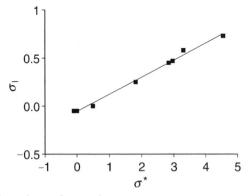

Figure 3   *Linear dependence of σ_I on σ^**

2.11 v_s is related to E_s by the equation:

$$v_s = -0.459E_s + 0.516$$

2.12 The data are governed by the equation:

$$pK_a^H - pK_a^X = 1.44\sigma_m + 0.042$$

and illustrate a linear relationship between pK_a of 4-substituted bicyclo-[222]octane-1-carboxylic acids and σ_m.

2.13 The Brønsted equation for the correlation for the mutarotation of glucose catalysed by general bases (**B**) is

$$\log k/M^{-1}\mathrm{sec}^{-1} = 0.42 pK_a^{BH} - 3.46.$$

The statistical allowance (Chapter 6) has been omitted. The β value of 0.42 is also the Leffler α and its magnitude indicates that the transition state structure involves about 40% of bond formation between transferring proton and the base.

2.14 The extended Brønsted correlation for the aminolysis of the 2,4,6-trinitrophenyl ester is linear with an equation:

$$\log k = 0.62 pK_a - 3.96$$

The other correlation (see Figure 4) is non-linear and is evidence for an intermediate (see Chapter 7). The data fit Equation (11) of Chapter 7 for the consecutive reaction.

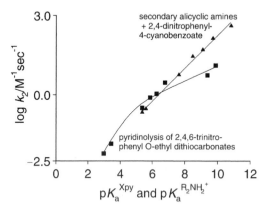

Figure 4   *Extended Brønsted correlations. The lines are theoretical*

2.15 The Swain–Scott dependence for the reaction of nucleophiles with α-*D*-glucosyl fluoride is:

$$\log k/M^{-1}\mathrm{sec}^{-1} = 0.185n - 1.56.$$

The slope ($s = 0.185$) is consistent with weak bond formation in the transition state compared with that in the reference reaction with methyl bromide where $s = 1.0$.

2.16 The solvolysis of α-D-glucosyl fluoride in aqueous methanol solutions follows the Grunwald–Winstein equation:

$$\log(k_S/k_{ss}) = 0.335\, Y_{ad} - 3.57$$

In this correlation the solvolysis of 1-adamantyl chloride is employed as the reference reaction (Y_{ad}). The slope m is consistent with a transition state with about 1/3 carbenium ion character of that of the reference reaction (where full carbenium ion character is assumed to be expressed).

Since there is an oxygen adjacent to the C_1 atom of the glucosyl species the result would also be consistent with complete charge development where 33.5% of the charge resides on C_1 and the rest on the ring oxygen.

2.17 The K_m data do not correlate well with Hansch's π parameters and it may be concluded that the catalysis does not involve a significantly apolar interaction between the amide backbone of the substrate and the active sub-site, ρ_1.[a] Reference to the original literature (reference 38 of Chapter 2) indicates that K_m correlates reasonably well with P_E, a parameter measuring polarisability; this is consistent with the postulate derived from structural data that there is hydrogen-bonding between the amide of the substrate and the binding sub-site of the active site.

2.18 The correlation of $pK_a^H - pK_a^X$ with σ′ (equivalent to the pK_a of 4-substituted bicyclo[2.2.2]octane carboxylic acid) is linear obeying the equation which could therefore be employed as a secondary definition of σ′:

$$\sigma' = (pK_a^H - pK_a^X)/0.73.$$

Chapter 3

Effective Charge Maps – Equilibria

3.1 Scheme 1 (Chapter 3) shows that aryl acetate esters and aryl thio-acetate esters have effective charges of +0.7 and +0.4 respectively on the O and S leaving atom. The effective charge map for the equilibrium can thus be written as in Scheme 3.

The effective charges on O and S cannot be directly compared because they are derived from different standard dissociation equilibria (phenols and thiophenols respectively).

---

[a] The term ρ_1 refers to a component of the active site of the enzyme complementary to the α-peptide bond of the substrate; it is not a Hammett slope.

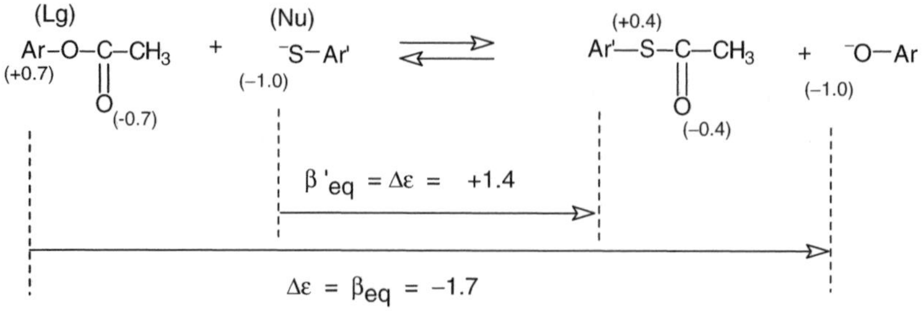

(Lg)   (Nu)   (+0.4)
$$Ar\text{–}O\text{–}\underset{\underset{O}{\|}}{C}\text{–}CH_3 \ (+0.7) \quad + \quad {}^-S\text{–}Ar' \ (-1.0) \quad \rightleftharpoons \quad Ar'\text{–}S\text{–}\underset{\underset{O}{\|}}{C}\text{–}CH_3 \ (-0.4) \quad + \quad {}^-O\text{–}Ar \ (-1.0)$$

$$\beta'_{eq} = \Delta\varepsilon = +1.4$$

$$\Delta\varepsilon = \beta_{eq} = -1.7$$

Scheme 3

3.2 Plotting β_{Lg} versus the pK_a of the attacking nucleophile gives a correlation $\beta_{Lg} = -0.111pK_{Nuc} + 1.62$. Extrapolation to pK_a 7.14 yields -0.83 for the β_{Lg} for the 4-nitrophenolate ion nucleophile. The extrapolation is not substantial and should therefore give a relatively accurate value of β_{Lg} which, when combined with β_{Nuc} (0.53), gives $\beta_{eq} = 1.36$. It should be appreciated that determining β_{Lg} experimentally is possible for attack of 4-nitrophenolate anion on substituted phenyl diphenylphosphate esters but would require a large investment of time, expensive equipment and isotopically labelled material.

3.3 A dissociation equilibrium where the acidic hydrogen is two atoms removed from the aromatic ring has a Hammett ρ value near to unity[b] due to the similar distances between the substituent and reaction centre in the reference dissociation.

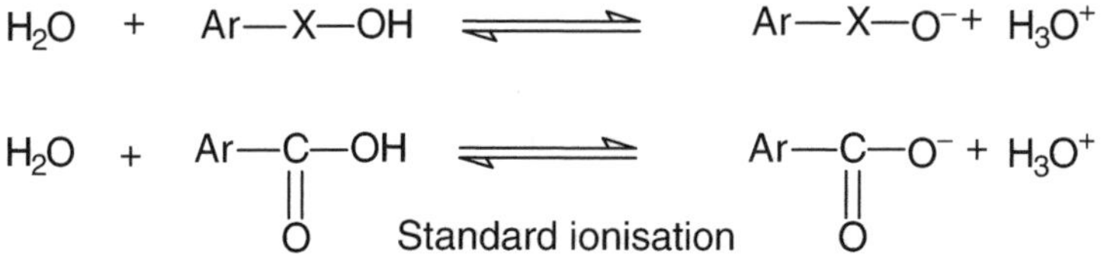

$$H_2O \ + \ Ar\text{–}X\text{–}OH \ \rightleftharpoons \ Ar\text{–}X\text{–}O^- + H_3O^+ \qquad (1)$$

$$H_2O \ + \ Ar\text{–}\underset{\underset{O}{\|}}{C}\text{–}OH \ \rightleftharpoons \ Ar\text{–}\underset{\underset{O}{\|}}{C}\text{–}O^- + H_3O^+ \qquad (2)$$

Standard ionisation

$ArB(OH)_2$ is a pseudo acid and the "dissociation" reaction involves the equilibrium:

$$2\,H_2O \ + \ Ar\text{–}B\overset{\displaystyle OH}{\underset{\displaystyle OH}{\diagup}} \ \rightleftharpoons \ Ar\text{–}\underset{\underset{OH}{|}}{\overset{\overset{OH}{|}}{B}}\text{–}OH \ + \ H_3O^+ \qquad (3)$$

The formal negative charge of the conjugate base resides on the atom (boron) *directly abutting* the aromatic ring and a positive Hammett ρ

[b] Table 1 in Appendix 4 shows a range of values between 0.7 and 1.4.

greater than 1 would therefore be expected. Table 1 in Appendix 4 shows a ρ value of about 2.3 for the dissociation of ArXH acids.

3.4 Perusal of the tables in Appendix 4 indicates that the pK_a values of anilinium ions and aryloxyacetic acids have Hammett equations:

$$pK_a^{\text{ArNH3}} = 2.89\sigma + 4.58 \quad\text{and}\quad pK_a^{\text{ArOCH2CO2H}} = 0.3\sigma + 3.17$$

Solving for equal pK_a values:

$$pK_a^{\text{ArNH}_3} - pK_a^{\text{ArOCH}_2\text{CO}_2\text{H}} = 0 = 2.59\sigma + 1.41$$

Thus

$$\sigma = -1.41/2.59 = -0.6$$

Data in Table 1 in Appendix 3 indicates that 4-EtNH– has a σ value of –0.61. The value of the pK_a at equality is calculated to be 1.23.

3.5

$$\rho_{eq} = \{(\varepsilon_p - \varepsilon_r)/ (\varepsilon'_{ps} - \varepsilon'_{rs})\}(d'_s/d)$$

$$\rho_s = \{(\varepsilon_{ps} - \varepsilon_{rs})/ (\varepsilon'_{ps} - \varepsilon'_{rs})\}(d'_s/d_s)$$

$$\rho_{eq}/\rho_s = \{(\varepsilon_p - \varepsilon_r)/ (\varepsilon_{ps} - \varepsilon_{rs})\}(d_s/d)$$

$$d_s = d, \varepsilon_p = -1 \quad\text{and}\quad \varepsilon_p = \varepsilon_{ps} \quad\text{and}\quad \varepsilon_{rs} = 0.$$

$$\rho_{eq}/\rho_s = 3.8/2.23 = 1.7 = (-1 - \varepsilon_r)/(-1 - 0)$$

Therefore

$$\varepsilon_r = 1.7 - 1 = +0.7 \quad\text{and}\quad \Delta\varepsilon = \varepsilon_p - \varepsilon_r = -1.7$$

Hence

$$\beta_{eq} = \Delta\varepsilon = -1.7$$

Negative charge development gives a positive Hammett ρ and a negative Brønsted β (Scheme 4). In an effective charge map negative β values measured from left to right indicate increase in negative effective charge. Deprotonation of substituted phenols in the standard dissociation increases negative charge and hence gives a negative β_s.

3.6 The value of β_{eq} is insensitive to R unless there is a gross change in polarity. Thus β_{eq} for R = C_2H_5– or –$(CH_3)_2CH$– would be approximately

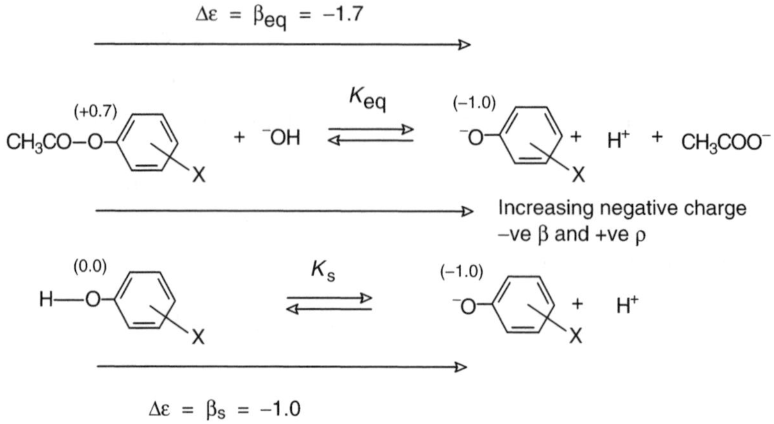

$$\Delta\varepsilon = \beta_{eq} = -1.7$$

(+0.7) K_{eq} (−1.0)

$CH_3CO-O-\text{(ring)}-X$ + ⁻OH $\rightleftharpoons$ ⁻O-(ring)-X + H^+ + CH_3COO^-

Increasing negative charge
−ve β and +ve ρ

(0.0) K_s (−1.0)

$H-O-\text{(ring)}-X$ $\rightleftharpoons$ ⁻O-(ring)-X + H^+

$$\Delta\varepsilon = \beta_s = -1.0$$

Scheme 4

the same as that for $R = CH_3-$ (+0.7). The β_{eq} for the "neutral" reaction can be obtained by difference from the effective charge map (Scheme 5).

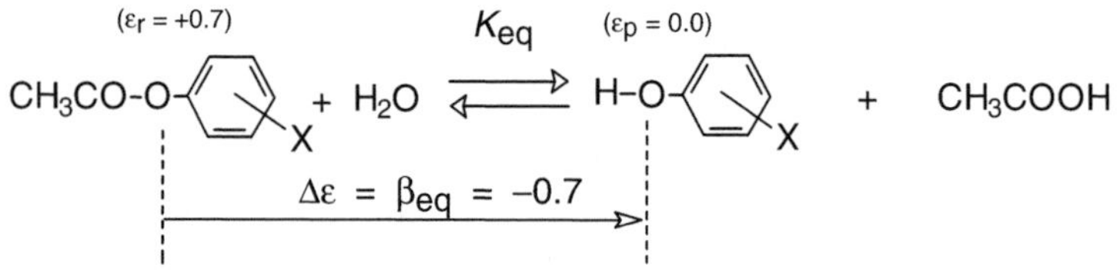

($\varepsilon_r = +0.7$) K_{eq} ($\varepsilon_p = 0.0$)

$CH_3CO-O-\text{(ring)}-X$ + H_2O $\rightleftharpoons$ $H-O-\text{(ring)}-X$ + CH_3COOH

$$\Delta\varepsilon = \beta_{eq} = -0.7$$

Scheme 5

3.7 The β_{eq} is independent of the nucleophile and reference to (Chapter 3, Scheme 1) shows that the effective charge on the pyridinium nitrogen is +1.6 in *N*-acetylpyridinium ions. The following effective charge map (Scheme 6) can be drawn:

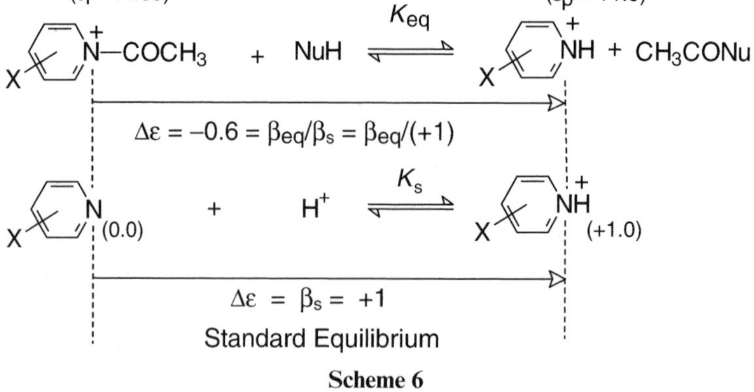

($\varepsilon_r = +1.60$) K_{eq} ($\varepsilon_p = +1.0$)

$X-\text{(ring)}-N^+-COCH_3$ + NuH $\rightleftharpoons$ $X-\text{(ring)}-N^+H$ + CH_3CONu

$$\Delta\varepsilon = -0.6 = \beta_{eq}/\beta_s = \beta_{eq}/(+1)$$

K_s

$X-\text{(ring)}-N$ (0.0) + H^+ $\rightleftharpoons$ $X-\text{(ring)}-N^+H$ (+1.0)

$$\Delta\varepsilon = \beta_s = +1$$

Standard Equilibrium

Scheme 6

$\Delta\varepsilon$ for the standard equilibrium is $\Delta\varepsilon = \beta_s = +1$ and for the transfer equilibrium is $\Delta\varepsilon = \beta_{eq}/\beta_s = -0.6$.

Thus:
$$\beta_{eq} = -0.6.$$

3.8 The effective charge on the sulfur atom in the displacement of an aryl-thio group from an alkyl donor is -0.16 units. The change in effective charge, $\Delta\varepsilon$, is therefore $+0.84$ from arylthio anion to thioether (Scheme 7).

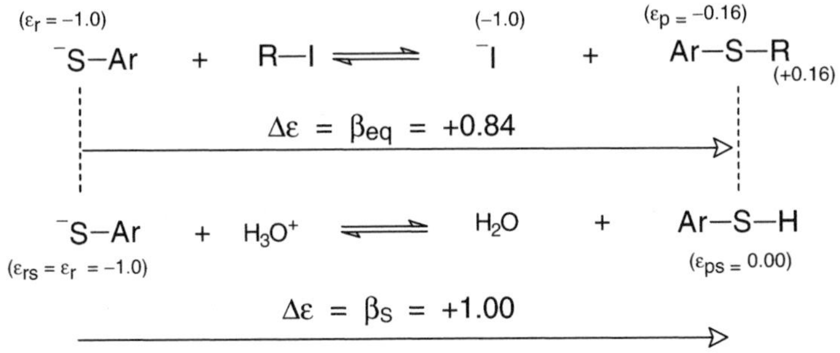

Scheme 7

The value of $\beta_{eq} = \Delta\varepsilon = (\varepsilon_p - \varepsilon_r)/(\varepsilon_{ps} - \varepsilon_{rs}) = +0.84$. Note that the concept of additivity predicts that the alkyl group, R–, has $+0.16$ units of effective charge and on this basis is a little more electropositive than the H– group.

The Hammett ρ_{eq} may be obtained from the ratio:

$$\rho_{eq}/\rho_s = +0.84$$

where $\rho_s = -2.26$. Therefore

$$\rho_{eq} = +0.84 . 2.23 = -1.88.$$

Although not needed for these calculations the value of β_s for the proto-nation of ⁻SAr is $+1.0$.

3.9 The charges on neutral and protonated pyridine nitrogens are defined as zero and $+1$. The change in effective charge on the nitrogen of the N-acetylpyridinium ion is $+1.6$. Using the protonation of substituted pyridines as the reference equilibrium ($\beta_s = +1$) the corresponding β_{eq} will therefore be $+1.6$.

The result can be formalised by use of the following equation where d for both reactions is the same (Scheme 8).

$$\beta_{eq}^{Xpy} = \Delta\varepsilon = (\varepsilon_p - \varepsilon_r)/(\varepsilon_{ps} - \varepsilon_{rs})(d/d) = (1.6 - 0)/(1 - 0) = +1.6.$$

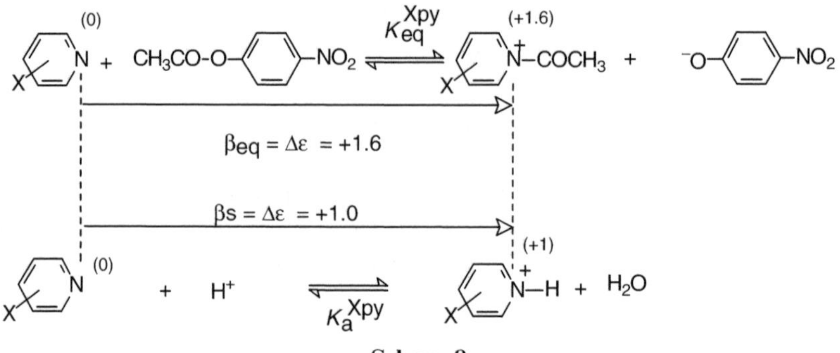

Scheme 8

3.10 Effective charge data show that the equilibrium for transfer of an acetyl group from a donor to aryloxide ions has a β_{eq} of +1.7 compared with that of the protonation of aryloxide ions ($\beta_s = +1$).

The effective charge on the aryl oxygen in the ester is +0.7 relative to the unit change in charge in the standard equilibrium (Scheme 9).

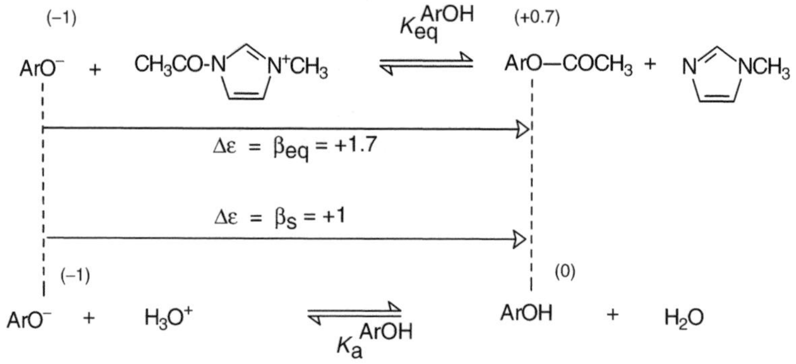

Scheme 9

3.11 The change in effective charge on the pyridine nitrogen ($\Delta\varepsilon_{Lg}$, measured from variation of X) is (−1.07). The change in effective charge on the aryl oxygen ($\Delta\varepsilon_{Nuc} = +1.36$) is measured using the variation in substituent in Ar. The two changes in effective charge cannot be compared directly because they are defined by *different* standard equilibria namely dissociation of substituted pyridinium ions and protonation of aryl oxide ions (Scheme 10).

Effective Charge Maps – Rates
3.12 The map of Hammett ρ values for the cyclisation of phenyl-substituted benzoylglycinate esters can be calculated from the three given

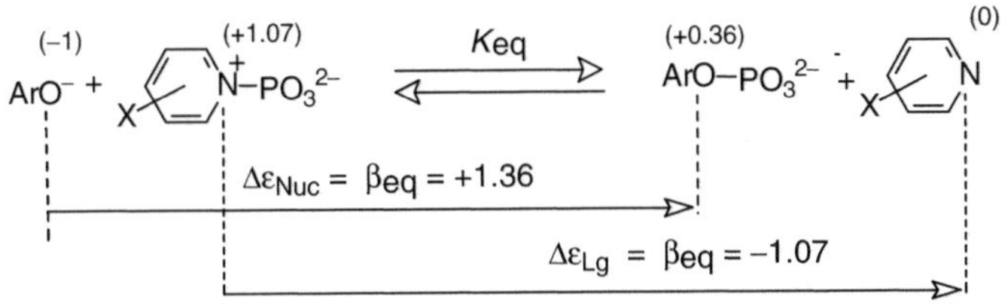

Scheme 10

ρ values and is shown below (Scheme 11). Bond formation is gauged by the Leffler index which is calculated to be 0.62.

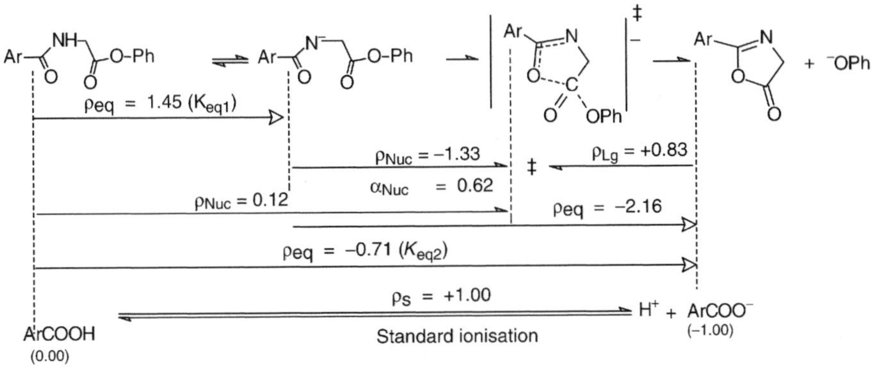

Scheme 11   *Map of Hammett ρ values for endocyclic bond formation*

Effective charges may be calculated for the substituent variation although they have little comparative meaning in this case. The ρ_{eq} of 1.45 together with $\rho_s = 1$ indicates that the amide conjugate base is more negative than the carboxylate ion. The overall equilibrium (K_{eq2}) has a ρ_{eq} of −0.71 indicating that the reaction centre interacting with the substituent in the oxazolinone product is more positive than that in the neutral amide. The reference equilibrium is not sufficiently similar to the reaction under investigation to enable any quantitative dissection of the effective charge; in the first instance the location of the computed effective charge is uncertain.

3.13   The map of Brønsted β values for the cyclisation are shown in Scheme 12. The α_{Lg} for the reaction of the conjugate base is 0.55. When the reference equilibrium for the effective charge of one substituent effect is not the same as that for the other (as in Problem 12) the bonding changes can *only* be compared when the β or ρ values are normalised as Leffler indices. In the case of cyclisation to the oxazolinone it is most convenient that bond fission is measured by a Brønsted relationship and bond formation by a Hammett equation. Both β and ρ can be normalised by division by β_{eq} and ρ_{eq} and the resulting α_{Lg} of 0.55 (from β values) and α_{Nuc} of 0.62 (from ρ values) indicates that bond formation is slightly more advanced than bond fission. The difference (0.55–0.62) indicates that there is hardly any charge development on the ester carbonyl function in the transition structure, which is *consistent* with a *synchronous* concerted mechanism.

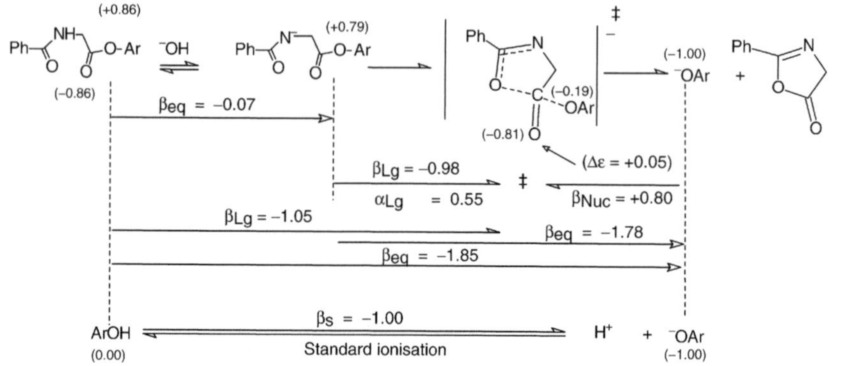

Scheme 12 *Map of effective charges for exocyclic bond fission*

3.14 The effective charge map for the sulfation of phenols with pyridine-N-sulfonate esters is illustrated in Schemes 13 and 14 for N–S bond fission and S–O bond formation. Effective charges are defined respectively by the dissociation reactions of substituted pyridinium ions and phenols. The pyridine nitrogen and the phenol oxygen suffer respectively changes in effective charge of -1.00 and $+0.23$ units from reactant to transition state structure.

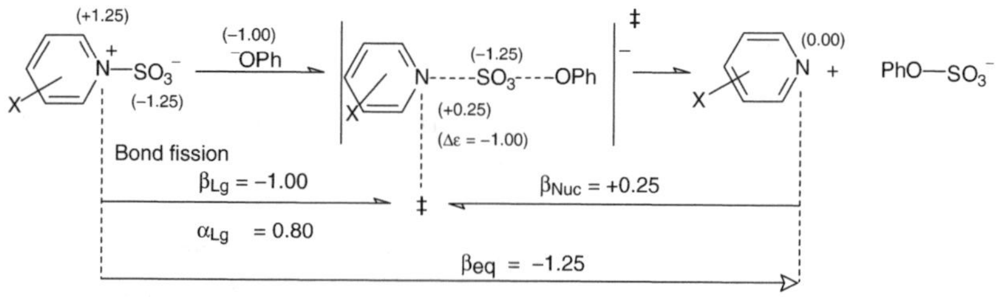

Scheme 13 *Substituent effect on N–S bond fission*

There is no change in effective charge on the $-SO_3-$ group from reactant to transition state structure as measured by variation in substituent on the pyridinium ion. The location of the residual effective charge of -1.25 is spread over the SO_3-O atoms.

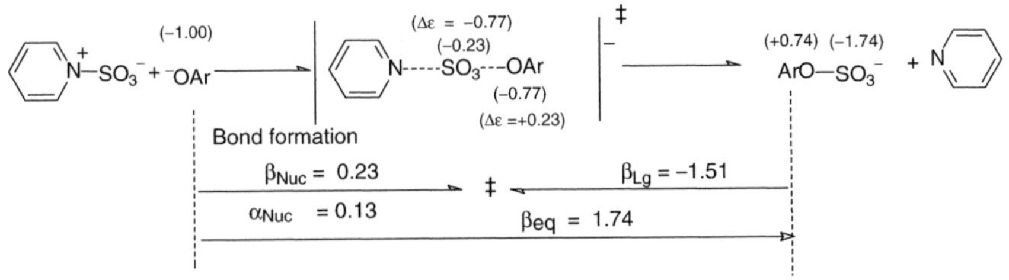

Scheme 14 *Substituent effect on S–O bond formation*

The increase in positive charge of 0.23 on the attacking aryl oxygen atom leads to an effective charge of -0.77 on this oxygen in the transition structure. This value cannot be utilised to decide the location of the -1.25 units determined from variation of substituent in the pyridinium ion because of the differing reference reaction for the effective charges.

The α_{Nuc} and α_{Lg} values indicate that there is a positive imbalance of $+0.67$ units of normalised effective charge on the SO_3 group of atoms, needed to supply electrons to support the charge during bond fission ($\alpha_{Lg} = 0.80$), in conjunction with a bond formation of $\alpha_{Nuc} = 0.13$. The displacement mechanism is thus a highly *asynchronous* concerted process with a transition state structure possessing substantial, neutral monomeric, sulfur trioxide character (Scheme 15).

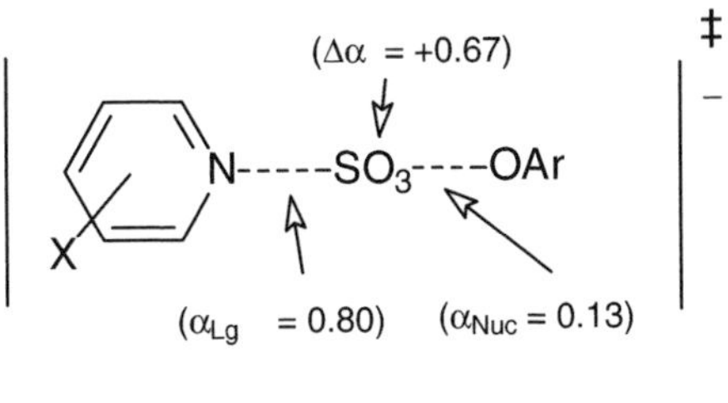

Transition state

Scheme 15 *Transition structure for sulfuryl group transfer*

3.15 The effective charge on the product $Xpy-PO_3H^-$ is derived by consideration of the map of Scheme 16 and the data yield the combined effective charge map as shown in Scheme 17:

Comparison of the α_{Nuc} and α_{Lg} indicates that the transition state of the concerted process (Scheme 18) is almost *synchronous*.

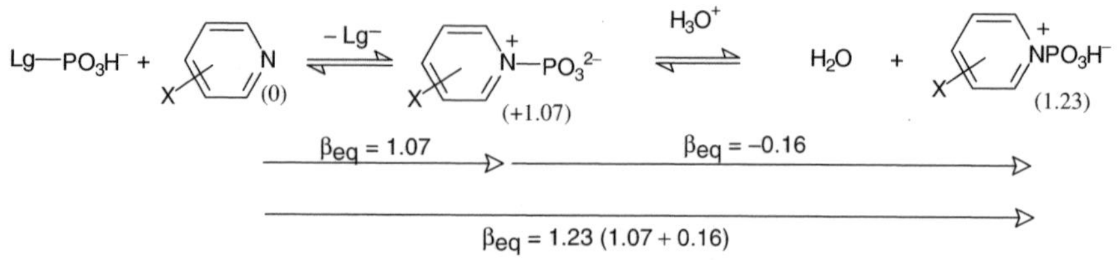

$Lg-PO_3H^-$ + (pyridine, X, (0)) $\xrightarrow{-Lg^-}$ (pyridinium, X, $N-PO_3^{2-}$, (+1.07)) $\xrightarrow{H_3O^+}$ H_2O + (pyridinium, X, NPO_3H^-, (1.23))

$\beta_{eq} = 1.07$

$\beta_{eq} = -0.16$

$\beta_{eq} = 1.23 \ (1.07 + 0.16)$

Scheme 16

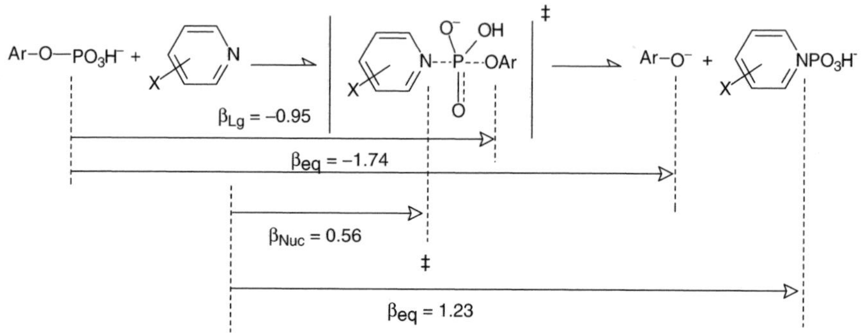

$Ar-O-PO_3H^-$ + (pyridine, X) $\longrightarrow$ (transition state: pyridinium $N---P$ with O^-, OH, $-OAr$, O) $\ddagger$ $\longrightarrow$ $Ar-O^-$ + (pyridinium, X, NPO_3H^-)

$\beta_{Lg} = -0.95$

$\beta_{eq} = -1.74$

$\beta_{Nuc} = 0.56$

$\ddagger$

$\beta_{eq} = 1.23$

Scheme 17

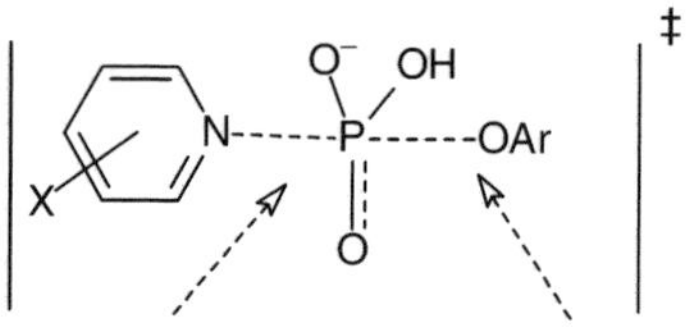

(transition state structure with O^-, OH, $N-----P------OAr$, O) $\ddagger$

$\alpha_{Nuc} = 0.56/1.23 = 0.46 \quad \alpha_{Lg} = -0.95/(-1.74) = 0.55$

Scheme 18 *Transition state for the phosphoryl transfer*

Chapter 4

4.1

Substituent	$\Sigma\sigma^*(calc)$	$\sigma^* \ (exp)$
$-CH(C_6H_5)_2$	$0 + 0.215 + 0.215 = 0.43$	0.41
$-CH_2C(CH_3)_3$	$0 + 0 - 3(0.115 \times 0.41) = -0.141$	-0.12
$-CH(OH)C_6H_5$	$0 + 0.215 + 0.555 = 0.77$	0.50
$-CHCl_2$	$0 + 2 \times 1.05 = 2.10$	1.94
$-CCl_3$	$3 \times 1.05 = 3.15$	2.65

Scheme 19

4.2

	pK_a (exp)	pK_a (calc)
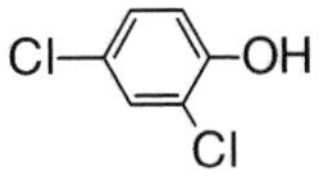 Cl—⟨⟩—OH with Cl	7.84	$9.38 - (9.95 - 8.48) = 7.91$
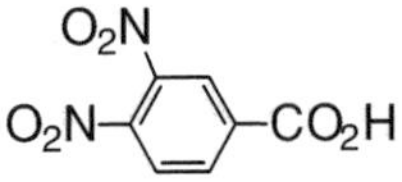 O$_2$N— ⟨⟩ —CO$_2$H with O$_2$N	2.82 ($\sigma_{exp.} = 1.38$)	$4.2 - 0.71 - 0.78 = 2.71$ ($\sigma_{calc.} = 1.49$)
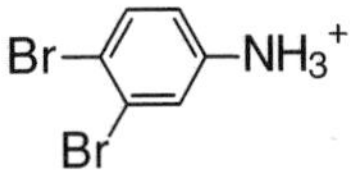 Br—⟨⟩—NH$_3^+$ with Br	2.34	$4.62 - 2(4.62 - 3.51) = 2.4$
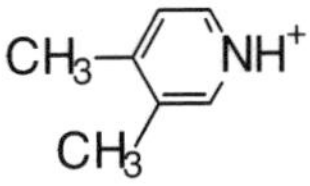 CH$_3$—⟨N⟩H$^+$ with CH$_3$	6.52	$5.17 - (5.17 - 5.68)$ $- (5.17 - 6.02) = 6.53$

Scheme 20

4.3 The plot of $[\log(k/k_H)]/\sigma_1$ versus σ_2/σ_1 is shown below (Figure 5) and does not exhibit a good fit, although the global fit using the bilinear equation $\log(k/k_H) = \sigma_1\rho_1 + \sigma_2\rho_2$ is excellent. The global fit may be illustrated by plotting $\log(k/k_H) - \rho_1\sigma_1$ against σ_2 (Figure 6).

The global analysis gives a result reasonably similar to that of the *original* article which employed the fit illustrated in Figure 5. (*global* $\rho_1 = -0.34$, $\rho_2 = 0.75$ *original* $\rho_1 = -0.52$ and $\rho_2 = 0.96$).

The effective charges on the carboxylate (see Scheme 21) may be obtained directly from the Hammett ρ_1 because in this case the

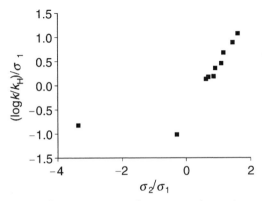

Figure 5 *Fit of data to Jaffé equation using the rearranged equation and linear fit*

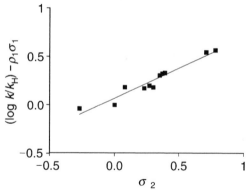

Figure 6   *Illustration of the global fit of data to the bilinear equation*

dissociation of benzoic acids is a suitable reference reaction for the proton transfer process.

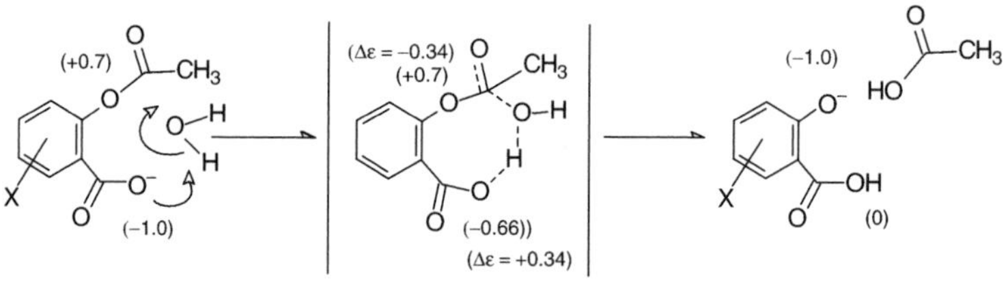

Scheme 21

The effective charge on the phenolic oxygen is obtained by use of the ρ_s for the dissociation of substituted phenols ($\beta_{Lg} = 0.75/{-}2.23 = -0.34$).

4.4   The pK_a values obey the Jaffé equation employing Hammett σ parameters for the 4-nitro substituents (Figure 7):

$$pK_{XY}/pK_{HH} = -1.92\sigma_X - 2.49\sigma_Y$$

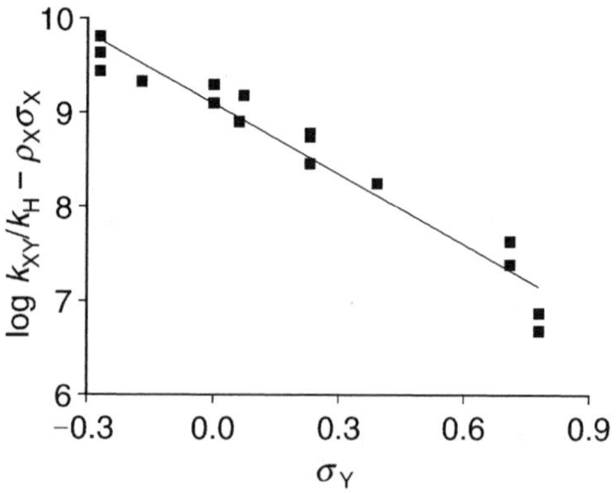

Figure 7   *Jaffé plot for the dissociation of sulfonamides with two transmission routes from substituent to ionising centre*

The high value of ρ_Y compared with ρ_X reflects the condition that the Y substituent is closer, by one atom, to the ionising NH than is the X substituent. The absence of a discernable resonance effect due to Y = 4-NO_2 indicates that a substantial amount of negative charge in the conjugate base does not reside on the nitrogen implying that the charge is spread over the SO_2 group. This would explain why the ρ_X value is not 0.4 of the ρ_Y value as predicted by the effect of an intervening atom.

4.5 The *para* substituents, which can withdraw electrons by resonance (such as CN or NO_2), stabilise the conjugate base (the aniline) and hence the pK_a values for these substituents will fall *below* the regression line for the *meta* series (□ in A in Figure 8).

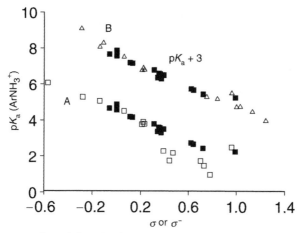

Figure 8 *Hammett plots of the pK_a of* meta- *and* para-*anilinium ions. A:* ■, meta *substituents;* □, para *substituents plotted against Hammett's σ; B:* ■, meta *substituents;* Δ, para *substituents plotted against σ^-. For the sake of clarity a constant increment of 3 is added to the pK_a ($ArNH_3^+$) in B to separate the lines*

When σ^- is used instead of σ the plot becomes linear (B in Figure 8).
4.6 The pK_R^+ values obey a Hammett σ correlation (Figure 9) for the *meta*-substituted compounds and the *para*-substituted congener (4-nitro-), which does not release electrons by resonance:

$$pK_R^+ = 3(-3.97)\sigma - 6.85$$

The *para* substituents (4NH₂, 4NMe₂ and 4MeO), which can donate electrons by resonance, stabilise the carbenium ion and hence the pK_R^+ for these substituents will lie *above* the regression line for the *meta* substituents.

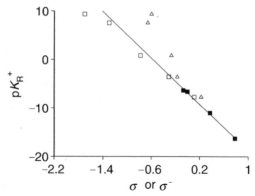

Figure 9   ■ *and* Δ *represent Hammett* σ *values for* meta *and* para *substituents respectively;* □ *represents the* σ⁺ *values of* para *substituents*

4.7   The data for the *meta*-substituted phenyl dimethylmethyl chlorides
fit the equation

$$\log(k/k_H) = 0.98F + 0.38R + 0.030$$

illustrated by Figure 10.

The relative values of the *r* (0.38) and *f* (0.98) coefficients are discussed
in the answer to Problem 8.

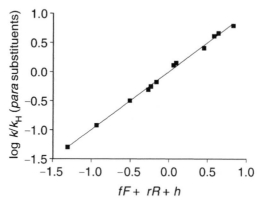

Figure 10   *Swain–Lupton plot for the solvolysis of* para-*substituted phenyl dimethylmethyl chlorides*

4.8   The data fit the Swain–Lupton equation for the solvolysis of the
para-substituted phenyl dimethylmethyl chlorides (Figure 10):

$$\log(k/k_H) = 0.64F + 1.84R + 0.084$$

The %*R* for the *meta* case is 27 (answer to Problem 7) and that for the
para case is 74 confirming that there is substantial resonance between

substituents and reaction centre in the *para* case. The resonance in the *meta* case is due to polar effect transmission through distortion of the π-electron cloud.

4.9 The Yukawa–Tsuno equation for the data is:

$$\log(k_X/k_H) = 3.51[\sigma + 0.64(\sigma^- - \sigma)]$$

The fit (Figures 11 and 12) indicates that there is substantial resonance transmission to the leaving oxygen in the transition structure by substituents in the *para* position which can withdraw electrons by resonance interaction. This is consistent with a *concerted* displacement mechanism at the silicon atom.

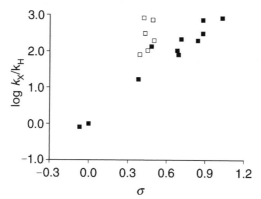

Figure 11 *Hammett correlations for alkaline hydrolysis of substituted phenyl triethyl silyl ethers. □, para substituents; ■, meta substituents*

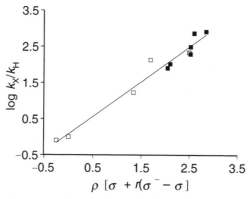

Figure 12 *Yukawa-Tsuno fits for alkaline hydrolysis of substituted phenyl triethyl silyl ethers. □, meta substituents; ■, para substituents*

4.10   *Calculated and experimental σ**

Substituent	$\sigma^*(calc)$	$\sigma^*(exp)$
$-CH_2CH_2Cl$,	$1.05 \times 0.41 = 0.43$	0.385
$-CH_2CH_2CF_3$	$0.92 \times 0.41 = 0.38$	0.32
$-CH_2CH_2CH_2CF_3$,	$0.92 \times (0.41)^2 = 0.155$	0.12
$-CH_2C_6H_5$,	$0.60 \times 0.41 = 0.246$	0.215
$-CH_2CH_2C_6H_5$	$0.60 \times (0.41)^2 = 0.10$	0.08
$-CH_2CH_2CH_2C_6H_5$	$0.60 \times (0.41)^3 = 0.041$	0.02

[a] See Table 1 in Appendix 3.

4.11 $0.47 \times 0.47 \times 1 = 0.22 \times 1 = 0.22$ (the observed value is 0.22)

4.12   The data fit the equation:

$$\log k_N = 1.28 E_N + 0.024 H_N - 6.51$$

The results indicate that in this case the E_N parameter is a good model and the dissociation of the conjugate acid is not a predominant factor in the nucleophilic substitution reaction. A correlation between $\log k_N$ and E_N could be tried as an exercise.

4.13

$$\Sigma\sigma = \sigma_{X1} + \sigma_{X2} + \sigma_{X3} + (0.22)(\sigma_{X4} + \sigma_{X5} + \sigma_{X6})$$

The attenuation factor for the $-CH_2CH_2-$ group is 0.22 (Table 1. Chapter 4).

4.14   The calculation is not perfect as the transmission factor is only assumed to be 0.4 for intervening groups of types $-CO-$ or $-NH-$. The acid in question has three atoms formally interposed between the oxygen and the hydrogen of a phenol. The β value will thus be $(-1)(0.4)^3 = -0.064$ which agrees with the value obtained experimentally (-0.07) (Problem 13, Chapter 3). The calculated value of β could be more negative because the attenuation factor of $-NH-$ is 0.69 (Chapter 4, Table 1) and that of the $>C=O$ group is possibly higher than 0.4.

4.15   The acid can formally be considered as composed of a pyridinium ion where a phosphorus and an oxygen atom are interposed between the nitrogen and the hydrogen. The value of β is thus given by $(-1)(0.4)^2 = 0.16$. Table 1 of Chapter 4 suggests that better attenuation factors could be employed (0.64 for oxygen) and possibly a value higher than 0.4 for the $-PO_2^-$ group.

4.16   The positive ρ of 1.45 indicates that negative charge accumulates in the product. The greater sensitivity than in the reference reaction is due to increased negative charge as the distances, d, between substituent and reaction centre are almost the same. The charge is likely to be localised on the oxygen in the conjugate base of the amide whereas it is spread equally over the oxygens in the carboxylate ion.

Chapter 5

5.1   The equation

$$\Delta G^{\ddagger}_{ij} = \omega^R + \gamma + \Delta G^{\circ}_{ij}/2 + \Delta G^{\circ}{}_{ij}{}^2/16\gamma$$

is differentiated at constant j to yield for the forward reaction ($\gamma = 0.5(\partial \Delta G^{\ddagger}_{ii} + \partial \Delta G^{\ddagger}_{jj})$):

$$\partial \Delta G^{\ddagger}_{ij} = 0.5\partial \Delta G^{\ddagger}_{ii} + 0.5\partial \Delta G^{\circ}_{ij} + \Delta G^{\circ}_{ij}/8\lambda(\partial \Delta G^{\circ}_{ij} - \Delta G^{\circ}_{ij}\partial \Delta G^{\ddagger}_{ii}/4\lambda)$$

If the squared term for the Marcus equation is neglected:

$$\beta_f = 0.5\beta_{ii} + 0.5\beta_{eq} \qquad \text{(forward reaction)}$$

For the return reaction:

$$\Delta G^{\ddagger}_{ij} = \omega^R + \gamma - \Delta G^{\circ}_{ij}/2 + \Delta G^{\circ}{}_{ij}{}^2/16\gamma$$

Differentiation at constant j yields:

$$\partial \Delta G^{\ddagger}_{ij} = 0.5\partial \Delta G^{\ddagger}_{ii} - 0.5\partial \Delta G^{\circ}_{ij} + \Delta G^{\circ}_{ij}/8\gamma(\partial \Delta G^{\circ}_{ij} - \Delta G^{\circ}_{ij}\partial \Delta G^{\ddagger}_{ii}/4\gamma)$$

If the squared term for the Marcus equation is neglected:

$$\beta_r = 0.5\beta_{ii} - 0.5\beta_{eq} \qquad \text{(return reaction)}$$

Subtracting the equation for β_r from that for β_f gives:

$$\beta_f - \beta_r = \beta_{eq}$$

(Note that the assumption that the squared term is zero is not necessary to derive this equation.)
   Summing the equations for β_r and β_f gives:

$$\beta_f + \beta_r = \beta_{ii}$$

5.2   The equation

$$\delta = \partial \log k_{ii}/\partial \log K_{ij} = \tau - 1$$

(K_{ij} is the equilibrium constant for constant j and varying i) may be written as

$$\delta = (\partial \log k_{ii}/\partial \log K_{ij})(\partial pK_a/\partial pK_a)$$

Substituting for $\partial \log K_{ij}/\partial pK_a = \beta_{eq}$ and $\partial \log k_{ii}/\partial pK_a = \beta_{ii}$ yields:

$$\delta = \{\partial \log k_{ii}/\partial pK_a\}\ \{\partial pK_a/\partial \log K_{ij}\} = \beta_{ii}/\beta_{eq}$$

Using the equation $\beta_{ii} = \beta_{Nuc} + \beta_{Lg}$ leads to:

$$\delta = (\beta_{Nuc} + \beta_{Lg})/\beta_{eq}$$

$$= \beta_{Nuc}/\beta_{eq} + \beta_{Lg}/\beta_{eq} = \alpha_{Nuc} + \alpha_{Lg}$$

Therefore

$$\tau - 1 = \alpha_{Nuc} + \alpha_{Nuc} - 1$$

Since $\alpha_{Nuc} - \alpha_{Lg} = 1$ (derived from $\beta_{eq} = \beta_{Nuc} - \beta_{Lg}$)

$$\tau - 1 = 2\alpha_{Lg} + 1$$

and thus

$$\tau = 2(\alpha_{Lg} + 1) \qquad \text{and} \qquad = \tau\ 2\alpha_{Nuc}$$

5.3   Substitute 7.14 (the pK_a of 4-nitrophenol) for pK_a^{ArOH}:

$$\log k_{ArO-} = 0.75 \times 7.14 - 7.28 = -1.93$$

$$k_{ArO-} = 1.19 \times 10^{-2}\ M^{-1}sec^{-1}$$

5.4   The data are analysed by a global technique (Appendix 1, Section A1.1.4.3). Alternatively the β_{Lg} and β_{Nuc} can be fitted separately to the linear equations.
Global analysis:

$$p_{xy} = 0.057; \qquad N = 0.430; \qquad L = -0.987$$

Separate fitting to Equations (30) and (31):

$$\beta_{Nuc} = 0.0605pK_a^{Lg} + 0.405$$

$$\beta_{Lg} = 0.0478pK_a^{Nuc} - 0.907.$$

This exercise illustrates the problem of fitting to Equations (30) and (31) separately. Summing β_{Nuc} and β_{Nuc} calculated from the equations obtained by global analysis (at $pK_a^{Lg} = pK_a^{Nuc}$) gives $\beta_{eq} = 1.42$.

5.5 Substituting into Equation (28) gives:

$$\beta_{eq} = 0.115pK_a^{NucH} - 1.63 - 0.115pK_a^{LgH} + 0.38 = 1.25$$

5.6 Differentiating Equation (32) yields:

$$\partial\beta_{ii} = 2p_{xy}\partial pK_a$$

$\tau - 1 = \delta = \partial logk_{ii}/\partial logK_{ij} = \beta_{ii}/\beta_{eq}$ (see answer to Problem 2)

Therefore

$$\partial\tau = \partial\beta_{ii}/\partial_{eq} = 2p_{xy}\partial pK_a/\beta_{eq}$$

and

$$\partial\tau/\beta pK_a = 2p_{xy}/\beta_{eq}$$

5.7 Substitute β_{ii} from $\tau - 1 = \beta_{ii}/\beta_{eq}$ into Equation (32) yields:

$$\tau = 1 + (2p_{xy}pK_a + N + L)/\beta_{eq}$$

$$(B = N + L)$$

5.8 The pK_a and $logk_B$ *are corrected statistically with* $p = 3$ *and* $q = 1$ throughout. The data can be force fit to a linear Brønsted equation:

$$logk_B = 0.711pK_a - 5.986$$

and to the binomial equation where $p_x = 0.054$, $\beta_o = 1.170$ and $C = 7.75$. The results are illustrated in Figure 13.

5.9 The Hammett ρ obeys a linear correlation $\rho = p_{xy}\sigma_Y + \rho_o$ where

$$p_{xy} = -0.744 \qquad and \qquad \rho_o = 0.976$$

5.10 The expression $\partial logk_X/\partial\sigma_x = \rho = \rho_o + 2m\sigma$ is integrated to give:

$$logk_X = \rho_o\sigma + m\sigma^2 + C$$

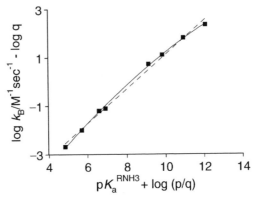

Figure 13   *Curvature in a free energy relationship due to a positive Hammond coefficient for the proton transfer from S,S-dimethyl (9-fluorenyl)sulfonium tetrafluoroborate. The dashed line is fit of the data to a linear Brønsted equation*

Set $\sigma = 0$ then $C = \log k_H$
and thus:

$$\log(k_X/k_H) = \rho_o\sigma + m\sigma^2$$

Chapter 6

6.1   The differentiated Marcus equation gives the value of Leffler's α:

$$\partial\Delta G^{\ddagger}/\partial\Delta G_o = 0.5(1 + \Delta G_o/4\Delta G_o^{\ddagger}) = \alpha$$

Inspection of the equation shows that:

$\alpha = 0$ when $\Delta G_o = -4\Delta G_o^{\ddagger}$; $\alpha = 1.0$ when $\Delta G_o = 4\Delta G_o^{\ddagger}$ and $\alpha = 0.5$ when $\Delta G_o = 0$.

The equation breaks down when $\Delta G_o < -4\Delta G_o^{\ddagger}$ when α becomes negative and when $\Delta G_o > 4\Delta G_o^{\ddagger}$ when α exceeds 1. This is not an explanation of Bordwell's anomaly because the breakdown criteria correspond to two chemically impossible structures where the cross-over points occur at bond lengths *less* than those in the relaxed states. The cross-over is illustrated for the case $\Delta G_o < -4\Delta G_o^{\ddagger}$ in Figure 14.

6.2   The value of ρ_{eq} for the dissociation of arylnitromethanes can be estimated from the average for the dissociation of groups adjacent to the benzene ring. Thus $\rho = 2.89, 2.24, 2.23$ for the dissociation of anilinium ions, thiophenols and phenols respectively. A ρ value (2.13) may also be calculated from that for the dissociation of substituted benzoic acids assuming an attenuation factor 0.47 for the $>C=O$ group. The average is 2.37 which is taken to be ρ_{eq} for the formation of the carbanion at (0,1).

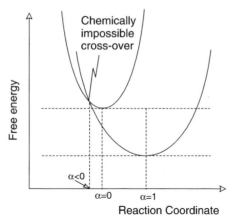

Figure 14 *Two-dimensional energy diagram illustrating the condition where the Marcus equation breaks down*

The value $\rho_f = 1.28$ combined with the estimated ρ_{eq} yields a Leffler α_f (1.28/2.37) of 0.54, consistent with development of −0.54 units of effective charge on the 1-carbon in the transition state of dissociation to pseudo acid. The charge on un-rehybridised carbon(1) (at 0,1) is defined as −1.

The value of ρ_{eq} for the equilibrium formation of pseudo base (at 1,1) is 0.83 and this corresponds to an effective charge on carbon(1) of −0.83/2.37 = −0.35.

6.3 The two relationships have the following parameters:
Brønsted plot for phenolate ion reaction

$$\log k_{PhO^-} = -0.3 pK_a^{ArOH} + 3.14 \qquad (r = 0.9007)$$

The relationship between reactivities of phenolate ion and 4-cyano-phenolate ion is:

$$\log k_{PhO^-} = 0.80 \log k_{4\text{-cyanophenolate}} + 1.34 \qquad (r = 0.9832).$$

The correlation coefficients indicate a much better relationship between the rates and this is explained by the closer similarity of the two systems being related compared with that for the Brønsted relationship. The graphs are illustrated in Figures 15 and 16:

6.4 When all the variant reagents have similar structures each will have identical p and q values (for example alkyl phosphate dianions all have $p = 1$ and $q = 3$). The slope, β, will be identical using both treatments but the intercepts will differ. The statistically uncorrected Brønsted graph is illustrated in Figure 17 for the base-catalysed decomposition of nitramide.

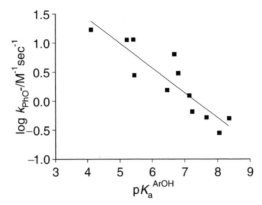

Figure 15 *Extended Brønsted correlation exhibits relatively substantial microscopic medium effects*

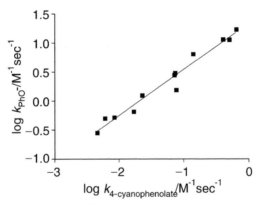

Figure 16 *Free energy correlation illustrating reduced microscopic medium effects*

The statistically corrected Brønsted line is also shown and it has a slightly better correlation coefficient than the uncorrected line. There is, however, little difference between the derived parameters even though both p and q values are not constant.

Uncorrected line:
  $\beta = 0.69 \pm 0.044$, intercept $= -4.39$ and $r = 0.9651$.
Corrected line:
  $\beta = 0.71 \pm 0.037$, intercept $= -4.70$ and $r = 0.9773$.

If the data were selected from the pK_a^{HB} range 8 to 11 a poorer correlation would have been obtained as well as a different slope.

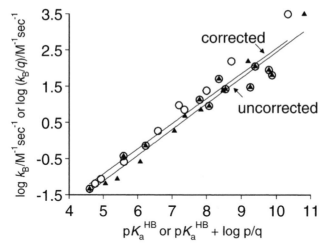

Figure 17 *Effect of statistical corrections on a Brønsted relationship. Closed triangles are statistically corrected; open circles are uncorrected. Lines are for least squares fits*

6.5 The random data created within the defined ranges are plotted as Figure 18 ($\log k_f$ versus $\log(k_f/k_r)$) and Figure 19 ($\log k_f$ versus $\log k_r$). Figure 18 suggests an *apparent* Leffler relationship which gets better as $\Delta \log k_r$ gets smaller but Figure 19 is diagnostic of a non-correlation.

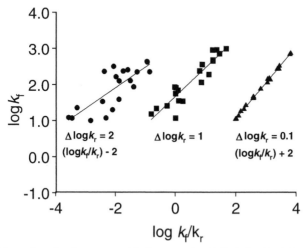

Figure 18 *Plots of randomly generated $\log k_f$ versus $\log(k_f/k_r)$ ($\log K_{eq}$) for varying ranges of $\log k_r$ at $\log k_f = 1.0$ to 3.0; correlation coefficients of the linear regressions are 0.7907, 0.9285 and 0.9987 for $\Delta \log k_r = 2.0$, 1.0 and 0.1 respectively. Note that the slopes of the linear regressions tend to unity as the range of $\log k_r$ decreases. To spread out the plots for clarity 2 has been subtracted from $\log(k_f/k_r)$ for $\Delta \log k_f = 2.0$ and 2 has been added to $\log(k_f/k_r)$ for $\Delta \log k_f = 0.1$*

5.0
4.0
3.0
2.0
1.0
0.0
log k_f
−2.0 −1.8 −1.6 −1.4 −1.2 −1.0
log k_r

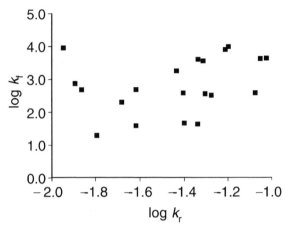

Figure 19 *Plot of* $\log k_f$ *versus* $\log k_r$ *for randomly generated numbers* $\Delta \log k_f = 2.0$ *and* $\Delta \log k_r = 0.1$

For a true free energy relationship the plot of Figure 19 should be linear with slope $-\beta/(1-\beta)$.

6.6

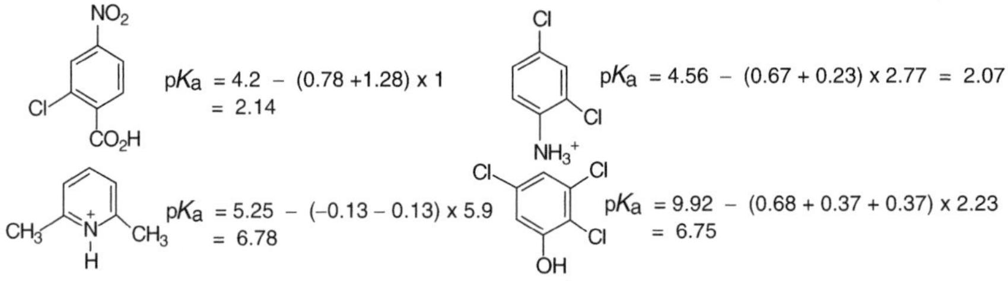

NO2
Cl
CO2H
pKa = 4.2 – (0.78 +1.28) x 1
= 2.14
Cl
Cl
NH3+
pKa = 4.56 – (0.67 + 0.23) x 2.77 = 2.07
CH3
N
H
CH3
pKa = 5.25 – (–0.13 – 0.13) x 5.9
= 6.78
Cl
Cl
Cl
OH
pKa = 9.92 – (0.68 + 0.37 + 0.37) x 2.23
= 6.75

Scheme 22

6.7 The graph (Figure 20) shows curvature and this may be quantified by fitting the data to a binomial equation:

$$\log k = a + b\Delta pK_a + c(\Delta pK_a)^2$$

The data do not fit an Eigen Equation (Equation 3 and Figure 2 of Chapter 6) as would be predicted for a change in rate-limiting step. The plot has "Marcus" curvature arising from changing slopes of the intersecting parabolas for the chemical proton transfer step (k_2 of Equation 1 of Chapter 6) as the free energy of the reaction changes. Note that in this case the change in slope is in accord with the reactivity–selectivity principle.

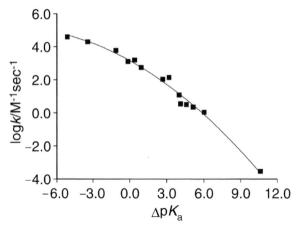

Figure 20 *Marcus curvature for proton transfer between carbon acid and heteroatom bases*

6.8 The rate of the forward proton transfer reaction can be deduced (Equation 1 of Chapter 5) using the steady-state assumption for reactive intermediates. At the steady state:

$$\partial[AH.B]/\partial t = [AH][B]k_1 + [A^-.HB]k_{-2} - (k_2 + k_{-1})[AH.B]$$

$$= 0$$

$$\partial[A^-.HB]/\partial t = [AH.B]k_2 - (k_{-2} + k_3)[A^-.HB] = 0$$

The two simultaneous equations can be solved for $[A^-.HB]$:
Thus:

$$(k_{-2} + k_3)[A^-.HB](k_2 + k_{-1}) = [AH][B]k_1 k_2 + [A^-.HB]k_{-2}k_2$$

and hence:

$$[A^-.HB] = [AH][B]k_1 k_2/((k_{-2} + k_3)(k_2 + k_{-1}) - k_{-2}k_2)$$

$$\text{Rate} = [A^-][HB]k_f$$

$$= [A^-.HB]k_3 = [A^-.HB]k_1 k_2 k_3/((k_{-2} + k_3)k_{-1} + k_2 k_3)$$

and

$$k_f = k_1 k_2 k_3/[(k_{-2} + k_3)k_{-1} + k_2 k_3] = k_1 k_2 k_3/D$$

Since $K_{eq} = k_f/k_r = k_1k_2k_3/k_{-1}k_{-2}k_{-3}$

$$k_r = k_{-1}k_{-2}k_{-3}/D \text{ (see text for } D)$$

When $k_2 > k_{-1}$

$$D = k_{-2}k_{-1} + k_2k_3 \quad \text{and} \quad k_f = k_1k_2k_3/((k_{-2}k_{-1} + k_2k_3)$$

$$k_f \equiv k_1/(k_{-1}K_2/k_3 + 1) \quad \text{where} \quad K_2 = k_{-2}/k_2$$

and

$$k_r = k_{-1}k_{-2}k_{-3}/(k_{-2}k_{-1} + k_2k_3) \equiv k_{-3}/[1 + k_3/(k_{-1}K_2)]$$

6.9 The fit to the Eigen equation yields $\log k_1/M^{-1}sec^{-1} = 9.03 \pm 0.16$, corresponding to a diffusional rate constant, and the pK_o parameter (8.1 ± 0.25) comes a little lower than the value determined by pH-titration.

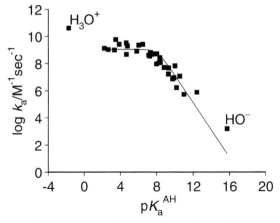

Figure 21   *Eigen plot for proton transfer to cyanide ion from AH^+*

The fit to an Eigen equation of the proton transfer step to the carbon base (CN^-) indicates that the k_2 step is not rate limiting and the cyanide ion behaves as if it were a heteroatom base. The anomalously large values of the rate constants for oxonium and hydroxide ions are consistent with their faster diffusion rates resulting from Grotthus translational-type effects.

6.10   The true position of the energy is a little lower than the intersection point because of resonance between the two structures. This lowering of energy can be considered constant throughout the range of ΔG_o and will not therefore affect the validity of the conclusions from the trigonometric arguments.

6.11 In the More O'Ferrall–Jencks diagram (Figure 22) the more electron withdrawing substituent stabilises the carbanion structure, the NW corner (0,1), and destabilises the SE corner (1,0). This skews the diagram and causes an increase in β because the transition structure moves towards the (0,1) coordinate as shown.

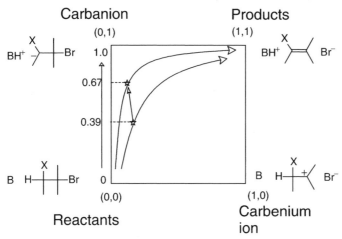

Figure 22 *More O'Ferrall–Jencks diagram for base-catalysed elimination*

6.12 Equation (20) can be verified in a number of equivalent ways one of which is:

$$\log k_f = \beta_f pK_a + C_f$$

$$\log k_r = \beta_r pK_a + C_r$$
$$K_{eq} = k_f/k_r$$

Therefore

$$\log K_{eq} = \log k_f - \log k_r$$

$$= \beta_f pK_a + C_f - \beta_r pK_a - C_r$$

$$\beta_{eq} = \partial \log K_{eq}/\partial pK_a = \beta_f - \beta_r$$

6.13 The Hammett ρ value of 2.23 translates to a Brønsted β of −2.23 because σ is essentially a pK_a of the substituted benzoic acid. As expected, electron withdrawing substituents in the acyl group of esters increase the rate of alkaline hydrolysis. The results indicate that the transition structure in ester hydrolysis is *more negatively* charged compared with the neutral ester than is the carboxylate anion compared with its undissociated acid in the reference reaction (which has a β defined as −1). An explanation is that the ionised tetrahedral intermediate (1) has more localised negative charge than the corresponding carboxylate anion (2) and therefore its stability has a greater dependence upon substituents.

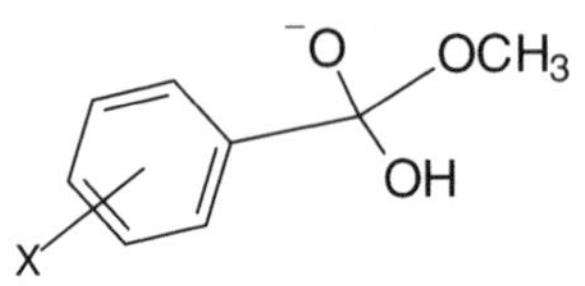

1 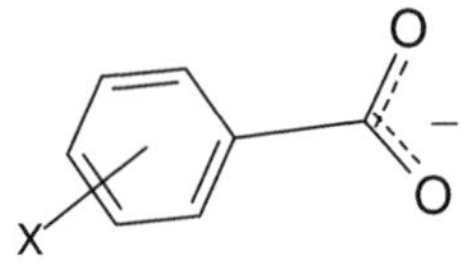   **2**

6.14

$$\Delta G^{\ddagger} = \Delta H^{\ddagger} - T\Delta S^{\ddagger}$$

$$\rho = \partial \Delta G^{\ddagger}/\partial \sigma = \partial \Delta H^{\ddagger}/\partial \sigma - T\partial \Delta S^{\ddagger}/\partial \sigma$$

$$\Delta H^{\ddagger} = T_i \Delta S^{\ddagger} + b$$

Therefore

$$\partial \Delta H^{\ddagger}/\partial \sigma = T_i \partial \Delta S^{\ddagger} \partial \sigma$$

Thus $\rho = \partial \Delta S^{\ddagger}/\partial \sigma (T_i - T)$ and, if $\partial \Delta S^{\ddagger}/\partial \sigma$ is temperature independent, $\rho = 0$ at $T = T_i$.

Chapter 7

7.1(a)

$$\log k_{\text{overall}} = \log(k_1 + k_2)$$

Substituting the Brønsted equations for k_1 and k_2

$$(\log k_1 = \beta_1 pK_a + C_1 \quad \text{and} \quad \log k_2 = \beta_2.pK_a + C_2)$$

into the equation for $\log k_{\text{overall}} = \log(k_1 + k_2)$ gives

$$\log k_{\text{overall}} = \log(10^{(\beta_1 pK_a + C_1)} + 10^{(\beta_2 pKa + C_2)}).$$

Figure 23 illustrates the three cases where a change in mechanism is indicated by non-linear free energy correlations. Break-points occur at $k_1 = k_2$ when

$$\log k_{overall} = 0.303 + \log k_1 = 0.303 + \log k_2$$

and at

$$pK_a = (C_2 - C_1)/(\beta_1 - \beta_2)$$

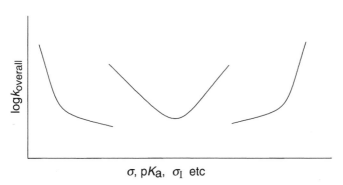

Figure 23   *Three non-linear free energy correlations indicating changes in mechanism*

7.1(b)   Equation (3) can be rearranged to give:

$$k_{overall} = k_1/(k_{-1}/k_2 + 1)$$

Hammett equations for each of the fundamental rate constants can be inserted:

$$\log k_1 = \rho_1\sigma + C_1$$

$$\log k_{-1} = \rho_{-1}\sigma + C_{-1}$$

$$\log k_2 = \rho_2\sigma + C_2$$

Hence

$$k_{overall} = (10^{\rho_1\sigma + C_1})/[(10^{\rho_{-1}\sigma + C_{-1}})/(10^{\rho_2\sigma + C_2}) + 1]$$

$$= (10^{\rho_1\sigma + C_1})/[10^{\{(\rho_{-1}-\rho_2)\sigma + C_{-1}-C_2\}} + 1]$$

Figure 24 illustrates the three cases of non-linear free energy correlations for a change in rate-limiting step. The break-points occur at $k_{-1} = k_2$ when

$$\log k_{\text{overall}} = \log k_1 - 0.303$$

and when

$$\sigma = (C_2 - C_{-1})/(\rho_{-1} - \rho_2)$$

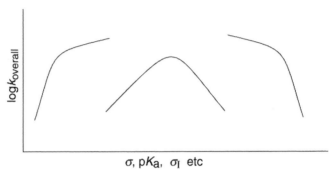

Figure 24 *Three non-linear free energy correlations indicating changes in rate-limiting step*

7.2 The Brønsted dependence of the hydrolysis (Figure 25) exhibits a shape typical of parallel mechanisms (see Problem 1a) with a break near $pK_a = 11$ due to a change in mechanism. The slope of the line at low pK_a values (~ -1.3) is significantly more negative than that for simple hydroxide ion attack at an ester function and is consistent with an E1cb mechanism (Scheme 23). The slope of the line for the high pK_a leaving groups is consistent with a mechanism involving attack of hydroxide at the ester function (Scheme 24 for example).

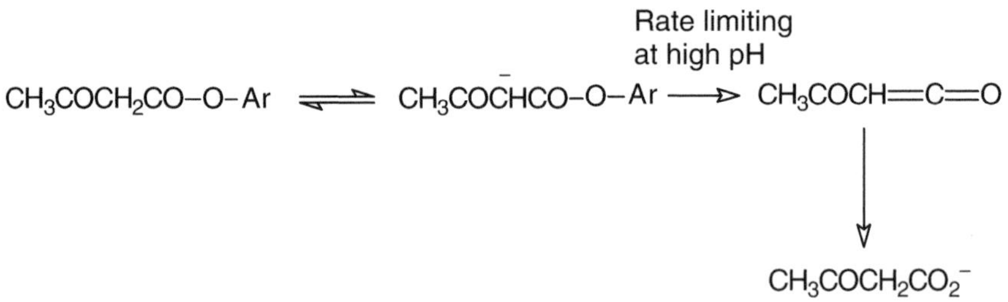

$$\text{CH}_3\text{COCH}_2\text{CO-O-Ar} \; \rightleftharpoons \; \text{CH}_3\text{COCHCO-O-Ar} \; \xrightarrow[\text{at high pH}]{\text{Rate limiting}} \; \text{CH}_3\text{COCH}{=}\text{C}{=}\text{O}$$

$$\downarrow$$

$$\text{CH}_3\text{COCH}_2\text{CO}_2^-$$

Scheme 23 *E1cb mechanism of ester hydrolysis*

The calculated alkaline hydrolysis rate constant for 4-nitrophenyl acetoacetate is obtained from ρ^*, the difference in σ^* between methyl and acetonyl ($\text{CH}_3\text{COCH}_2-$) and the rate constant for 4-nitrophenyl acetate: *antilog* $[2.5 \times (0.62 - 0) \times 0.978] = 33$ M^{-1} sec^{-1}); the large difference between observed (1.32×10^5 M^{-1} sec^{-1}) and calculated rate

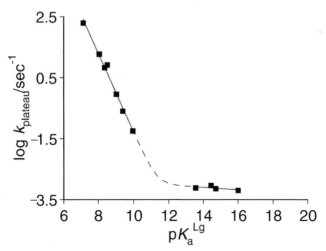

Figure 25 *Brønsted plot for the hydrolysis of acetoacetate esters as a function of the pK of the conjugate acid of the leaving group. The dashed line is the global fit to parallel mechanisms (see equation in Answer to Problem 1a); the full lines are linear Brønsted correlations for the individual segments*

constants is additional evidence that the mechanism of the acetoacetate hydrolysis is not the same as that for hydroxide ion attack on 4-nitro-phenyl acetate.

7.3 The alkaline hydrolysis of aryl acetates has an effective charge map (Scheme 24) derived from the given data:

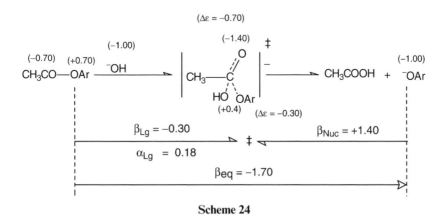

Scheme 24

7.4 The attack of phenolate ions on phenyl esters can be computed from the data to have the following reaction map (Scheme 25). Only two β values are required to construct this map if the reaction under investigation is a quasi-symmetrical reaction.

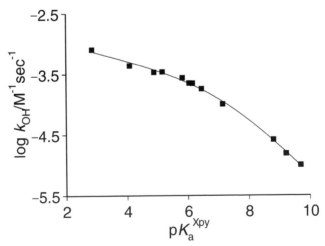

The change in charge on attacking phenolate ion to the transition state (+0.75) is not balanced by the change in charge on the leaving phenyl oxygen (−0.95); a change in charge on the components of the CH_3CO- group (+0.2) is needed to conserve the overall effective charge. Alternatively, α_{Nuc} (corresponding to bond formation) is smaller than α_{Lg} (corresponding to bond fission) consistent with more advanced bond fission than formation and a build-up of positive charge on the CH_3CO group in the transition structure.

7.5 The Brønsted-type graph is non-linear as shown in Figure 26:

Figure 26 *Brønsted plot for the hydroxide ion catalysed elimination reaction of N(2-cyanoethyl)pyridinium ions (Scheme 26). Line is theoretical for a change in rate limiting step*

The data fit Equation (4) (Chapter 7) modified for the Brønsted equation:

$$\log k = -0.531 \text{p}K_a + 0.18 - \log(1+10^{-0.402\text{p}K_a + 2.90})$$

The free energy relationship is diagnostic of a change in rate-limiting step (near $\text{p}K_a = 7$) and the simplest mechanism fitting these results is the E1cb mechanism of Scheme 26.

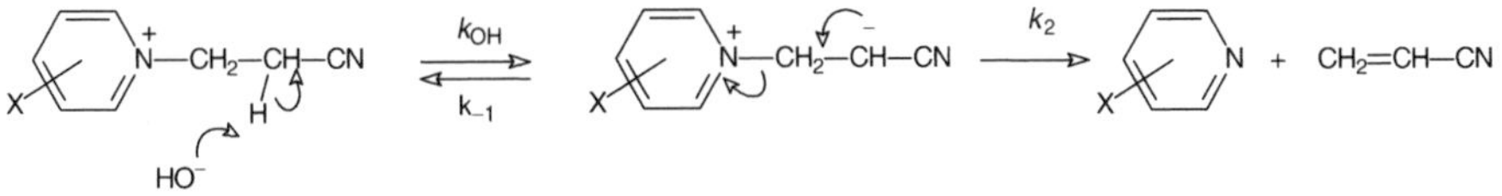

Scheme 26

7.6 The general acid-catalysed halogenation of acetone fits the mechanism of Scheme 27.

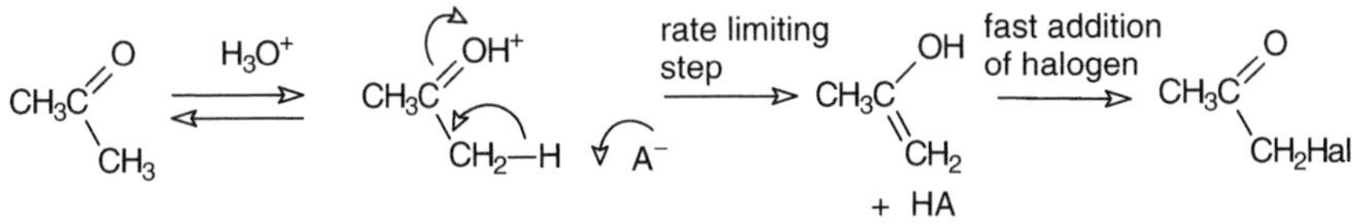

Scheme 27

The reactivities of ammonium ion-catalysed reactions are several orders of magnitude larger than those predicted from the Brønsted Equation for the general acid-catalysed process. This result indicates a different mechanism and the favoured one involves the formation of an iminium ion (Scheme 28) intermediate which is present in substantial proportions at neutral pH.

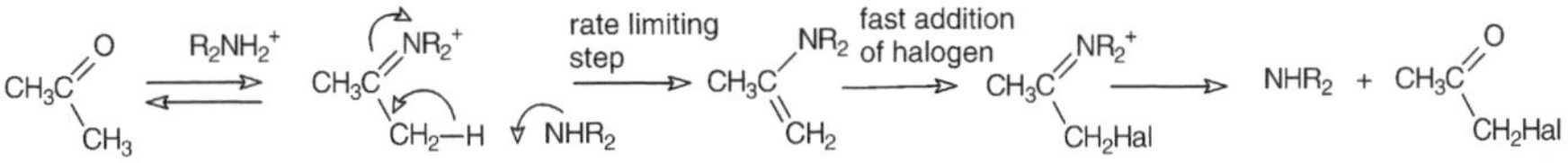

Scheme 28

7.7 The Brønsted graph is linear from $\text{p}K_a$ 5 to 10 straddling the $\text{p}K_a$ of 7.14 for 4-nitrophenol. The reaction is a quasi-symmetrical displacement with an identity reaction when the nucleophile is 4-nitrophenolate ion. A concerted displacement mechanism is consistent with the absence of a break at $\text{p}K_a = 7.14$.

7.8 The map of Brønsted β values is shown in Scheme 29. When the pK_a of the conjugate acid of the attacking phenolate ion is smaller than that of 4-nitrophenol, the k_2 step would be rate limiting (in a stepwise mechanism) so that the value of $\beta_{eq(1)} < 0.95$.

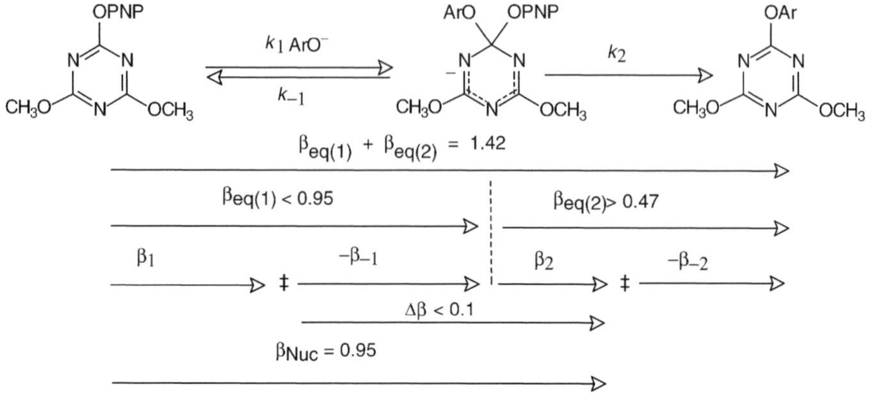

Scheme 29

Since the overall $\beta_{eq} = 1.42$ it follows that the value of $\beta_{eq(2)} > 0.47$ in the putative stepwise scheme. This value is relatively large and would not be expected for a k_2 step where *no substantial* bonding change is occurring in the Ar–O bond.

7.9 The value $\Delta\beta$ corresponds to the difference in effective charge between the two transition structures of the putative stepwise process. A concerted process will thus possess a $\Delta\beta$ of zero but owing to the incursion of microscopic medium effects the Brønsted relationship, although linear, may appear to be slightly curved; the curvature, measured by $\Delta\beta$, will essentially be dependent on the number of available data points. The criterion of significance is the comparison of the $\Delta\beta$ value with the overall β_{eq} for the displacement. In this case the β_{eq} is 1.42 so that the difference in charge on the two transition states for the putative stepwise process (0.1/1.42) is about 7% of the overall difference in charge between reactants and products. The charge difference between one of the transition states and the intermediate in the putative stepwise mechanism must be 3.5% of the total change. This difference is con-sidered too small to accommodate a significant barrier for a stable intermediate.

In the pyridinolysis of *N*-triazinyl-pyridinium ions, the $\Delta\beta$ is 0.82 in a total change in charge of 1.25; the charge difference between the putative intermediate and transition states is 0.65 in an overall change of 1.21, a substantial value which requires the existence of an intermediate.

7.10

	pKa(exp.)	pKa(calc.)
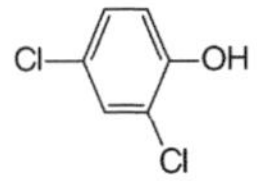 Cl—⬡—OH (with Cl)	7.84	$9.38 - (9.95 - 8.48) = 7.91$
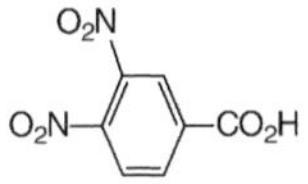 O_2N, O_2N—⬡—CO_2H	2.82	$4.2 - 0.71 - 0.78 = 2.71$
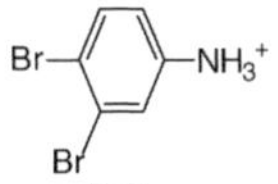 Br—⬡—NH_3^+ (with Br)	2.34	$3.91 - (4.62 - 3.51) = 2.80$
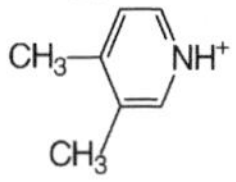 CH_3—⬡—NH^+ (with CH_3)	6.52	$6.02 - (5.17 - 5.68) = 6.53$

Scheme 30 *Values of pK$_a$ calculated by additivity*

7.11 The β_{eq} values are recorded in Scheme 31.

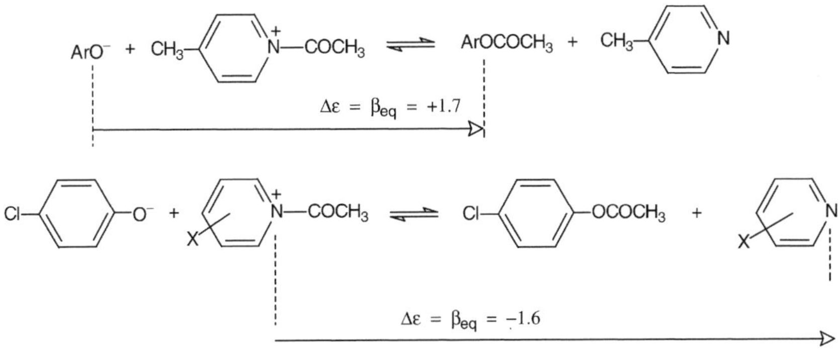

$$\Delta\varepsilon = \beta_{eq} = +1.7$$

$$\Delta\varepsilon = \beta_{eq} = -1.6$$

Scheme 31

The equilibrium constant for formation of 4-chlorophenyl acetate from 4-chlorophenolate ion and acetyl 4-methylpyridinium ion is given by:

$$\log K_{eq} = 1.7 \times (9.38 - 7.14) + \log 3.37 \times 10^6 = 10.34.$$

The equilibrium constant for reaction of 4-chlorophenolate ion with acetyl 4-dimethylaminopyridinium ion is given by:

$$\log K_{eq} = -1.6 \times (9.68 - 6.14) + 10.34 = 4.68$$

7.12 The Brønsted relationship is linear over the range of the measurements (Figure 27). The equation is $\log k_2 = -1.98 \, pK_a^{Lg} + 24.07$. The value of k_2 becomes $10^{13} \, sec^{-1}$ when the $pK_a^{Lg} = 5.6$ and is consistent with a change from an E1cb mechanism over the range of the measurements to a concerted E2 mechanism when the pK_a^{Lg} is less than 5.6.

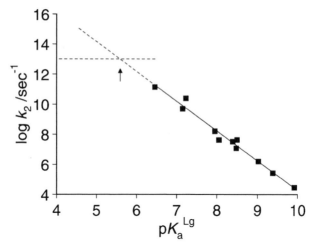

Figure 27 *Decomposition of the anion of substituted phenylmethane sulfonate esters. Arrow indicates the pK$_a$ of the leaving phenol where the lifetime of the intermediate becomes too small to support its existence*

7.13 The overall equation for the reaction catalysed by HA, H_3O^+ and H_2O is:

$$k_{overall} = k_{HA} \times 0.1 + k_{H_2O} \times 55.5 + k_{H_2O} \times 10^{-5}$$

For the case where $\alpha = -1.0$

$$\log k_{HA} = -1 \times 5 + C$$

$$\log k_{H_2O} = -1 \times 15.7 + C$$

$$\log k_{H_3O} = -1 \times (-1.7) + C$$

Substituting into the overall equation gives:

$$k_{overall} = 10^C (10^{-5} \times 0.1 + 10^{-15.7} \times 55.5 + 10^{1.7} \times 10^{-5})$$

$$= 10^C [10^{-6} \{HA\} + 1.11 \times 10^{-14} \{H_2O\} + 5.01 \times 10^{-4} \{H_3O^+\}]$$

Therefore reaction flux has:[c]

$$0.2\% \{HA\} + 2.2 \times 10^{-9}\% \{H_2O\} + 99.8\% \{H_3O^+\}$$

For case where $\alpha = -0.1$

$$\log k_{HA} = -0.1 \times 5 + C$$

$$\log k_{H_2O} = -0.1 \times 15.7 + C$$

$$\log k_{H_2O} = -0.1 \times (-1.7) + C$$

Substituting into the overall equation gives:

$$k_{overall} = 10^C(10^{-0.5} \times 0.1 + 10^{-1.57} \times 55.5 + 10^{0.17} \times 10^{-5})$$

$$= 10^C(0.0316 + 1.494 + 1.48 \times 10^{-5})$$

Therefore reaction flux has:

$$2.1\% \{HA\} + 97.9\%\{H_2O\} + 0.001\%\{H_3O^+\}$$

At $\alpha = 0.1$ some 98% of the total reaction flux is taken by the solvent water acting as the acid; at $\alpha = 1.0$ the oxonium ion takes 99.8% of the total reaction flux. At intermediate α values the general acid takes a substantial proportion of the reaction flux. The consequences of these results are that extreme values of α are more difficult to measure because the solvent species take most of the reaction, resulting in only small changes in rate constant for variation in general acid concentration. Similar arguments can be advanced for general base catalysis in competition with water or hydroxide ion.

7.14 The reverse of the reaction of aryloxide ions with acetylketene is the alkaline hydrolysis of aryl acetoacetates which has the mechanism given in Scheme 32.

The general approach is similar to that for the reaction of phenolate ions with fluorenylketene (Scheme 3, Chapter 7). The β_{Lg} of -1.29 refers to the change in charge from conjugate base to transition structure of

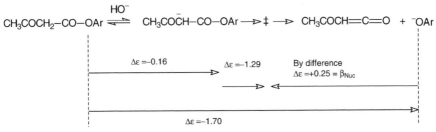

Scheme 32 *ElcB mechanism for the alkaline hydrolysis of substituted phenyl acetoacetate esters*

the rate-limiting step. The value of β for the ionisation of the α-proton (-0.16) is estimated from the value of -1 for substituted phenols allowing for two-atom attenuation of 0.4. The value of β_{Nuc} (0.25) for attack of phenolate ions on the ketene is deduced by subtraction $(+1.70 - 0.16 - 1.29)$.

The value of β_{Nuc} is less than that found for the reaction of phenolate ions with the fluorenyl ketene (Scheme 3, Chapter 7) and for attack on isocyanic acid (Scheme 3, Chapter 3). The observed values of β for the ionisation of the carbamate (Chapter 3) and fluorenyl esters are more negative than that calculated for the acetoacetate case.

7.15

(a) $\log P_{obs} = 0.26$

   $\log P_{calc} = 2f(CH_3) + f(CH) + f(NH_2) + 2f_b + f_{gbr}$

   $= 2(0.89) + 0.43 + (-1.54) + 2(-0.12) + (-0.22)$

   $= 0.21$

(b) $\log P_{obs} = 2.92$

   $\log P_{calc} = f(Ph) + f(CH_2) + f(Br) + f_b$

   $= 1.90 + 0.66 + 0.20 + (-0.12)$

   $= 2.64$

(c) $\log P_{obs} = 0.37$

   $\log P_{calc} = 3f(CH_3) + f(C) + f(OH) + 3f_b + 2f_{gbr}$

   $= 3(0.89) + 0.20 + (-1.64) + 3.(-0.12) + 2(-0.22)$

   $= 0.43$

(d) $\log P_{obs} = 3.44$

   $\log P_{calc} = 6f CH_2 + 5f_b(\text{single bond in ring})$

   $= 6 \times 0 \times 66 + 5(-0.09)$

   $= 3.96 - 0.45 = 3.51$

The f_b component allows for the contribution due to the extra bonds after the first and f_{gbr} allows for the contribution due to a group branching fragment. The examples illustrate very simple systems where fragments such as phenyl or methylene possess "global" fragmentation factors. Much more sophisticated protocols break the molecule into smaller fragments (see Further Reading in Chapter 7) and these form the basis of software programs.

7.16

(a) acetoacetic acid $CH_3COCH_2CO_2H^*$

pK_a of aliphatic carboxylic acid	4.80
ΔpK_a for a CH_3CO-	−1.10
pK_a^{calc}	3.70
pK_a^{exp}	3.58

(b) citric acid $*HO_2CC(OH)(CH_2CO_2H)_2$

pK_a of aliphatic carboxylic acid	4.80
ΔpK_a for $\beta -CO_2H$	$(-1.37) \times 0.4 \times 2$
ΔpK_a for $a -OH$	−0.90
pK_a^{calc}	3.70
pK_a^{exp}	3.58

(c) 2-ammoniocycloheptanol (RNH_3^+ acid)

pK_a of aliphatic primary ammonium ion	10.77
ΔpK_a for $\beta -OH$	1.10
pK_a^{calc}	9.67
pK_a^{exp}	9.25

(d) 2-aminoethylammonium ion
$H_2NCH_2CH_2NH_3^+$

pK_a of aliphatic primary ammonium ion	10.77
ΔpK_a for $\beta -NH_2$	−0.80
pK_a^{calc}	9.97
pK_a^{exp}	9.98

7.17 The graph fits the equation $\log k_{ij} = 0.414 \log K_{ij} - 1.89$ and is illustrated in Figure 28.

The relationship is linear over a good range of $\log K_{ij}$ on both sides of the predicted breakpoint where $\log K_{ij} = 0$. This is evidence for the concerted transfer of the hydride between the donor electrophiles (Scheme 33) and that the hydride ion is not an intermediate even in a pre-association mechanism.

A stepwise mechanism where an addition intermediate is formed (Scheme 34) can be excluded because it would also predict a break in the free energy plot (Figure 28).

7.18 The reaction of Equation (5) has the rate law

$$k_{overall} = k_1/(k_H[H_3O^+]/k_2 + 1)$$

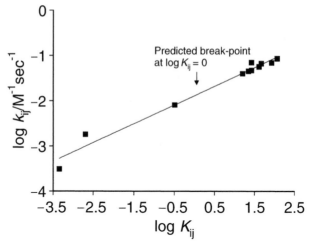

Figure 28 *Plot of logk_{ij} versus logK_{ij} for the transfer of hydride from the dihydro derivatives of 1-alkyl-3-cyano and 3-aminocarbonylquinolinium ions to 1-methyl acridinium ion*

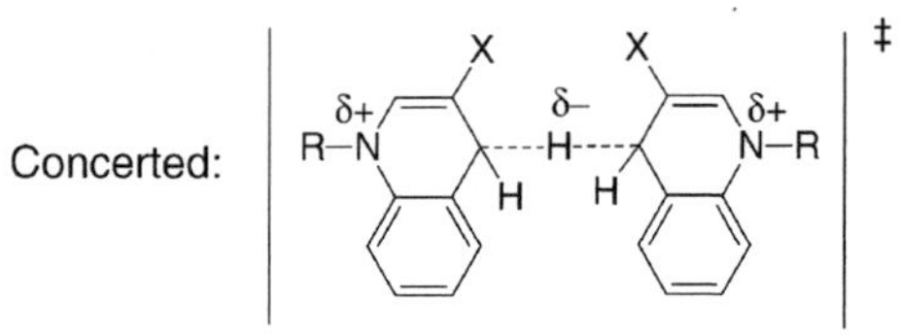

Concerted:

$$E_1\text{—H} \underset{}{\overset{[E_2^+].k_a}{\rightleftarrows}} [E_1\text{—H . } E_2^+] \underset{}{\overset{k_1}{\rightleftarrows}} [E_1^+.H^-.E_2^+] \underset{}{\overset{k_2}{\rightleftarrows}} [E_2\text{—H . }E_1^+] \overset{k_d}{\rightleftarrows} E_2\text{—H + } E_1^+$$

Stepwise:

Scheme 33 *Putative mechanisms involving a concerted pathway and a hydride ion intermediate in an encounter complex; k_a and k_d are association and dissociation diffusion processes respectively*

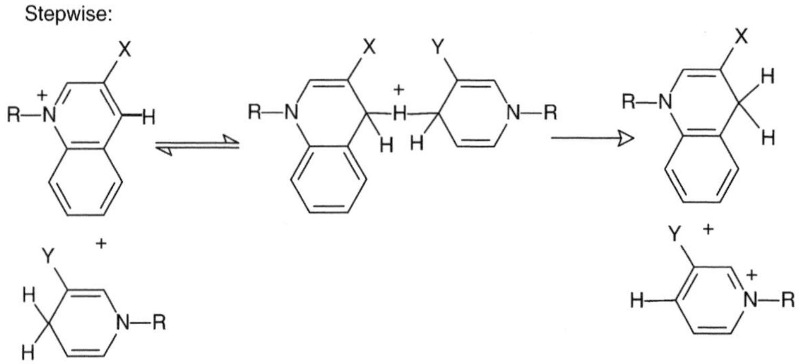

Stepwise:

Scheme 34 *Addition intermediate in putative stepwise hydride transfer*

Substituting the parameters for the pH-range 1 to 10 gives a non-linear pH-dependence following that of the titration of an acid with $pK_a = 6$. The pH-dependence is essentially a free energy relationship and the direction of the break indicates a change in rate limiting step.

7.19 The data for the acid-catalysed formation of carbinolamine from methoxyamine and 4-methoxybenzaldehyde fit the Eigen equation and follow a plot similar to that shown in Figure 20. The parameters are:

$$k_{max} = 251 \text{ M}^{-2} \text{ sec}^{-1} \quad \text{and} \quad pK_o = 8.68.$$

The fit to the Eigen equation is consistent with a mechanism involving a stepwise proton transfer to the oxyanion of the zwitterion (Scheme 35). The parameter pK_o is close to pK_a calculated for the protonation of the zwitterion (8.97).

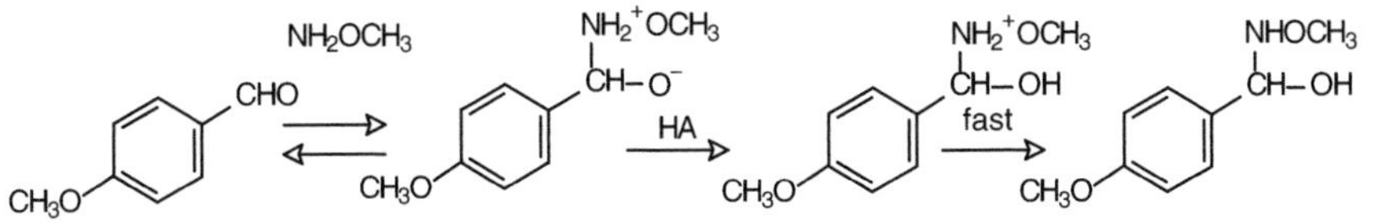

Scheme 35

The calculation of the pK_a of the conjugate acid of the zwitterion is carried out as follows: starting with an estimated pK_a of 9.98 for $CH_3N^+H_2CH_2OH$, H– is replaced by Ph– to decrease the pK_a by -0.20. Substitution of CH_3O- ($\sigma_1 = 0.25$) for CH_3- on nitrogen gives a ΔpK_a of $-8.4 \times 0.25 \times 0.5 = -1.05$ and correcting for the 4-CH_3O substituent on the phenyl ring increases the pK_a by $+0.24$.

$CH_3N^+H_2CH_2OH$	9.98
H– by Ph–	-0.20
CH_3- on nitrogen by CH_3O-	-1.05
CH_3O- for H– in Ph–	$+0.24$
Calculated pK_a	8.97

7.20 The equilibrium constant for the transfer process can be derived from the sum of the two equations and is $K_{eq} = K_1/K_2$

thus

$$\log K_{eq} = \log K_1 - \log K_2$$

Substituting the Brønsted equations

$$\log K_1 = 1.36 p K_a^{ArOH} + C \quad \text{and} \quad \log K_2 = 1.36 p K_a^{Ar'OH} + C$$

gives

$$\log K_{eq} = 1.36(pK_a^{ArOH} - pK_a^{Ar'OH})$$

$$= 1.36(pK_a^{acceptor-H} - pK_a^{donor-H})$$

7.21 Substituted phenolate anions react with 4-nitrophenyl diphenyl-phosphinate in a *quasi-symmetrical* reaction so that the effective charges can be determined for both forming and breaking bonds from the two quantities β_{Nuc} and β_{eq} without any special experiments made on the leaving group substituents. The transition structure for the reaction, which has a concerted mechanism, has:

$$\alpha_{Nuc} = 0.46/1.25 = 0.37 \qquad \text{and} \qquad \alpha_{Lg} = -0.79/(-1.25) = 0.63$$

Thus bond formation is 37% of its full potential and bond fission has proceeded to 63%. There is thus an imbalance in bond formation which is reflected by the phosphinoyl group's becoming more positive than in its reactant or product state (Scheme 36).

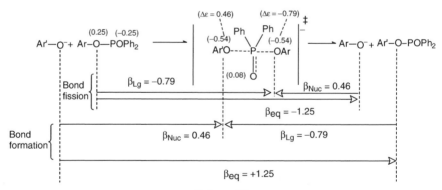

Scheme 36

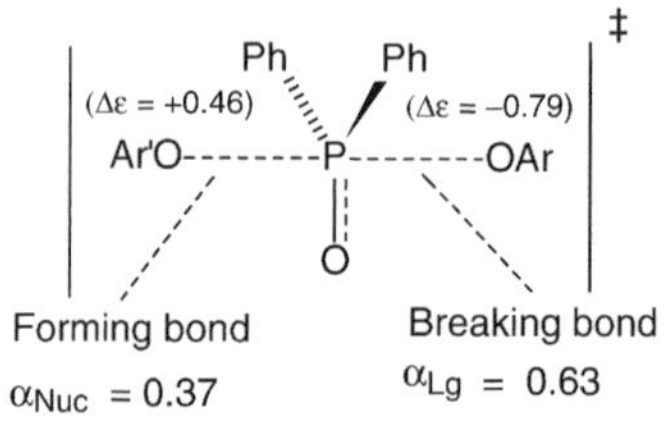

Scheme 37

Since bond fission and formation both involve an ArO–P bond change, the same reference dissociation equilibrium is employed (the dissociation of phenols) and hence the effective charges in the reaction can be directly compared. The transition state structure results from an increase of +0.33 units on the phosphinyl group, a gain of +0.46 units in the attacking oxygen and a gain of −0.79 units in the leaving oxygen. This is consistent with imbalance and a reaction coordinate in the 1,0 (SE) quadrant of the More O'Ferrall–Jencks diagram (analogous to Figure 9 of Chapter 5).

7.22 The Leffler index α_{Nuc}, 0.21 (Scheme 38), indicates that bond formation is only 21% complete in the transition structure (Scheme 39). It is not possible to determine the percentage of bond fission from this data. If, however, the reaction were assumed to be *quasi-symmetrical*, then the results would indicate that bond fission is well advanced in the transition state.

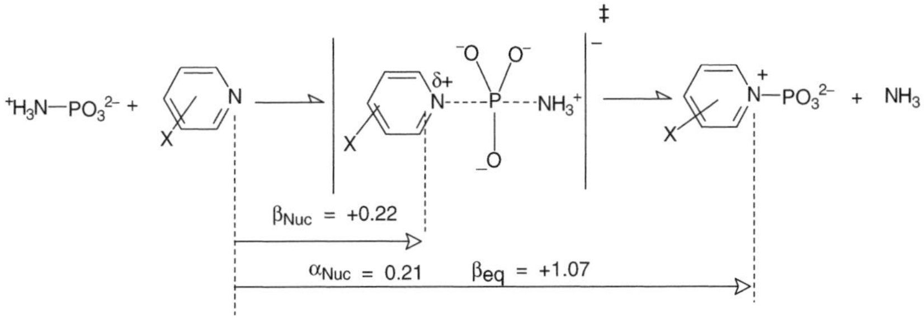

Scheme 38

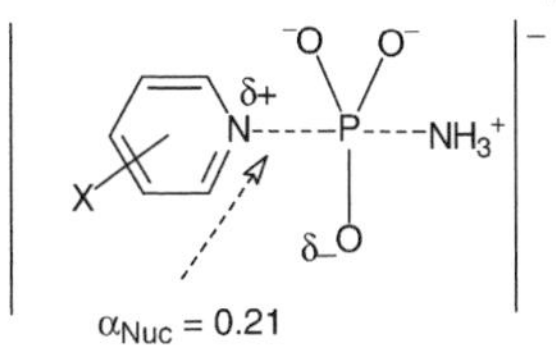

Scheme 39

APPENDIX 3

Tables of Structure Parameters

The structure parameters are collected from various sources as indicated. The present tables are meant to provide a convenient source for the reader rather than a critical assessment. Each value can vary depending on its source and if the constants are to be used for critical work they should be checked against the literature which is quoted in the foot-notes. Exner[a] provides an excellent critical and extensive compilation of structure parameters.

[a] From O. Exner, *Correlation Analysis in Chemistry: Recent Advances*, N.B. Chapman and J. Shorter (eds.), Plenum Press, New York, 1978, p 439.

Table 1 *Hammett, Taft, Charton, σ_p^+, σ_p^-, F, R and MR constants for some substituents[b]*

Formula[c]	Structure	MR[d]	$\sigma_p^{+\,e}$	$\sigma_p^{-\,e}$	F[f]	R[f]	$\sigma_I{}^a$	$\sigma_{meta}{}^g$	$\sigma_{para}{}^g$	$\sigma^{*\,g}$
Br	–Br	8.88	0.15	0.25	0.45	–0.22	0.45	0.39	0.23	2.84
					0.72	*–0.18*				
Cl	–Cl	6.03	0.11	0.19	0.42	–0.19	0.47	0.37	0.23	2.96
						–0.24				
F	–F	0.92	–0.07	–0.03	0.45	–0.39	0.52	0.34	0.06	3.21
H	–H	1.03	0.00	0.00	0.00	0.00	0.00	0.00	0.00	0.49
					0.00	*0.00*				
I	–I	13.94	0.14	0.27	0.42	–0.24	0.39	0.35	0.18	2.46
					0.65	*–0.12*				
N3	–N3	10.20		0.11	0.48	–0.40	0.42	0.27	0.15	2.62
O	–O⁻		–2.30	–0.82	–0.26	–0.55	–0.16	–0.47	–0.81	–2.78
S	–S⁻		–2.62		0.03	–1.24		–0.36		
BF2	–BF2				0.26	0.22	0.16	0.32	0.48	
Br3Ge	–GeBr3				0.61	0.12	0.59	0.66	0.73	3.7
Br3Si	–SiBr3				0.44	0.13	0.39	0.48	0.57	2.4
Cl3Ge	–GeCl3		0.57		0.65	0.14	0.63	0.71	0.79	3.9
Cl3Si	–SiCl3		0.57		0.44	0.12	0.39	0.48	0.56	2.4
F3Ge	–GeF3				0.76	0.21	0.74	0.85	0.97	4.6
F3S	–SF3				0.63	0.17	0.60	0.70	0.80	3.75
F3Si	–SiF3				0.47	0.22	0.42	0.54	0.66	2.62
GeH3	–GeH3				0.03	–0.02		0.00	0.01	–0.7
HO	–OH	2.85	–0.92	–0.37	0.33	–0.70	0.24	0.12	–0.37	1.34
					0.46	*–1.89*				
HS	–SH	9.22	–0.03		0.30	–0.15	0.26	0.25	0.15	1.68
					0.52	*–0.26*				
H2N	–NH2	5.42	–1.30	–0.15	0.08	–0.74	0.19	–0.16	–0.66	0.62
					0.38	*–2.52*				

(continued overleaf)

Table 1 (*continued*)

Formula[c]	Structure	MR[d]	σ_p^{+e}	σ_p^{-e}	F[f]	R[f]	σ_I[a]	σ_{meta}[g]	σ_{para}[g]	σ^{*}[g]
H$_3$N	–NH$_3^+$			–0.56	0.92	–0.32	0.60	0.67	0.53	3.76
H$_3$N$_2$	–NHNH$_2$	8.44			0.22	–0.77	0.14	–0.02	–0.55	0.40
H$_3$Si	–SiH$_3$		0.14		0.06	0.04	0.09	0.05	0.10	3.71
IO$_2$	–IO$_2$	63.51			0.61	0.17		0.70	0.76	
					0.99	*0.99*				
NO	–NO	5.20		1.63	0.49	0.42	0.33			2.08
NO$_2$	–NO$_2$	7.36		1.24	0.65	0.13	0.67	0.72	0.78	4.25
					1.00	*1.00*				
NO$_3$	–ONO$_2$			0.08	0.48	0.22	0.62	0.55	0.70	3.86
O$_2$S	–SO$_2^-$				0.03	–0.08		–0.02	–0.05	
O$_3$P	–PO$_3^{2-}$			–0.16				–0.02		
O$_3$S	–SO$_3^-$			0.58	0.29	0.06	0.13	0.05	0.09	0.81
					–0.05	*0.53*				
AsHO$_3$	–AsO$_3$H			0.46	0.04	–0.06	0.01	0.00	–0.02	0.14
BH$_2$O$_2$	–B(OH)$_2$	11.04		0.45	–0.03	0.15	–0.08	–0.01	0.12	0.95
ClOS	–SOCl				1.16	–0.05	0.68	0.75	0.82	4.25
ClO$_2$S	–SO$_2$Cl		0.38		0.70	0.20	0.80	0.92	1.04	5.0
Cl$_2$OP	–POCl$_2$		0.33		0.63	0.17	0.65	0.78	0.90	4.1
Cl$_3$PS	–PSCl$_2$				0.67	0.16	0.59	0.70	0.80	3.7
FOS	–SOF			1.54	0.72	0.19	0.66	0.74	0.83	4.12
FO$_2$S	–SO$_2$F	8.65			0.74	0.15	0.75	–0.89	–0.99	–4.7
F$_2$OP	–POF$_2$							0.81	0.89	
HO$_2$P	–PO(OH)								0.14	
HO$_2$S	–SO(OH)				0.01	–0.08		–0.04	–0.07	
HO$_3$P	–PO$_3$H				0.19	0.07		0.25	0.17	1.41
					–0.19	*–0.32*				
HO$_3$S	–SO$_2$OH							0.55		
HO$_4$P	–OPO$_3$H				0.41	–0.41		0.29	0.00	

Formula	Substituent									
H₂NO	-NHOH	7.22			0.11	-0.45		-0.04	-0.34	0.30
H₂O₃P	-PO(OH)₂							0.36	0.42	
H₃NO	-ONH₃⁺			0.94			0.47	0.53		2.92
H₂NO₂S	-SO₂NH₂	12.28			0.49 *(0.55)*	0.21 *(1.07)*	0.44	0.46	0.57	2.61
N₂	-N₂⁺				1.58 *(2.36)*	0.33 *(2.81)*		1.76	1.91	
CBr₃	-CBr₃				0.28	0.01	0.26	0.28	0.29	2.43
CCl₃	-CCl₃				0.38	0.09	0.41	0.40	0.46	2.65
CF₃	-CF₃	5.02		0.65	0.38 *(0.64)*	0.16 *(0.76)*	0.42	0.46	0.53	2.61
CH₃	-CH₃	5.65	-0.31	-0.17	0.01 *(-0.01)*	-0.18 *(-0.41)*	-0.10	-0.06	-0.14	0.00
CN	-CN	6.33		0.88	0.51 *(0.90)*	0.15 *(0.71)*	0.58	0.56	0.66	3.30
CN₇	-NC				0.47	0.02	0.67	0.48	0.49	
	5-azido-1-tetrazolyl[h]				0.53	0.01		0.54	0.54	
CO₂	-COO⁻	6.05	-0.02	0.31	-0.10 *(-0.27)*	0.10 *(0.40)*	-0.17	-0.10	0.00	-1.06
CClN₄	5-Cl-1-tetrazolyl[h]	23.60		0.70	0.58	0.03	0.38	0.80	0.61	
CClO	-COCl			1.24	0.46	0.15	0.51	0.53	0.69	2.37
CCl₃O	-OCCl₃				0.46	-0.11	0.39	0.43	0.35	3.19
CFO	-COF			0.27	0.48	0.22	0.55	0.55	0.70	2.44
CF₃O	-OCF₃	7.86			0.39	-0.04	0.44	0.36	0.33	
CF₃S	-SCF₃	13.81		0.64	0.36	0.14	0.30	0.38	0.50	2.75
CHBr₂	-CHBr₂				0.31	0.01	0.31	0.31	0.32	1.96
CHCl₂	-CHCl₂				0.312	0.01	0.32	0.31	0.32	1.94
CHF₂	-CHF₂				0.29	0.03	0.26	0.32	0.35	2.05
CHI₂	-CHI₂				0.27	-0.01	0.21	0.26	0.26	1.62
CHN₂	-NHCN	10.14		0.28	-0.22	0.37		0.60		
CHN₄	1-tetrazolyl[h]	18.33		0.57	0.52	-0.02			0.57	3.4

(continued overleaf)

Table 1 (*continued*)

$Formula^c$	$Structure$	MR^d	$\sigma_p^{+\,e}$	$\sigma_p^{-\,e}$	F^f	R^f	σ_I^a	σ_{meta}^g	σ_{para}^g	$\sigma^{*\,g}$
CHO	–CHO	6.88		1.03	0.33	0.09	0.25	0.36	0.44	2.15
CHO₂	–COOH	6.93		0.77	0.34	0.11	0.39	0.35	0.44	2.08
					0.44	*0.66*				
CH₂Br	–CH₂Br	13.39	0.02		0.14	0.00	0.18	0.12	0.14	1.00
CH₂Cl	–CH₂Cl	10.49	–0.01		0.13	–0.01	0.16	0.11	0.12	1.05
CH₂F	–CH₂F				0.15	–0.04	0.18	0.11	0.10	1.10
CH₂I	–CH₂I	18.60		0.08	0.12	–0.01	0.16	0.10	0.11	0.85
CH₃O	–CH₂OH	7.19	–0.04		0.03	–0.03	0.11	0.01	0.01	0.54
					0.54	*–1.68*				
CH₃O₂	–OCH₃	7.87	–0.78	–0.26	0.29	–0.56	0.30	0.11	–0.27	1.81
	–CH(OH)₂						0.22			1.37
CH₃S	–CH₂SH						0.10	0.08		0.62
	–SCH₃	13.82	–0.60	0.06	0.23	–0.23	0.25	0.15	0.00	1.56
					0.68	*–1.30*				
CH₃S₂	–SSCH₃	17.03			0.27	–0.14		0.22	0.13	0.95
CH₃Se	–SeCH₃				0.16	–0.16		0.1	0.00	
					0.28	*–0.39*				
CH₄N	–CH₂NH₂	10.33			0.04	–0.15	0.08	–0.03	–0.11	0.50
	–NHCH₃		–1.81		0.03	–0.73	0.18	–0.30	–0.84	–0.81
CH₅N	–CH₂NH₃⁺				0.59	–0.06	–0.36	0.32	0.29	2.24
	–NH₂CH₃⁺						0.60	0.96		3.74
CNO	–OCN				0.69	–0.15	0.80	0.67	0.54	5.0
	–NCO		–0.19		0.31	–0.12	0.36	0.27	0.19	2.25
CNS	–NCS	17.24		0.34	0.51	–0.13	0.42	0.48	0.38	2.62
	–SCN	13.40		0.59	0.49	0.03	0.58	0.41	0.52	3.43
CN₃O₆	–C(NO₂)₃				0.65	0.17	0.73	0.72	0.82	4.6
CHN₄O	5-OH-1-tetrazolyl[h]				0.41	–0.08		0.39	0.33	

Formula	Substituent									
CHN₄S	5-SH-1-tetrazolyl[h]				0.44	-0.01	0.41	0.45	0.45	2.62
	1,2,3,4-Thia-triazol-5-yl-amino[i]				0.35	-0.16	0.37	0.30	0.19	
CH₂ClO	-OCH₂Cl				0.33	-0.25	0.27	0.25	0.08	2.56
CH₂FO	-OCH₂F				0.29	-0.27	0.27	0.20	0.02	2.31
CH₂FS	-SCH₂F				0.25	-0.05	0.32	0.23	0.20	1.69
CH₂NO	-CONH₂	9.81		0.61	0.26	0.10	0.24	0.28	0.31	1.68
	-NHCHO	10.31			0.28	-0.28		0.19	0.00	1.62
	-CH=NOH	10.28			0.28	-0.18		0.22	0.10	
CH₃Br₂Si	-SiBr₂(CH₃)		0.08						0.29	1.5
CH₃Cl₂Si	-SiCl₂(CH₃)		0.23		0.29	0.10		0.31	0.39	
CH₃F₂Si	-SiF₂(CH₃)				0.32	-0.09		0.29	0.23	
CH₃N₂O	-NHCONH₂	13.72			0.09	-0.33	0.21	-0.03	-0.24	1.31
CH₃N₂S	-NHCSNH₂	22.19			0.26	-0.1	0.29	0.22	0.16	1.8
CH₃OS	-SOCH₃	13.70			0.24	-0.07	0.25	0.21	0.17	1.56
	-S(O)CH₃			0.73	0.52 *(0.80)*	-0.03 *(0.45)*	0.50	0.52	0.49	2.88
CH₃O₂S	-SO₂CH₃	13.49		1.13	0.53 *(0.80)*	0.19 *(0.85)*	0.59	0.64	0.73	3.68
CH₃O₃S	-S(O)OCH₃				0.47	0.07	0.45	0.50	0.54	2.84
	-OS(O)CH₃			0.89	0.43	0.02		0.44	0.45	3.62
	-OSO₂CH₃	16.99	0.16		0.40	-0.04		0.39	0.36	2.07
CH₂NOS	-SCONH₂	18.17			0.28	-0.25	0.33	0.34	0.03	
CH₄NOS	-NHSO₂CH₃						0.42	0.20	0.39	3.1
CHF₃NO₂S	-NHSO₂CF₃						0.49	0.44	0.52	2.56
C₂F₅	-CF₂CF₃	9.23	0.18	0.69			0.41	0.50	0.23	2.18
C₂H	-C≡CH	9.55		0.53	0.22	0.01	0.35	0.20	0.23	0.56
C₂H₃	-CH=CH₂	10.99	-0.16	0.69	0.13	-0.17	0.09	0.08	-0.08	-0.10
C₂H₅	-C₂H₅	10.30			0.00 *(-0.02)*	-0.15 *(-0.44)*	-0.05	-0.07	-0.15	
C₂F₃O	-COCF₃			1.09	0.63		0.47		0.80	3.7

(continued overleaf)

Table 1 (*continued*)

Formula[c]	Structure	MR[d]	σ_p^{+e}	σ_p^{-e}	F[f]	R[f]	σ_I^a	σ_{meta}^g	σ_{para}^g	σ^{*g}
C₂F₅O	-OCF₂CF₃						0.65	0.48	0.28	4.06
C₂F₃O₂	-OCOCF₃			0.61			0.16	0.56	0.46	1.00
C₂HCl₂	-CH=CCl₂						0.16	0.11		2.00
C₂HS	-SC≡CH						0.32	0.26	0.19	2.00
C₂H₂Cl₃	-CH₂CCl₃						0.12	0.06		0.75
C₂H₂F₃	-CH₂CF₃	10.11	0.16	0.14	0.15	-0.06	0.14	0.16	0.14	0.92
C₂H₂N	-CH₂CN		-0.53	0.11	0.17	0.01	0.18	0.16	0.18	1.30
C₂H₂O₂	-CH₂COO⁻			-0.16			0.01			-0.06
C₂H₃O	-COCH₃	11.18		0.84	0.33 / 0.50	0.17 / 0.90	0.29	0.38	0.50	1.65
C₂H₃O₂	-OCOCH₃	12.47	-0.19		0.42 / 0.70	-0.11 / -0.04	0.41	0.39	0.31	2.56
	-CH₂COOH		-0.01	0.05			0.17		-0.07	1.05
	-COOCH₃	12.87		0.75	0.34	0.11	0.34	0.32	0.39	2.00
C₂H₃S₂	-SC(S)CH₃						0.48			3.00
C₂H₄Cl	-CH₂CH₂Cl									0.385
C₂H₅O	-OC₂H₅	12.47	-0.81	-0.28	0.26 / 0.61	-0.50 / -1.72	0.27	0.10	-0.24	1.68
	-CH₂OCH₃	12.07	-0.05		0.13	-0.12	0.11	-0.10	0.03	0.52
	-CH₂CH₂OH						0.060			
C₂H₅S	-SC₂H₅	18.42			0.26	-0.23	0.25	0.23	0.03	1.56
C₂H₆N	-N(CH₃)₂	15.55	-1.70	-0.12	0.15 / 0.69	-0.98 / -3.81	0.10	-0.15	-0.83	0.32
C₂H₆S	-NHC₂H₅	14.98			-0.04	-0.57		-0.24	-0.61	-0.62
	-S(CH₃)₂⁺			0.83	0.98 / 1.62	-0.08 / 0.52	0.90	1.00	0.90	5.09
C₂H₇N	-CH₂CH₂NH₃⁺							0.23		4.36
	-NH(CH₃)₂⁺						0.70	0.84	0.17	3.74
	-NH₂C₂H₅⁺							0.96		

Formula	Group									
C₂F₃OS	–SCOCF₃									3.19
C₂H₄Cl	–CH₂CH₂Cl									0.385
C₂H₃OS	–SCOCH₃	18.42		0.46	0.37 *(0.53)*	0.07 *(0.68)*	0.51	0.48	0.46	2.29
C₂H₄NO	–NHCOCH₃	14.93		–0.46	0.31 *(0.77)*	–0.31 *(–1.43)*	0.41	0.39	0.44	1.40
	–CH₂CONH₂				0.08	–0.01	0.28	0.21	0.00	0.31
	–CONHCH₃				0.35	–0.01	0.05	0.06	0.07	
C₂H₄NO₂	–CH₂CH₂NO₂	14.57			0.29	0.05	0.29	0.35	0.36	
C₂H₄NS	–CSNHCH₃				0.30	–0.18		0.30	0.34	0.50
	–NHCSCH₃	23.40						0.24	0.12	
C₂H₅O₂S	–SO₂C₂H₅	18.14	0.72		0.59	0.18	0.60	0.66	0.77	3.74
	–CH₂SO₂CH₃							0.18	0.50	1.32
C₂H₆OP	–PO(CH₃)₂				0.40	0.10	0.35	0.43	0.55	2.81
C₂H₆O₃P	–PO(OCH₃)₂	21.87			0.37	0.16	0.24	0.42	0.04	
C₂H₆O₄P	–OPO(OCH₃)₂							0.04		
C₂H₃ClNO	–NHCOCH₂Cl	19.77			0.27	–0.30	0.36	0.17	–0.03	2.06
C₂H₆ClNP	–P(Cl)N(CH₃)₂			0.88	0.31	0.25	0.42	0.38	0.56	
C₂H₆NO₂S	–SO₂N(CH₃)₂			0.68	0.44	0.21	0.48	0.51	0.65	2.62
C₃F₇	–CF(CF₃)₂		0.86				0.21	0.37	0.53	3.00
C₃H₃	–C≡CCH₃				0.29	–0.26	0.13	0.10	0.12	1.20
	–CH₂C≡CH							0.07		0.81
C₃H₅	–cyclo-C₃H₅	13.53	–0.41	–0.09	0.02	–0.23		–0.07	–0.21	
	–CH=CHCH₃									0.36
C₃H₇	–CH(CH₃)CH₃	14.96	–0.28	–0.16	0.04	–0.19	–0.02	–0.07	–0.15	–0.19
	–CH₂CH₂CH₃	14.96	–0.29	–0.06	0.01	–0.14	–0.02	–0.05	–0.15	–0.115
C₃HF₆	–CH(CF₃)₂									1.32
C₃H₃O	–CH=CHCHO				0.29	–0.16	0.21	0.24	0.13	1.00
C₃H₃O₂	–CH=CHCOOH									1.00
C₃H₄F₃	–CH₂CH₂CF₃							0.14	0.90	0.32

(continued overleaf)

Table 1 (*continued*)

Formula[c]	Structure	MR^d	σ_p^{+e}	σ_p^{-e}	F^f	R^f	σ_I^a	σ_{meta}^g	σ_{para}^g	σ^{*g}
C₃H₄O	–CH₂CH₂COO⁻				0.07	–0.02	0.16	0.04	0.05	0.02
C₃H₅O₂	–CH₂OCOCH₃	16.52			0.02	–0.09	0.34	–0.03	–0.07	1.00
	–CH₂CH₂COOH				0.34	0.11		0.37	0.45	0.35
	–COOC₂H₅	17.47		0.64	*0.47*	*0.67*				2.26
C₃H₇O	–CH₂COOCH₃	16.72		0.07	–0.06	–0.11	0.17	0.13		1.06
	–CH₂OC₂H₅									0.58
	–CH₂CH(OH)-CH₃				0.34	–0.79	–0.01	–0.12		–0.06
	–OCH(CH₃)CH₃	17.06	–0.85		*0.90*	*–2.88*	0.26	0.05	–0.45	1.62
	–OCH₂CH₂CH₃	17.06	–0.83		0.26 / *0.63*	–0.51 / *–1.77*	0.27	0.10	–0.25	1.68
C₃H₇S	–SCH(CH₃)CH₃	23.07			0.30	–0.23	0.25	0.23	0.07	1.56
	–SCH₂CH₂CH₃						0.24	0.22		1.49
C₃H₉Ge	–Ge(CH₃)₃				0.03	–0.03	0.00	0.00	0.00	
C₃H₉N	–CH₂NH(CH₃)⁺		0.50		0.39	0.04		0.40	0.43	
	–NH₂CH₂CH₂-CH₃⁺						0.60	0.71		3.74
	–N(CH₃)₃⁺	21.20	0.41	0.77	0.86 / *1.54*	–0.04 / *(0.00)*	0.73	0.88	0.82	4.55
C₃H₉P	–P(CH₃)₃⁺				0.71	0.02	0.40	0.50	0.80	2.5
C₃H₉Si	–Si(CH₃)₃	24.96	0.02		0.01 / *–0.10*	–0.08 / *0.16*	–0.13	–0.04	–0.07	–0.81
C₃H₉Sn	–Sn(CH₃)₃		–0.12				0.00	0.0	0.0	
C₃H₆NO	–CH₂NHCOCH₃	21.18			0.03	–0.03	0.07	–0.04	–0.05	0.43
C₃H₆NO₂	–NHCOC₂H₅							0.23		1.56
	–OCON(CH₃)₂									2.87
	–NHCOOC₂H₅						0.26	0.07	–0.15	1.99

Formula	Substituent									
$C_3H_7N_2O$	$-NHCONHC_2H_5$							0.04	-0.26	
$C_3H_7N_2S$	$-NHCSNHC_2H_5$	31.66						0.30	0.07	
C_4F_9	$-C(CF_3)_3$			0.71			0.35	0.35	0.52	0.12
$C_4H_6F_3$	$-CH_2CH_2CH_2CF_3$						0.05			0.31
C_4H_7	$-CH{=}CHC_2H_5$						0.00	-0.04		0.13
	$-CH_2CH{=}CHCH_3$						0.03	-0.04		0.19
	$-CH{=}C(CH_3)_2$							-0.06		
C_4H_9	$-(CH_2)_3CH_3$	19.61	-0.29	-0.12	-0.01	-0.15	-0.04	-0.07	-0.16	-0.13
	$-CH(CH_3)C_2H_5$	19.62	-0.26	-0.13	-0.02	-0.10	-0.03	-0.08	-0.19	-0.125
	$-C(CH_3)_3$				-0.02	-0.18	-0.07	-0.10	-0.20	-0.30
	$-CH_2CH(CH_3)CH_3$				*-0.11*	*-0.29*	-0.03		-0.12	-0.21
C_4N_3	$-C(CN)_3$			0.01	0.92	0.04	0.98	-0.7	0.99	6.1
$C_4H_2F_7$	$-CH_2C_3F_7$						0.14	0.98	0.87	0.25
C_4H_3O	2-furyl		-0.39		0.1	-0.08	0.17	0.08	0.02	1.31
C_4H_3S	2-thienyl	24.04	-0.43	0.21	0.13	-0.08	0.21	0.06	0.05	0.62
	3-thienyl	24.04		0.13	0.08	-0.10	0.27	0.09	-0.02	
C_4H_4N	1-pyrrolyl	21.85			0.50	-0.13	0.26	0.03	0.37	1.68
C_4H_9O	$-O(CH_2)_3CH_3$	21.66			0.29	-0.61	-0.04	0.47	-0.32	1.62
	$-OCH(CH_3)C_2H_5$				*0.72*	*-2.16*	-0.03	-0.05		-0.25
	$-CH_2C(OH)(CH_3)_2$						0.23	0.25		
$C_4H_9O_3$	$-C(OCH_3)_3$							-0.16		
C_4H_9S	$-S(CH_2)_3CH_3$							-0.03	-0.04	1.44
	$-SCH(CH_3)C_2H_5$							0.21		1.32
$C_4H_{10}N$	$-NH(CH_2)_3CH_3$	24.26	-2.07	-0.43	-0.11	-0.06	0.16	-0.34	-0.51	-1.08
	$-N(C_2H_5)_2$	24.85			0.01	-0.05	0.60	-0.15	-0.53	3.74
$C_4H_{11}N$	$-NH_2(CH_2)_3CH_3^+$				-0.21	-0.30		0.71		
	$-NH_2C(CH_3)_3^+$				0.01	-0.73		0.71		

(continued overleaf)

Table 1 (*continued*)

Formula[c]	Structure	MR[d]	σ_p^+[e]	σ_p^-[e]	F[f]	R[f]	σ_I[a]	σ_{meta}[g]	σ_{para}[g]	σ^{*}[g]
	$-CH_2CH_2NH(CH_3)_2^+$				0.29	−0.15		0.24	0.14	3.60
	$-NH_2CH_2CH(CH_3)_2^+$				0.38	0.06	0.60	0.40	0.44	1.90
	$-CH_2N(CH_3)_3^+$		0.50				0.30			
$C_4H_{11}Si$	$-CH_2Si(CH_3)_3$	29.61	−0.66		−0.09 *−0.19*	−0.12 *−0.32*	−0.05	−0.16	−0.21	−0.26
C_4H_8NO	-morpholino[j]							−0.10	−0.68	0.69
$C_4H_8NO_2$	$-NHCH_2COOC_2H_5$									
$C_4H_{10}O_3P$	$-P(O)(OC_2H_5)_2$	31.16	0.54	0.84			0.35	0.49	0.57	3.02
$C_4H_{11}NO_2$	$-NH(CH_2CH_2OH)_2^+$									4.43
C_5H_{11}	$-(CH_2)_2CH(CH_3)_2$				0.03	−0.14	−0.02	−0.13	−0.23	−0.165
	$-CH_2C(CH_3)_3$	24.26	−0.31		−0.01	−0.14	−0.04	−0.08	−0.12	−0.23
	$-(CH_2)_4CH_3$			−0.19					−0.15	−0.225
	$-C(CH_3)_2C_2H_5$								−0.19	
	$-CH(C_2H_5)_2$									
C_5H_4N	2-pyridyl			0.55	0.40	−0.23	0.20	0.33	0.17	
C_5H_5S	2-thenyl[k]						0.05	−0.04		0.31
C_5H_9O	$-COC(CH_3)_3$				0.26	0.06		0.27	0.33	
$C_5H_{10}N$	-piperidino[l]				0.19	−0.06		−0.12	−0.12	
$C_5H_{13}N$	$-CH_2CH_2NMe_3^+$				0.27	−0.03			−0.01	
C_6C_{15}	$-C_6C_{15}$				0.27	−0.06	0.25	0.16	−0.03	
C_6F_5	$-C_6F_5$			0.43		0.00	0.31	0.24	0.24	1.56
C_6H_5	$-C_6H_5$	25.36	−0.18	0.02	0.12 *0.55*	−0.13 *1.07*	0.12	0.06	−0.01	0.60
C_6H_{11}	-cyclohexyl		−0.29	−0.14	0.03	−0.18	−0.02	−0.14	−0.22	−0.15
C_6H_{13}	$-(CH_2)_5CH_3$						−0.06	−0.16		−0.25
	$-CH(CH_3)C(CH_3)_3$									−0.28

Formula	Substituent									
C₆H₅N₂	–N=N–C₆H₅	31.31	–0.19	0.45	0.30	0.09	0.19	0.29	0.31	1.87
C₆H₅O	–OC₆H₅	27.68	–0.50	–0.10	0.37	–0.40	0.34	0.25	–0.32	2.43
					0.76	*–1.29*				
C₆H₅S	–SC₆H₅		–0.55	0.18	0.30	–0.23	0.30	0.17	0.13	1.87
C₆H₅Se	–SeC₆H₅		–0.47	–0.29			0.37		0.13	2.30
C₆H₆N	–NHC₆H₅	30.04					–		–0.45	
C₆H₅OS	–S(O)C₆H₅			0.76	0.22	–0.78	0.52	–0.12	0.46	3.24
C₆H₅O₂S	–SO₂C₆H₅	33.20		1.21	0.51	–0.07	0.57	0.51	0.70	3.25
C₆H₅O₃S	–S(O)₂OC₆H₅			1.11	0.58	–0.10	0.55	0.62	0.51	
	–OSO₂C₆H₅	36.70					0.58		0.33	3.62
C₆H₁₄O₃P	–P(O)(OC₃H₇)₂				0.37	–0.04	0.32	0.36	0.50	1.99
C₆H₁₅O₃Si	–Si(OC₂H₅)₃		0.17		0.33	0.17	0.04	0.38	0.08	0.215
C₆H₆NO₂S	–SO₂NHC₆H₅	37.88			0.03	0.05	0.27	0.02	0.65	2.2
	–NHSO₂C₆H₅				0.51	0.14		0.56	0.01	2.57
C₇H₇	–CH₂C₆H₅	30.01	–0.98	–0.09	0.24	–0.23		0.16	–0.09	0.50
C₇H₅O	–COC₆H₅	30.33	–0.28	0.83	–0.04	–0.05		–0.08	0.46	0.85
C₇H₅O₂	–OCOC₆H₅	32.33			0.31	0.12		0.36	0.13	1.56
	–COOC₆H₅	32.31			0.26	–0.13		0.21	0.44	1.56
C₇H₆N	–N=CHC₆H₅	33.01			0.34	0.10		0.37	–0.55	1.68
	–CH=NC₆H₅	33.01			0.14	–0.69		–0.08	0.42	
C₇H₇O	–OCH₂C₆H₅	32.19	–0.07		0.33	0.09	0.08	0.35	–0.41	
	–CH(OH)C₆H₅						0.16	0.00	–0.03	
	–CH₂OC₆H₅						0.25	0.04	0.05	
C₇H₇S	–SCH₂C₆H₅	34.64			0.05	–0.08	0.27	0.23	0.41	
C₇H₆NO	–CONHC₆H₅		–0.60		0.08	–0.01	0.25	0.23	0.08	
	–NHCOC₆H₅							0.22		
C₇H₁₃	–CH₂*cyclo*C₆H₁₁									–0.06
C₉H₁₁	–CH(C₂H₅)C₆H₅									0.04
	–CH₂CH₂CH₂C₆H₅									0.02
C₈H₅	–C≡CC₆H₅	33.21	–0.03	0.30	0.15	0.01	0.22	0.14	0.16	1.35
C₈H₇	–CH=CHC₆H₅	34.17	–1.00	0.13	0.10	–0.17	0.07	0.03	–0.07	0.41

(continued overleaf)

Table 1 (*continued*)

Formula[c]	Structure	MR^d	$\sigma_p^{+\,e}$	$\sigma_p^{-\,e}$	F^f	R^f	σ_I^a	σ_{meta}^g	σ_{para}^g	σ^{*g}
C_8H_9	$-CH_2CH_2C_6H_5$		-0.28	-0.12	-0.01	-0.11	-0.01	-0.07	-0.12	0.08
	$-CH(CH_3)C_6H_5$						0.06	-0.03		0.11
C_8H_6N	3-indolyl						0.01	-0.12		-0.06
$C_{10}H_7$	(α-naphthyl)						0.12	0.06		0.75
	β-naphthyl						0.12	0.06		0.75
$C_{10}H_{15}$	-adamantyl[m]	59.29	-0.27	-0.14	-0.07	-0.06		-0.12	-0.24	
$C_{12}H_{10}OP$	$-PO(C_6H_5)_2$		0.49	0.68	0.32	0.21	0.27	0.40	0.54	1.71
$C_{12}H_{10}PS$	$-PS(C_6H_5)_2$			0.73	0.23	0.24	0.40	0.29	0.47	
$C_{13}H_{11}$	$-CH(C_6H_5)_2$		-0.19		0.01	-0.06	0.07	-0.03	-0.04	0.405
$C_{18}H_{15}Si$	$-Si(C_6H_5)_3$		0.12	0.29	-0.04	0.14	0.13	-0.03	0.10	
$C_{19}H_{15}$	$-C(C_6H_5)_3$		-0.21		0.01	0.01	0.10	-0.01	0.02	~0.4
$C_{19}H_{15}S$	$-SC(C_6H_5)_3$						0.12	0.05		0.69
C_n	general									
OR	$-OR$		-0.83				0.27	0.1	-0.35	1.69
SR	$-SR$			0.74			0.25	0.18	0.05	1.56
HNR	$-NHR$						0.17	-0.29	-0.56	
H_2NR	$-NH_2R^+$						0.60			3.75
CO_2R	$-COOR$						0.31	0.35	0.44	1.94
O_2PR_2	$-P(OR)_2$						0.09	0.12	0.15	
O_2SR	$-SO_2R$						0.57	0.64	0.73	3.56
O_3PR_2	$-PO(OR)_2$						0.28	0.38	0.50	
O_3SR	$-SO_2OR$						0.50	0.71	0.90	3.12
CHO_2R_2	$-CH(OR)_2$						-0.02	-0.04	-0.05	
CHS_2R_2	$-CH(SR)_2$						0.15			1.12
CH_2OR	$-CH_2OR$						0.10	0.02	0.02	0.94

Formula	Substituent					
C₂H₂O₂R	-CH₂COOR	0.17				1.12
C₂H₂O₃R	-OCH₂COOR	–			-0.18	
C₃H₂O₂R	-CH=CH-COOR	0.18	0.19	0.03		1.12
CHNO₂R	-NHCOOR	0.29	0.07	-0.15		
CH₂O₂SR	-CH₂SO₂R	0.21	0.15	0.17		1.12
CH₃NO₂R	-NHCH₂COOR	0.26	-0.10	-0.68		

(additional entries: 0.16, 0.01)

[a] σ_I from O. Exner, *Correlation Analysis in Chemistry: Recent Advances*, N.B. Chapman and J. Shorter (eds.), Plenum Press, New York, 1978, p 439.

[b] Parameters obtained mainly from footnotes a to d; other databases which can be consulted for parameters are:
F.E. Norrington *et al.*, *J. Med. Chem.*, 1975, **18**, 604; D.H. McDaniel and H.C. Brown, *J. Org. Chem.*, 1958, **23**, 420; M. Charton, *Progr. Phys. Org. Chem.*, 1981, **13**, 119; M. Charton, *J. Org. Chem.*, 1964, **29**, 1222; H.H. Jaffé, *Chem. Rev.*, 1953, **53**, 191.

[c] The molecular formula sequence is that adopted by Perrin (footnote g) where the formulae are blocked together initially by their carbon number and then by the number of elements.

[d] MR (Molar refractive index) from C. Hansch and A. Leo, *Substituent Constants for Correlation Analysis in Chemistry and Biology*, John Wiley, New York, 1979; C. Hansch *et al.*, *J. Med. Chem.*, 1973, **16**, 1207.

[e] σ_p^+ and σ_p^- from C. Hansch *et al.*, *Chem. Rev.*, 1991, **91**, 165.

[f] F and R parameters are those from C. Hansch *et al.*, *Chem. Rev.*, 1991, **91**, 165; they are defined from different databases than are those in Swain's papers. Hansch's F and R are derived with the following equations:

$$F = \sigma_I = 1.297.\sigma_m - 0.385.\sigma_p + 0.033$$
$$R = \sigma_p - F$$

The quoted F and R parameters were derived by Hansch from σ_m, σ_p and σ_I values different from those given here. The *italicised F and R* parameters are those defined by C.G. Swain, *J. Am. Chem. Soc.*, 1983, **105**, 492.

[g] σ_m, σ_p, and σ^* from D.D. Perrin *et al.*, *pK$_a$ Prediction for Organic Acids and Bases*, Chapman & Hall, London, 1981.

h

R

R = N₃, OH, Cl, H, SH

i (NH)

j (N— , O)

k (CH₂— , S)

l (N—)

m

Table 2 *Charton's v_x (corrected atomic radii) and Taft–Pavelich steric parameters*[a]

Substituent	E_S	v_x	Substituent	E_S	v_x
Me	(0.00)	0.52	Bu₂CH		1.56
i-Bu	−0.93	0.98	*n*-C₅H₁₁	−0.40	0.68
t-Bu	−1.54	1.24	(i-PrCH₂)₂CH		1.70
CCl₃	−2.06	1.38	(t-BuCH₂)₂CH	−3.18	2.03
CBr₃	−2.43	1.56	*i*-PrCHEt		2.11
CI₃	−0.90	1.79	*t*-BuCHMe	−3.33	2.11
CF₃	−1.16	0.91	*t*-BuCH₂CHMe	−1.85	1.41
Me₃Si		1.40	H-	1.24	(0.00)
F		0.27	*t*-BuCMe₂	−3.90	2.43
Cl		0.55	*t*-BuCH₂CMe₂	−2.57	1.74
Br		0.65	Et₃C	−3.80	2.38
I		0.78	PhCH₂	−0.38	0.70
Ph	−2.48	1.66	PhCH₂CH₂	−0.38	0.70
Ph(CH₂)₃	−0.45	0.70	CHPh₂	−1.43	1.25
Et	−0.07	0.56	Ph(CH₂)₄	−1.76	0.70
Pr	−0.36	0.68	PhMeCH	−1.19	0.99
n-Bu	−0.39	0.68	PhEtCH	−1.50	1.18
BuCH₂	−0.40	0.68	Ph₂CH	−1.43	1.25
BuCH₂CH₂	−0.30	0.73	ClCH₂	−0.24	0.60
Bu(CH₂)₃		0.73	BrCH₂	−0.27	0.64
Bu(CH₂)₄	−0.33	0.68	Cl₂CH	−1.54	0.81
i-PrCH₂	−0.93	0.98	CH₂CH₂Cl	−0.90	
t-BuCH₂	−1.74	1.34	Br₂CH	−1.86	0.89
EtMeCHCH₂		1.00			
i-PrCH₂CH₂	−0.35	0.68	Br₂CMe		1.46
t-BuCH₂CH₂	−0.34	0.70	BrMe₂C		1.39
t-BuCHEtCH₂CH₂		1.01	FCH₂	−0.24	0.62
cyclo-C₆H₁₁CH₂	−0.98	0.97	ICH₂	−0.37	0.67
cyclo-C₆H₁₁CH₂CH₂		0.70	CNCH₂	−0.94	0.89
cyclo-C₆H₁₁(CH₂)₃		0.71	I₂CH		0.97
CH₂SMe	−0.34		CHMeBu-*t*	−3.33	2.11
CH₂OPh	−0.33		CMe₂Et	−1.8	
CH₂CH₂OMe	−0.77		*n*-C₈H₁₇	−0.33	0.68
CH(CH₂CHMe₂)	−2.47	1.70			
CPh₃	−4.68		CHMeEt	−1.13	1.02
cyclo-C₄H₇	−0.06		CH₂CH₂CHMe₂	−0.35	0.68
cyclo-C₅H₉	−0.51		CH(CH₂Bu-t)₂	−3.18	2.03
cyclo-C₆H₁₁	−0.79	0.87	CHMeCH₂Bu-*t*	−1.85	1.41
cyclo-C₇H₁₃	−1.10		CMe(Bu-*t*)CH₂Bu-*t*	−4.00	
i-Pr	−0.47	0.76	F₂CH	−0.67	0.68

[a] Parameters are from O. Exner, in *Correlation Analysis in Chemistry: Recent Advances*, N.B. Chapman and J. Shorter (eds.), Plenum Press, New York, 1978, p. 439. The E_s parameter is sometimes standardised on $E_s(\text{H}) = 0$; the figures quoted above refer to E_s standardised on $E_s(\text{Me}) = 0$. The two scales are related by a constant increment: $E_s(\text{H}) = E_s(\text{Me}) -; 1.24$. Thus E_s for hydrogen on the Me scale is 1.24.

Table 3 *Polar and steric substituent constants for ortho groups*

Group	E_S	σ_o^*
Methoxy	0.99	−0.22
Ethoxy	0.90	n/a
Fluoro	0.49	0.41
Chloro	0.18	0.37
Bromo	0.00	0.38
Methyl	(0.00)	(0.00)
Iodo	−0.2	0.38
Nitro	−0.75	0.97
Phenyl	−0.90	n/a

n/a = not available
Parameters from M. Charton, *Progr. Phys. Org. Chem.*, 1971, **8**, 235.

Table 4 *Ritchie N_+ parameters*

Nucleophile	Solvent	N_+^a	Nucleophile	Solvent	N_+^a
H_2O	water	0.73	Piperidine	water	7.26^b
NH_3	water	3.89		DMSO	8.68^b
n-BuNH$_2$	methanol	6.16	OH$^-$	water	4.75^c
EtNH$_2$	water	5.28	MeO$^-$	methanol	7.51
$H_2NCH_2CH_2NH_2$	water	5.44	HC≡CCH$_2$O$^-$	water	5.84
EtO$_2$CCH$_2$NH$_2$	water	4.40	CF$_3$CH$_2$O$^-$	water	5.06
	methanol	5.13	CN$^-$	water	4.12
	DMSO	6.54		DMF	9.44
$H_2NCH_2CH_2NH_2$	water	4.35		DMSO	8.64
$^-O_2CCH_2NH_2$	water	5.36	N$_3^-$	water	7.54
Glycylglycinate	water	4.69		methanol	8.78
MeOCH$_2$CH$_2$NH$_2$	water	5.07	ClO$^-$	water	7.41
	DMSO	7.56	HOO$^-$	water	8.52
n-PrNH$_2$	DMSO	7.88	SO$_3^-$	water	8.01
CF$_3$CH$_2$NH$_2$	water	3.45	BH$_4^-$	water	6.95
	DMSO	4.86	C$_6$H$_5$S$^-$	water	9.10
H_2NNH_2	water	6.01		methanol	10.41
	methanol	6.89	EtS$^-$	water	8.76
	DMSO	8.17	HOCH$_2$CH$_2$S$^-$	water	8.87
HONH$_2$	water	5.05		DMSO	12.71
MeONH$_2$	water	4.37	MeO$_2$CCH$_2$S$^-$	water	8.89
	DMSO	6.38		DMSO	12.71
C$_6$H$_5$NHNH$_2$	water	4.77	n-PrS$^-$	water	8.93
$H_2NCONHNH_2$	water	3.73	$^-O_2CCH_2S^-$	water	9.09
	DMSO	6.22			
Morpholine	water	6.78^b	Morpholine	DMSO	8.56^b

[a] Based on reactions of malachite green or tri-4-anisylmethyl cation. Data from C.D. Ritchie, *Can. J. Chem.*, 1986, **64**, 2239.
[b] Calculated from N_+ for the pyronin-Y cation.
[c] All values relative to $N_+ = 4.75$ for HO$^-$ in water.

Table 5 *Swain–Scott, Pearson and Edwards parameters for nucleophilic reagents*

Nucleophile	n_{MeI}	E_n	pK_a	Nucleophile	n_{MeI}	E_n	pK_a
$MeOH$	(0.00)	(0)	−1.7	$(C_2H_5)_3As$	6.90	<2.61	
CO	<2.0			$(C_6H_5)_3P$	7.00	2.73	
MeO^-	6.29	1.65	15.7	$(n\text{-}C_4H_9)_3P$	8.69	8.43	
F^-	2.7	−0.27	3.45	$(C_2H_5)_3P$	8.72		8.69
Cl^-	4.37	1.24	−5.7	$MeCOO^-$	4.3	0.95	4.75
NH_3	5.50	1.36	9.25	$C_6H_5COO^-$	4.5		4.19
Imidazole	4.97		7.10	$C_6H_5O^-$	5.75	1.46	9.89
Piperidine	7.30		11.21	$C_6H_5N(Me)_2$	5.64		5.06
Aniline	5.70	1.78	4.58	2,6-Dimethyl-pyridine	3.51		
Pyridine	5.23	1.20	5.23	2-Picoline	4.7		6.48
NO_2^-	5.35	1.73	3.37	Pyrrolidine	7.23		11.27
$(C_6H_5CH_2)_2S$	4.84			N,N-Dimethyl-cyclo-hexylamine	6.73		
N_3^-	5.78	1.58	4.74	$(C_2H_5)_3N$	6.66		10.70
NH_2OH	6.60		5.82	$(C_2H_5)_2NH$	7.0		11.0
NH_2NH_2	6.61		7.93	$C_6H_5Se^-$	10.7		
C_6H_5SH	5.70			$(C_6H_5)_3SiO^-$	6.2		
Br^-	5.79	1.51	−7.7	$C_6H_5SO_2NCl^-$	6.8		3.0
$(C_2H_5)_2S$	5.34			$C_6H_5SO_2NH^-$	5.1		8.5
$((C_2H_5)_2N)_3P$	8.54			Phthalimide anion	5.4		7.4
$(Me)_2S$	5.54		−5.3	HS^-	8	2.10	7.8
$(MeO)_2PO$	7.00			SO_4^{2-}	3.5	0.59	2.0
$(CH_2)_5S$	5.42			NO_3^-	1.5	0.29	−1.3
$(CH_2)_4S$	5.66		−4.8	$(C_6H_5)_3Ge^-$	12		
$SnCl_3^-$	3.84			$(C_6H_5)_3Sn^-$	11.5		
I^-	7.42	2.06	−10.7	$(C_6H_5)_3Pb^-$	8		
$(C_6H_5CH_2)_2Se$	5.23			$Re(CO)_5^-$	8		
$(Me)_2Se$	6.32			$Mn(CO)_5^-$	5.5		
SCN^-	6.70	1.83	−0.7	$Co(CO)_4^-$	3.5		
SO_3^{2-}	8.53	2.57	7.26	HO_2^-	7.8		
C_6H_5NC	<2.0			$C_6H_5S^-$	9.92	2.9	6.52
$(C_6H_5)_3Sb$	<2.0			$SC(NH_2)_2$	7.27	2.18	−0.96
$(C_6H_5)_3As$	4.77			$(MeO)_3P$	5.2		
$SeCN^-$	7.85			$S_2O_3^{2-}$	8.95	2.52	1.9
CN^-	6.70	2.79	9.3				

Data from R.G. Pearson, *J. Am. Chem. Soc.*, 1968, **90**, 319.

Values of solvent ionizing power *Y* for ... 2-adamantyl tosylates at 25°C

Composition/%	Solvent	Y_{G-W}[a]	Y_{OTs}[b]	N_{OTs}[b]
100	EtOH	-2.03	-1.96	0.06
90		-0.75	-0.77	0.07
80		0.00	0.00	0.00
70		0.60	0.47	-0.05
60			0.92	-0.08
50		1.12	1.29	-0.09
40		1.66	1.97	-0.23
30			2.84	-0.35
20			3.32	-0.34
10			3.78	-0.41
100	water	3.49	4.1	-0.44
100	MeOH	-1.09	-0.92	-0.04
90		-1.09	-0.05	-0.05
80		0.38	0.47	-0.08
70		0.96	1.02	-0.13
60		1.49	1.52	-0.19
50		1.97	2.00	-0.21
40			2.43	-0.31
30			2.97	-0.35
20			3.39	-0.41
10			3.78	
95	acetone	-1.86	-2.95	-0.39
90		-0.67	-1.99	-0.42
80		0.13	-0.94	-0.42
70		0.80	0.07	-0.41
60		1.40	0.66	-0.39
50			1.26	-0.38
40			1.85	-0.38
30			2.50	-0.40
20			3.05	-0.38
10			3.58	-0.41

Composition/%	Solvent	Y_{G-W}[a]	Y_{OTs}[b]	N_{OTs}[b]
90	dioxan		-2.41	-0.51
80			-1.30	-0.29
100	MeCN		-3.21	
50			1.2	
35			1.8	
30			1.9	
25			2.5	
20			2.7	
10			3.6	
100	(Me)$_2$CHOH	-2.73	-2.83	0.12
100	(Me)$_3$COH		-3.74	
100	CF$_3$CH$_2$OH		1.77	-3.07
97			1.83	-2.79
85			1.92	-2.01
70			2.00	-1.20
50			2.14	-0.93
100	(CF$_3$)$_2$CHOH(HFIP)		3.82	
100	MeCO$_2$H(A)	-1.64	-0.90	-2.28
90	A/F		0.70	
75			1.62	
50			2.34	
100	HCO$_2$H(F)	2.05	3.04	-2.35
100	CF$_3$CO$_2$H(TFA)	1.84	4.57	-5.56
80	TFA/EtOH	0.41	0.98	-1.72
60			0.21	-1.01
40			-0.44	-0.55
60	H$_2$SO$_4$/H$_2$O		5.29	-0.71
40			4.67	-1.20
20			4.39	-2.02
100	HCONHMe		-1.70	
100	MeCONHMe		-2.69	
100	HCON(Me)$_2$		-4.14	
100	MeCON(Me)$_2$		-4.99	

[a] Parameters from J.E. Leffler and E. Grunwald, John Wiley, New York, 1963.   [b] T.W. Bentley and G. Llewellyn, *Progr. Phys. Org. Chem.*, 1990, **17**, 121.

Table 7 *Some Leo–Hansch* π *parameters for use in calculating logP values*

Substituent	ArOCH$_2$CO$_2$H	ArOH	Substituent	ArOCH$_2$CO$_2$H	ArOH
H	0	0	3-COOH	−0.15	0.04
2-F	0.01	0.25	4-COOH		0.12
3-F	0.13	0.47	2-OMe	−0.33	
4-F	0.15	0.31	3-OMe	0.12	0.12
2-Cl	0.59	0.69	4-OMe	−0.04	−0.12
3-Cl	0.76	1.04	3-OCF$_3$	1.21	
4-Cl	0.70	0.93	3-OCH$_2$COOH		−0.70
2-COMe	0.01		4-OCH$_2$COOH		−0.10
3-COMe	−0.28	−0.07	3-NME$_2$		0.10
4-COMe	−0.37	−0.11	3-NH$_2$		−1.29
3-CN	−0.30	−0.24	4-NH$_2$		−1.63
4-CN	0.32	0.14	2-NO$_2$	−0.23	0.33
2-Br	0.75	0.89	3-NO$_2$	0.11	0.54
3-Br	0.94	1.17	4-NO$_2$	0.24	0.50
4-Br	1.02	1.13	3-NHCOPh	−0.72	
2-I	0.92	1.19	3-NHCONH$_2$	−1.01	
3-I	1.15	1.47	3-NHCOMe	−0.79	
4-I	1.26	1.45	3-*n*Bu	1.90	
2-Me	0.68		3-*t*Bu	1.68	
3-Me	0.51	0.56	4-*Cyclo*hexyl	2.51	
4-Me	0.52	0.48	3,4[CH$_2$]$_3$	1.04	
2-Et	1.22		3,4[CH]$_4$	1.24	
3-Et	0.97	0.94	3,4[CH$_2$]$_4$t	1.39	
3-*n*Pr	1.43		3-OH	−0.49	−0.66
3-*i*Pr	1.30		4-OH	−0.61	−0.87
4-*i*Pr	1.40				
4-*sec*Bu	1.82		4N = NPh	1.71	
4-*Cyclo*pent	2.14		3-SMe	0.62	
3-Ph	1.89		3-SO$_2$CF$_3$	0.93	
			3-CH$_2$-COOH		−0.61
3-CF$_3$	1.07	1.49	3-CH$_2$OH		−1.02
3-SCF$_3$	1.58		4-CH$_2$OH		−1.26

Data taken from: T. Fujita *et al.*, *J. Am. Chem. Soc.*, 1964, **86**, 5175.

Table 8 *Fragmental constants, f, for calculating logP*

Calculation step	f	Calculation step	f
CH_2	0.66	Single bond in chain	$-0.12(f_b)$
H	0.23	CH	0.65
Single carbon (C)	0.20	Branching in C chain	$-0.13(f_{cbr})$
Single bond in ring	$-0.09(f_b)$	Branching at a group	$-0.22 \ (f_{gbr})$
Phenyl group	1.90	Me	0.89
Aliphatic groups:			
COOH	-1.09	NH_2	-1.54
CN	-1.28	$CONH_2$	-2.18
CI	0.06	I	0.60
NH	-2.11	OH	-1.64
COR_2	-1.90	$-O^-$	-1.81
F	-0.38	Br	0.20
NO_2	-1.26	$-S^-$	-0.79
Aromatic groups:			
COOH	-0.03	NH_2	-1.00
CN	-0.34	$CONH_2$	-1.26
CI	0.94	I	1.35
NH	-1.03	OH	-0.40
CO	-0.32	$-O^-$	-0.57
F	0.37	Br	1.09
NR_2	-1.17	$-CONR_2$	-2.82
$-CO-OR$	-0.56	CONHR	-1.81
NO_2	-0.02	$-S^-$	0.03

Data taken from A.J. Leo *et al.*, *J. Med. Chem.*, 1975, **18**, 865.
R. Rekker and R. Mannhold, *Calculations of Drug Lipophilicity*, VCH, Weinheim, 1992;
H. Kubinyi, *Prog. Drug Res.*, 1979, **23**, 97; C. Hansch and A.J. Leo, *Substituent Constants for Correlation Analysis in Chemistry and Biology*, Wiley-Interscience, New York, 1979, pp. 18–43.

Table 9a *Fragmental constants for calculation of pK_a values for carboxylic acids*

Substituent	$-\Delta pK_a$	Substituent	$-\Delta pK_a$	Substituent	$-\Delta pK_a$
Substituents at α-carbon.[a]					
$-Br$	1.85	$-C\equiv CH$	1.43	$-CH_2OH$	0.26
$-Cl$	1.92	$-CH=CH_2$	0.41	$-OMe$	1.20
$-F$	2.08	$-COCF_3$	2.39	$-SMe$	1.04
$-I$	1.61	$-CH_2CN$	0.88	$-SCN$	2.22
$-N_3$	1.71	$-COMe$	1.10	$-CONH_2$	1.12
$-OH$	0.90	$-OCOMe$	1.41	$-NHCHO$	1.08
$-SH$	1.12	$-COOMe$	1.32	$-NHCONH_2$	0.89
$-NH_2$	0.45	$-NH(Me)_2^+$	2.81	$-SO_2Me$	2.38
$-NH_3^+$	2.43	$-NHCOMe$	0.94	$-SCONH_2$	1.36
$-NO_2$	2.58	$-N(Me)_3^+$	2.93	$-COC_6H_5$	1.45
$-ONO_2$	2.49	$-Si(Me)_3$	-0.45	$-SCH_2C_6H_5$	1.04
$-SO_3^-$	0.57	$-NHCOC_2H_5$	1.04	$-CONHC_6H_5$	1.04
$-CCl_3$	1.73	$-NHCOOC_2H_5$	1.31	$-N(COMe)C_6H_5$	0.92
$-CF_3$	1.70	$-CH=CHC_2H_5$	0.26	α-naphthyl	0.53
$-CN$	2.14	$-CH_2COOC_2H_5$	0.58	β-naphthyl	0.53
$-COO^-$	-0.61	$-CH=CHC_2H_5$		$-CH=CHR_n$	0.25
$-COOH$	1.37	$-C_6H_5$	0.44	$-OCOR_n$	~ 1.70
$-CH_2Br$	0.69	$-OC_6H_5$	1.59	$-O$-*cyclo*-C_6H_{11}	1.20
$-CH_2Cl$	0.65	$-SC_6H_5$	1.24	$-S$-*cyclo*-C_6H_{11}	1.28
$-CH_2F$	0.75	$-CH_2I$	0.69		

Data taken from D.D. Perrin *et al.*, *pK_a Prediction for Organic Acids and Bases*, Chapman & Hall, London, 1981.
[a] The pK_a of the starting carboxylic acid is taken as 4.80; the attenuation across a $-CH_2-$ group is taken as 0.40.

Table 9b *Fragmental constants for calcutation of pK$_a$ values for ammonium ions*

Substituent	$-\Delta pK_a^a(\alpha^b)$	$-\Delta pK_a^b(\beta^b)$	Substituent	$-\Delta pK_a^a(\alpha^b)$	$-\Delta pK_a^b(\beta^b)$
–F		1.9	–COCF$_3$		3.5
–OH		1.1	–COMe		1.7
–SH		1.3	–COOMe		1.3
–NH$_2$		0.8	–NHCOMe		1.5
–NH$_3^+$		3.6	–N(Me)$_3^+$		4.2
–NO$_2$		3.8	–Si(Me)$_3$		–0.4
–ONO$_2$		3.6	–NHCOC$_2$H$_5$		1.6
–SO$_3^-$		1.0	–NHCOOC$_2$H$_5$		2.0
–CCl$_3$		2.6	2-pyridyl	2.20	1.14
–CF$_3$		2.6	3-pyridyl	2.70	
–CN	5.8	3.2	–C$_6$H$_5$	1.4	0.8
–COO$^-$	0.8^c	–0.2	–COC$_6$H$_5$		2.2
	–0.1^d		–CH=CR$_2$	~1.0	~0.5
–OMe		1.2	–C≡CR	~2.0	~1.0
–SMe		1.6	–NHR	~1.7	~0.9
–CONH$_2$	2.8	1.7	–NR$_2$	~1.7	~0.9
–NHCHO		1.7	–OR		1.2
–NHCONH$_2$		1.4	–COR		1.6
–SO$_2$Me		3.5	–COOR	3.0	1.3
–SCONH$_2$		2.1	–OCOR		~1.7
–C≡CH		1.9	–SR	~3.5	1.4

Data taken from D.D. Perrin *et al.*, *pK$_a$ Prediction for Organic Acids and Bases*, Chapman & Hall, London, 1981.
^{a} The starting pK$_a$ values of primary, secondary and tertiary amines are taken to be 10.77, 11.15 and 10.5 respectively. ΔpK_a values of +0.2 and –0.2 are added for ring formation and N-methylation respectively. A –CH$_2$– group possesses an attenuation factor of approximately 0.4.
^{b}

$$\alpha = R-\overset{|}{\underset{|}{C}}-NH^+ \qquad \beta = R-\overset{|}{\underset{|}{C}}-\overset{|}{\underset{|}{C}}-NH^+$$

^{c} Adjacent to primary or secondary amine.
^{d} Adjacent to tertiary amine.

Table 10 *Hammett σ_{ortho} constants for use in systems relating to benzoic acids, phenols, anilines or pyridines.*

Substituent	ArCOOH	ArOH	ArNH$_2$	Pyridines
–Br	1.35	0.70	0.71	0.58
–Cl	1.28	0.68	0.67	0.79
–F	0.93	0.54	0.47	0.96
–I	1.34	0.63	0.70	0.58
–OH	1.22	0.04	–0.09	0.76
–NH$_2$		0.03	0.00	–0.27
–NH$_3^+$	2.15		1.23	
–NO$_2$	1.99	1.40	1.72	
–Me	0.29	–0.13	0.10	–0.13
–CN	1.06	1.32	1.26	0.93
–COO$^-$	–0.91		–0.13	–0.05
–CHO		0.75		0.25
–COOH	0.95		0.88	0.51
–CH$_2$OH		0.04		
–OMe	0.12	0.00	0.00	0.34
–SMe	0.52	0.30		0.28
–CH$_2$NH$_3^+$		0.41		
–CONH$_2$	0.45	0.72		0.53
–SOMe		1.04		
–C$_2$H$_5$	0.41	–0.09	0.05	–0.13
–OCH$_2$COO$^-$	–0.27			
–COMe	0.07			
–OCOMe	–0.37			0.52
–COOMe	0.63		0.84	0.51
–OC$_2$H$_5$	–0.01	–0.08	0.02	
–N(Me)$_2$		–0.36		
–CH$_2$CH$_2$NH$_3^+$		0.28		0.25
–N(Me)$_3^+$		1.07		
–Ph	0.74	0.00	0.28	0.13
–OPh	0.67			
–SPh			0.72	
–CH$_2$Ph				0.02
–COPh	0.65			

Data largely from D.D. Perrin *et al.*, *pK$_a$ Prediction for Organic Acids and Bases*, Chapman & Hall, London, 1981.

Table 11 *Reichardt* E_T^N *parameters*

Solvent	E_T^N	Solvent	E_T^N	Solvent	E_T^N
Gas phase	−0.111	2-Phenyl ethanol	0.580	$CH_3CONHCH_3$	0.657
$CH_3CH(CH_3)C_2H_5$	0.006	2-Methoxyethanol	0.657	CH_3NO_2	0.481
n-Pentane	0.009	2,2,2-Triflaoroethanol	0.898	$C_2H_5NO_2$	0.398
Cyclohexane	0.006	1-Propanol	0.617	Piperidine	0.148
CH_2Cl_2	0.309	2-Propanol	0.546	Morpholine	0.318
CH_2Br_2	0.269	1-Butanol	0.586	Aniline	0.420
CH_2I_2	0.179	*tert*-Butanol	0.389	*N*-methylaniline	0.364
$CHCl_3$	0.259	Diethyl ether	0.117	*N,N*-dimethylaniline	0.179
C_2H_5Br	0.213	1,2-Dimethoxyethane	0.231	CS_2	0.065
1,1-Dichloroethane	0.269	Furan	0,164	H_2O	1.00
1,2-Dichloroethane	0.327	Tetrahydrofuran	0.207	$SP(N(CH_3)_2)_3$	0.272
Benzene	0.111	Tetrahydropyran	0.170	$(CH_3)_2SO$	0.444
Toluene	0.099	Dioxane	0.164	1-Methylpyrrolidin-	
				2-one	0.355
Fluorobenzene	0.194	Anisole	0.198	CH_3CN	0.460
Bromobenzene	0.182	Acetone	0.355	CCl_3CN	0.287
Pyridine	0.302	Cyclohexanone	0.281	*tert*-Butylamine	0.179
Methanol	0.762	Acetophenone	0.306	1,2-Diaminoethane	0.349
Benzyl alcohol	0.608	HC_2H	0.728	Diethylamine	0.145
Ethanol	0.654	CH_3CO_2H	0.648	Pyrrolidine	0.259
Ethanol/water 80%	0.710	$(CH_3CO)_2O$	0.407	$OP(OCH_3)_3$	0.398
HCO_2CH_3	0.346	$(CH_3O)_2CO$	0.232	$OP(OC_2H_5)_3$	0.324
$HCO_2C_2H_5$	0.315	$(C_2H_5O)_2CO$	0.185	$OP(OnC_3H_7)_3$	0.302
$CH_3CO_2CH_3$	0.253	$HCONH_2$	0.775	$OP(N(CH_3)_2)_3$	0.315
$CCl_3CO_2CH_3$	0.275	$HCONHCH_3$	0.722	Sulpholane	0.410
$CH_3CO_2C_2H_5$	0.228	$HCON(CH_3)_2$	0.382	$HCON(CH_3)C_6H_5$	0.312
$CH_3CON(CH_3)_2$	0.377	$HCSN(CH_3)_2$	0.410	$CH_3CON(C_2H_5)_2$	0.330

Data from C. Reichardt, *Chem. Rev.*, 1994, **94**, 2319; $E_T^{water} = 0.00$.

Some Linear Free Energy Parameters

Table 1 *Some Hammett relationships for dissociation constants* (pK_a^{HA})

Acid	$-\rho^a$	$pK_{intercept}^b$	Acid	$-\rho^a$	$pK_{intercept}^b$
Carboxylic acids:			2-Naphthylamines[5]	2.81	4.35
ArCOOH(water)	1.00	4.20	*anti*-ArCH=NOH	0.86	10.70
ArCOOH(40% EtOH/water)	1.67	4.87	ArOH	2.23	9.92
ArCOOH(70% EtOH/water)	1.74	6.17	$ArCOCH_2PPh_3^+$		
ArCOOH (EtOH)	1.96	7.21	(80% EtOH/H_2O)	2.40	6.0
ArCOOH (CH_3OH)	1.54	6.51	$ArCOCH_2CONHPh$		
ArCOOH(50% butylcellosolve/			(water/dioxane)	0.79	9.4
H_2O)	1.42	5.63	$ArCOCH_2COCH_3$	1.72	8.53
Butylcellosolve/(H_2)			ArSH (48.9% EtOH/H_2O)	2.24	7.67
2-Hydroxybenzoic acids	1.10	4.00	$ArNH_3^+$	2.77	4.56
$ArCH_2COOH$	0.47	4.30	$ArCH_2NH_3^+$	1.06	9.39
$ArCH_2CH_2COOH$	0.22	4.55	$ArNHNH_3^+$	1.17	5.19
trans-ArCH=CHCOOH	0.47	4.45	ArNHPh	4.07	22.4
4-ArC_6H_4COOH			$ArN^+H_2CH_2CO_2^-$	3.06	2.24
(50% butylcellosolve/H_2O)	0.48	5.64	$ArN^+H_2CH_2CO_2H$	3.02	22.03
ArC≡CCOOH (35% dioxane/			$ArS(Me)CH_2COPh^+$	1.4	7.32
H_2O)	0.80	3.26	$ArCH_2NO_2$	0.83	6.88
$ArOCH_2COOH$	0.30	3.17	$ArCH_2NO_2$ (50% MeOH)	1.22	7.93
$ArSCH_2COOH$	0.32	3.38	$ArCH(Me)NO_2$	1.03	7.39
$ArSeCH_2COOH$	0.35	3.75	$ArCH(Me)NO_2$		
$ArSOCH_2COOH$	0.17	2.73	(50% dioxane/H_2O)	1.62	10.30
$ArSO_2CH_2COOH$	0.25	2.51	ArC(OH)=CHOMe	1.10	8.24
$ArN^+H_2CH_2CO_2H$	0.125	22.06	$ArCOCH_2S(Me)Ph^+$	2.0	7.32
$ArNHCH_2CO_2H$	0.323	2.99	$ArCH(NO_2)_2$	1.47	3.89
			$ArCH_2CH(Me)NO_2$		
Other acids:			(50% MeOH/H_2O)	0.40	9.13
$ArB(OH)_2$ (25% EtOH/H_2O)	2.15	9.70	substituted fluorenes		
$ArPO(OH)_2$	0.76	1.84	(water/DMSO)[2]	6.3	22.1
$ArPO_2(OH)^-$	0.95	6.96	2-nitrophenols	2.16	6.89
$ArAsO_2(OH)^-$	0.87	8.49	$ArSO_2NH_2$	0.88	10.02
$ArAsO(OH)_2$	1.05	3.54	$ArSO_2NHPh$	1.16	8.31
$ArSeO_2H$	0.90	4.74	$ArNHSO_2Ph$	1.74	8.31
			ArC$^+$(OH)Me	2.6	−6.00
Conjugate acids of the bases:			$ArC(OH)_2^+$	1.2	−7.26
Pyridines	5.9	5.25	$ArNHMe_2^+$	3.46	5.13
Pyrimidines	5.9	1.23	ArCONHOH	1.02	8.78
Quinolines[3]	5.9	4.90	$ArCH(CF_3)OH$	0.995	11.91
Isoquinolines[4]	5.9	5.4	$ArC(CF_3)(OH)_2$	1.07	9.94
1-Naphthylamines[3]	2.81	3.9			

Parameters (25 °C) mostly from D.D. Perrin *et al.*, *pK*$_a$ *Prediction for Organic Acids & Bases*, Chapman and Hall, London, 1981.

[a] In the case of multi-substitution Ss is employed.
[b] Calculated value at $\sigma = 0$.
[c] dmso = dimethylsulfoxide.
[d] 2,3 & 4 substituted.
[e] 1,3 & 4 substituted.
[f] 3 & 4 substituted.

Table 2 *Some Taft and Charton relationships for dissociation constants* $(pK_a^{HA})^1$

Acid	ρ^a	$pK_{intercept}^b$	Acid	ρ^a	$pK_{intercept}^b$
Taft Relationships			$CH_3COCHRCOOEt$	−3.44	12.59
RNH_3^+	−3.14	10.15	$RCH(NO_2)_2H$	−2.23	5.35
$RR_1NH_2^+$	−3.33	10.59	$R_1R_2C(OH)_2$	−1.32	14.19
$RR_1R_2NH^+$	−3.30	9.61	$RNHC(NH_2)_2^+$	−3.60	14.0
			$RNHNH_3^+$	−2.80	7.8
RPH_3^+	−2.64	2.46	2-substituted benzoic		
$R_2PH_2^+$	−2.61	3.59	acids	−1.79	4.20
R_3PH^+	−2.67	7.85	2-substituted anilines	−2.90	4.58
$RCOOH$	−1.62	4.66	*Charton Relationships*	ρ_I	$pK_{intercept}^b$
RCH_2COOH	−0.67	4.76	$X\text{-}CH_2\text{-}NH_3^+$	−10.02	10.39
RCH_2OH	−1.32	15.7	$X\text{-}CH_2NH_2CH_3^+$	−10.53	10.93
ROH_2^+	−2.36	−2.18	$X\text{-}CH_2NH(CH_3)_2^+$	−11.01	10.03
RSH	−3.5	10.22	$X\text{-}CH_2CH_2NH_3^+$	−4.71	10.54
RCH_2SH	−1.47	10.54	$X_1X_2X_3COH$	−8.23	15.88
$RCH(SR')OH$	−1.32	$11.1+0.16pK_a^{R'SH}$	XCH_2COOH	−1.00	4.66
$RCH(OH)_2$	−1.32	13.54			

Parameters for ρ^* taken mostly from D.D. Perrin *et al.*, *pK_a Prediction for Organic Acids and Bases*, Chapman & Hall, London, 1981; for r_I from P.J. Taylor, *J. Chem. Soc., Perkin Trans. 2*, 1993, 1423 and R.W. Taft, *J. Am. Chem. Soc.*, 1953, **75**, 4231.
[a] In the case of multi-substitution $\Sigma\sigma^*$ or $\Sigma\sigma_I$ is employed.
[b] Calculated value at σ^* or $\sigma_I = 0$.

Table 3 *Some linear free energy relationships for rates*

Reactions	T/°C	ρ or (β)	$logk_H$
ArPhCHCl + EtOH → ArPhCHOEt + HCl (EtOH)	25	−5.09	−2.06
ArNMe$_2$ + MeI → ArNMe$_3^+$ + I$^-$ (90% acetone)	35	−3.3	−3.37
ArH + NO$_2^+$ → ArNO$_2$ + H$^+$ (acetone)	18	−5.93	0.10
ArCO$_2$Me + $^-$OH → ArCO$_2^-$ + MeOH (60% acetone)	25	2.23	−2.08
ArCO$_2$H + H$^+$ + CH$_3$OH → ArCOOCH$_3$ + H$_3$O$^+$ (CH$_3$OH)	25	−0.23	−3.84
ArSO$_2$OEt + H$_2$O → ArSO$_3^-$ + EtOH (30%EtOH)	25	1.17	5.25
ArSO$_2$OEt + $^-$OEt → ArSO$_3^-$ + EtOEt (EtOH)	35	1.37	2.88
ArO$^-$ + (CH$_2$CH$_2$O) → ArOCH$_2$CH$_2$O$^-$ (98% EtOH)	70.4	−0.95	−4.25
ArCHO + HCN → ArCH(OH)CN (95% EtOH)	20	2.33	−7.70
ArCH$_2$Cl + H$_2$O → ArCH$_2$OH + Cl$^-$(48% EtOH)	30	−2.18	−5.61
ArO$^-$ + EtI → ArOEt + I$^-$ (EtOH)	42.5	−0.99	−3.96
ArNH$_2$ + HCO$_2$H → ArNHCHO (67% pyridine)	100	−1.22	−4.12
ArCOCl + H$_2$O → ArCO$_2$H + Cl (95% acetone)$^-$	25	1.78	−4.20
ArB(OH)$_2$ + Br$_2$ → ArBr (20% acetic acid)	25	−4.50	−3.21
ArCONH$_2$ + H$^+$ → ArCO$_2$H + NH$_4^+$ (60% EtOH)	52.4	−0.48	−5.61
ArCONH$_2$ + $^-$OH →ArCO$_2^-$ + NH$_3$ (60% EtOH)	52.8	1.36	−5.19
2,4-Dinitrophenylarylether + CH$_3$O$^-$ → ArOCH$_3$ + 2,4-Dinitrophenol (MeOH)	20	1.45	−2.32
PhCOCl + ArNH$_2$ → PhCONHAr + Cl$^-$ (benzene)	25	−2.78	−2.89
2,4-dinitrofluorobenzene + ArNH$_2$ → 2,dinitrophenylarylamine (99.8% EtOH)	20	−4.24	−2.28
ArNCO + CH$_3$OH → ArNHCOOCH$_3$ (di-*n*–butylether)	20	2.46	−2.26
ArCOCH$_3$ + Br$_2$ → ArCOCH$_2$Br + Br$^-$ (75% acetic acid + 0.5M HCl)	25	−0.46	−2.41
2ArCHO + $^-$OH → ArCH$_2$OH + ArCO$_2^-$ (50% MeOH)	40	3.63	−5.06
ArOCOCH$_3$ + $^-$OH →ArO$^-$ + CH$_3$CO$_2^-$	30	−0.31	0.55
ArOCOCH$^-$COCH$_3$ + H$_2$O→ ArOH + CH$_3$COCH$_2$CO$_2^-$	30	(−1.29)	−1.23
ArOCONHPh + $^-$OH→ ArO$^-$ + PhNH$_2$ + CO$_2$	25	(−1.34)	1.87
ArOCONH$_2$ + $^-$OH→ ArO$^-$ + NH$_3$ + CO$_2$	25	(−1.15)	2.32
ArOCONHCH$_3$ + $^-$OH→ ArO$^-$ + CH$_3$NH$_2$ + CO$_2$	25	(−1.1)	0.53
ArOPO(OPh)$_2$ + $^-$OH→ ArO$^-$ + (PhO)$_2$PO$_2^-$	25	(−0.55)	−1.86
ArCH=NCl + MeO$^-$ → ArCN (92.5% EtOH)	0	2.24	−1.76
Ph$_2$CHCH$_2$-O-SO$_2$C$_6$H$_4$Y + MeO$^-$ → Ph$_2$C=CH$_2$	50	1.09	−2.44
ArCHCH$_3$CH$_2$-O-SO$_2$C$_6$H$_4$4CH$_3$ + Bu^{t}O$^-$ → ArC-CH$_3$=CH$_2$ (Bu^{t}OH)	50	2.19	−3.63
XC$_6$H$_4$CH$_2$CH$_2$-O-C$_6$H$_4$Y + Bu^{t}O$^-$ → XC$_6$H$_4$CH=CH$_2$ (Bu^{t}OH)	40	2.34(σ_x) 1.08(σ_Y)	−2.22 ($logk_{HH}$)
ArCO-O-a–chymotrypsin hydrolysis (water at pH 8.50)	25	1.73	−3.48
ArCN + HO$^-$ → ArCO$_2$H (EtOH)	82	2.13	−1.00
ArCO$_2$H + CH$_3$OH +H$^+$ → ArCO$_2$CH$_3$ (CH$_3$OH)	25	−0.23	−3.84
ArCH$_2$CO$_2$C$_2$H$_5$ + HO$^-$ → ArCH$_2$CO$_2^-$ (88% C$_2$H$_5$OH)	30	0.82	−1.81
ArCH$_2$CH$_2$CO$_2$C$_2$H$_5$ + HO$^-$ → ArCH$_2$CH$_2$CO$_2^-$ (88% C$_2$H$_5$OH)	30	0.49	−2.20
ArCH=CHCO$_2$C$_2$H$_5$ + HO$^-$ → ArCH=CHCO$_2^-$ (88% C$_2$H$_5$OH)	30	1.33	−2.75
ArCH=N-O-COCH$_3$ + H$^+$ → ArCHO (80% acetone)	25	−0.13	−4.65

	T/°C	ρ or (β)	$logk_H$
(MeOH)	25	3.94	−6.39

Data largely from H.H. Jaffé, *Chem. Rev.*, 1953, **53**, 191.

Table 4 *Hammett and Taft relationships for some equilibria*

Reaction	Solvent	$\rho\ (\rho^*)$	$\log K_H$ $(\log K_{Me})$	$T/^\circ C$
$ArNH_2 + HCO_2H \rightleftharpoons HCONHAr + H_2O$	67% pyridine	-1.43	0.630	100
$ArCHO + HCN \rightleftharpoons ArCH(OH)CN + H_2O$	95% EtOH	-1.49	-2.14	20
$ArCHO + HO^- \rightleftharpoons ArCH(OH)(O^-)$	aqueous	2.84	-0.97	25
$ArCOCH_3 + H_2 \rightleftharpoons ArCH(OH)CH_3$	aqueous	1.63	-3.913	25
aminotetrazole tautomerism (ArN / $ArHN$)	glycol	0.98	0.05	197
$2\,ArNH_2 + Ag^+ \rightleftharpoons Ag(ArNH_2)_2^+$	59% EtOH/H_2O	-1.45	3.18	25
$ArO-$ triazine + $PhO^- \rightleftharpoons PhO-$ triazine + ArO^-	aqueous	-1.48	0.00	25
pyridinium triazine exchange	aqueous	-1.25	0.00	25
O_2N–sultone + $ArO^- \rightleftharpoons O_2N$–$SO_2$–$OAr$	aqueous	-1.8	2.10	25
$ArCONH-CH_2-CO-OPh \rightleftharpoons Ar-$oxazolone $+ PhO^-$	aqueous	-0.71	0.79	25
$RCHO + HO^- \rightleftharpoons RCH(OH)(O^-)$	aqueous	$\rho^* = 3.02$	-0.32	25
$RCHO + RS^- \rightleftharpoons RCH(SR)(O^-)$	aqueous	$\rho^* = 2.97$	-9.7 $+0.84 pK_a^{RSH}$	25
$RCHO + RSH \rightleftharpoons RCH(SR)(OH)$	aqueous	$\rho^* = 1.65$	-1.41 (independent of pK_a^{RSH})	25
$ArCHO + H_2O \rightleftharpoons ArCH(OH)(OH)$ (activity = 1)	aqueous	$+1.77$	-2.17	25
$RCHO + H_2O \rightleftharpoons RCH(OH)(OH)$ (activity = 1)	aqueous	$+1.70$	-0.78	25

Data mainly from H.H. Jaffé *Chem. Rev.*, 1953, **53**, 191; R.W. Taft and I.C. Lewis, *J. Am. Chem. Soc.*, 1959, **81**, 5343. In this table the equilibrium constant is defined as $K = $ [Products]/[Reactants].

Table 5 *Linear free energy relationships for some transport and complexation processes*

Reaction	Equation
Hydrolysis of 4-nitrophenyl esters by serum albumin	$\log k = 0.95\log P + 3.5E_s - 0.47$
Inhibition of chymotrypsin:	
with aromatic acids	$\log K_i = 0.94\log P - 0.96\pi K + 3.66$
with PhCOR	$\log K_i = -0.31\pi - 2.53$
with ArOH	$\log K_i = -0.95\log P + 1.88$
with hydrocarbons	$\log K_i = -1.47\log P + 1.2$
Reaction of chymotrypsin:	
with PhCONHCH$_2$COOR	$\log K_m = -0.41\pi - 0.4Es + 0.71$
with Ph[CH$_2$]$_2$COOR	$\log K_m = -0.21\pi - 3.16$
Binding of serum albumin:	
with barbiturates	$-\log C = 0.58\log P + 2.39$
with phenols	$-\log C = 0.68\pi + 3.48$
Binding of bovine haemoglobin with non-polar compounds	$-\log C = 0.67\log P + 1.96$
Partition coefficient between *n*-octanol and water	
aryloxyacetic acids[a]	$\log P = \pi + 1.21$
phenols[b]	$\log P = \pi + 1.46$

[a] Taking π values from the phenoxyacetic acid in Table 7 in Appendix 3.
[b] Taking π values from the phenol in Table 7 in Appendix 3.

Table 6 *Some applications of the Yukawa–Tsuno equation*

Reaction	ρ	r
Rates		
Hydrolysis of ArCMe$_2$Cl	−4.52	1.00
Brominolysis of ArB(OH)$_2$	−3.84	2.29
Methanolysis of Ar$_2$CHCl	−4.02	1.23
Bromination of ArH	−5.28	1.15
Nitration of ArH	−6.38	0.90
Ethanolysis of Ar$_3$CCl	−2.52	0.88
Decomposition of ArCOCHN$_2$	−0.82	0.56
Beckmann rearrangement of ArC(Me)=NOH	−1.98	0.43
Semicarbazone formation from ArCHO	1.35	0.40
Rearrangement of ArCH(OH)CH=CHMe to ArCH=CHCH(OH)Me	−4.06	0.40
Decomposition of Ar$_2$CN$_2$ by benzoic acid	−1.57	0.19
Water + ArOSiEt$_3$	2.84	0.95
Hydroxide ion + ArOSiEt$_3$	3.52	0.46
Hydroxide ion + CH$_3$COOAr	1.69	0.2
Hydroxide ion + PhCOOAr	1.87	0.2
ArCH(CH$_3$)–O–CO–Cl → ArCH(CH$_3$)Cl + CO$_2$ (dioxan, 70°C)	−3.06	1.17
ArCOCHN$_2$ → ArCOCH$_2$OH (75% acetic acid, 25°C)	−1.22	0.603
Equilibria:		
Basicity of ArN=NPh	−2.29	0.85
PhArCHOH $\rightleftharpoons$ Ph$_2$CH$^+$ + H$_2$O (H$_2$SO$_4$)	−5.24	1.49
Ar$_2$CHOH $\rightleftharpoons$ Ar$_2$CH$^+$ + H$_2$O (H$_2$SO$_4$)	−9.11	1.18
Ar$_3$CCl $\rightleftharpoons$ Ar$_3$C$^+$ + Cl$^-$ (H$_2$SO$_4$)	−4.05	0.101

Data mostly from Y. Yukawa and Y. Tsuno, *Bull. Chem. Soc. Japan*, 1959, **32**, 971.

Table 7 *Table of f and r values for some reactions*

Reaction	f	r	Resonance/%
Dissociations			
σ_m	0.57	0.14	20
σ_p	0.49	0.30	38
σ_p^-	0.46	0.66	59
σ' X—⬡—CO₂H	0.64	0.01	1
σ'' X—∿—CO₂H	0.48	(0.00)	(0.00)
4-Substituted picolinium ions			
(MeOH)	1.39	0.25	15
(CH₃CN)	1.70	0.31	16
Dimethylformamide	1.35	0.03	2
3-Substituted pyridines (water)	3.36	0.81	19
4-Substituted pyridines (water)	2.56	1.59	38
X—△—CO₂H	0.71	0.03	4
X—▱—CO₂H	0.48	0.02	3
(dibenzobicyclo CO₂H / X)	0.73	0.09	10
(naphthalene 1,4 CO₂H / X)	0.50	0.38	43
(naphthalene CO₂H / X)	0.53	0.10	16
Rates			
σ_p^+	0.34	0.54	62
σ_m^+	0.52	0.15	23
$CH_3O^- + ArCl$	2.63	2.43	48
$ArCHO + NADH$	0.87	1.27	60
Mandelate racemase + substituted mandelates	0.97	0.78	44
σ^0	0.55	0.24	30

Data from C. G. Swain *et al.*, *J. Am. Chem. Soc.*, 1983, **105**, 492.

[a] The parameters σ_m, σ_p, and σ_p^- are for the dissociation constants of *m*-, *p*-substituted benzoic acids and of *p*-substituted phenols. σ_m^+ and σ_p^+ are for the solvolyses of *m*- and *p*-substituted phenyldimethyl carbinyl chloride in 90% acetone–water. σ^0 is for the reaction of hydroxide ion with the ethyl esters of substituted phenylacetic acids.

Subject Index

α, Brønsted equation, 27
α, Leffler equation, 27
a, reaction parameter (Swain solvent equation), 94
A, acity parameter (Swain solvent equation), 94
Acetaldehyde, hydrate, dehydration, acid catalysed, 28
  addition of thiols, general acid catalysis, level lines, 115
Acetone, general acid catalysis of halogenation, 190
  iodination, base catalysed, 28
Acetophenones, addition of bisulphite ion, 87
Acidity and Basicity functions, 42
Acity parameter, A, 95
Active complex, 4
Acylketenes, calculation of β_{eq} values for reaction of phenoxide ions with, 193
1-Adamantyl chloride, solvolysis, 38
Additivity, inductive effects, 78
  effective charge, 65,67
Ambiguity, resolution of kinetic, 181
Anilines, pK_as, 99
Anti-Hammond effect, 137
Arrhenius plots, 148
Aryl acetoacetate esters, hydrolysis, 189
2-Arylethylquinuclidinium ions, level lines of elimination reaction, 117
Aryl monophosphate esters, reaction with pyridines, 72
Aspartate amino transferase, without lysine-258, 187
Aspirin hydrolysis, application of Jaffé equation, 98
Assembly, reactant molecules, 3
Attenuation, table of factors, 76
  polar effect, 76

β_{eq}, equation for, 121
  equation derived from Marcus theory, 124
  estimation for aryl diphenylphosphate, 124
  estimation and calculation of, 178
  examples of calculations of, 192
β_{ii}, equation for, 121
  equation derived from Marcus theory, 124
b, reaction parameter (Swain solvent equation), 94
B, basity parameter (Swain solvent equation), 94
Balance, degree of bond order, 162, 163
Base catalysis, decomposition of nitramide, 150
Basity parameter, B, 95
Bema Hapothle, 109
Benzene diazonium ions, reaction with Cl⁻, 91
Bond order, degree of balance, 162
  effective charge, 161
  synchronous mechanism, 162
Bond, order (η), 120
  Lewis, 3
Bordwell's anomaly, 138
Breakpoint, convex, 165
  concave, 167
  criteria for in quasi-symmetrical reaction technique, 173
  free energy relationship, concave upwards, 167
  prediction for stepwise quasi-symmetrical reactions, 170
1-Bromo-2-benzamidoethane, solvolysis, 38
Brønsted equation, 27
  extended (Class II), 30
  statistical treatment, 144

t-Butyl chloride, solvolysis, 37

Calculation, β_{eq} values for reaction of phenoxide ions with acylketenes, 193
   β_{eq} values, examples of, 192
   dissociation constants, 174
   equilibrium constants, 174
   log P values, examples of, 193
   partition coefficients (Log P), 180
   physico-chemical constants, 174
   pK_a values, comparison of techniques, 175, 176
   pK_a values, examples of, 191, 193
   rate constants, 179
Carbamate esters, alkaline hydrolysis, 68
Carbenium ion formation, equilibrium, pK_R^+, 100
Carbon acids, dissociation, 132
2-Carboxyphenylsulphate, hydrolysis, 81
Catalytic triad hypothesis, test, 142
Change in rate limiting step, and non-linear free energy relationship, 164
Charton's equation, 25
Chymotrypsin, complexation with inhibitors, 41
ClogP®, 181
Complex, active, 4
   encounter, 4
   reaction, 4
Concave upwards breakpoint, free energy relationship, 167
Concept, similarity, 2
Concerted mechanism, technique of Quasi-symmetrical reactions, 169
   open, 163
   demonstration, 169
Conservation of effective charge, maps, 160
Convex upwards break, 165
Cordes-Thornton coefficient, p_{xy}, 109
   magnitude, 118
   p_{xy}, 110
Coupling between bonds, 107
Coupling coefficients, shapes of energy surfaces, 110

Cross interaction, 107
Cross-coupling, 115
Curtin-Hammett principle, 183

Deviant points, 140
   criteria for parallel mechanism, 142
Dewar-Grisdale equation, 88
Diffusion controlled limit, reaction rate, 34
Dimroth, E_T scale, 39
Disparity index, 120
Dissociation constants, calculation, 174
Dissociation of carbon acids, 132
Dissociative processes, quasi-symmetrical reaction technique, 172
Distance, effect on charge, 57

η, bond order, 120
E_N, Table of standard electrode potential, 92, 274
E_s (ortho), Table of values, 273
E_s (Taft), Table of values, 272
E_T scale, Dimroth and Reichardt, 39
   Table of values, 281
Edwards equation, 92
Effective charge maps, 67
   conservation of charge, 160
   equilibria, 69
   rates, 71
Effective charge, additivity in equilibria, 65
   additivity in transition structures, 67
   bond order balance, 161
   conservation, 161
   equilibria, 55
   from Hammett slopes, 206
   measurement in equilibria, 58
   rates, 66
   table of values, 62
Eigen equations, 130, 131
   example of free energy plot, 194
Encounter complex, 4
Energy surfaces, model equations, 111
   quadratic equation, 112

quartic equation, 112
  shapes, 110
Enzfitter®, 202
Enzyme mechanisms, free energy
  relationships, 184
Equilibrium constants, calculation,
  174
Ethyl bromide, solvolysis, 38

f, Fragmentational constants for
  LogP, Table of, 277
f, Swain-Lupton reaction constant,
  89
  Table of parameters, 289
f, attenuation factor, 76
F, Swain-Lupton Field constant,
  Table of parameters, 259
  measurement, 89
F, field effect, 75
^{19}F-NMR spectroscopy, 41
FigP®, 202
Fluorenes, detritiation by amines,
  125
Fluorenoyl esters, hydrolysis, 161
Fragmentation method, calculation
  of partition coefficients (Log P),
  180
Free energy relationships, classes, 6
  definition, 6
  non-linear, 129
  non-linear due to change in
    mechanism, 166
  non-linear due to change in rate
    limiting step, 164
  origin, Bell's theory, 7
  origin, Leffler and Grunwald
    theory, 9
  as statistical artifacts?, 145
  Class I, definition, 7
  Class II, definition, 7
  and enzyme mechanisms, 184
  for complexation, Table of
    equations, 287
  for rate constants, Table of
    equations, 285
  Class I, derivation of, 132
Front-side nucleophilic aliphatic
  substitution, 118
Fundamental chemical process, 110

g_1 and g_2, level line slope, 114
General acid catalysis, definition of,
  30
General acid or specific acid-general
  base catalysis, resolution, 182
General base catalysis, definition of,
  30
General base or nucleophilic
  catalysis, resolution, 181
Glucose, mutarotation of, 48
Glucosyl fluoride, solvolysis, 49
Grob's equation, σ_I^Q, 24
Grunwald-Winstein equation, 36
  extended, 94

H_N, for Edwards' equation, 92
Hammett and Taft relationships for
  equilibria, Table of equations, 286
Hammett equation, definition of σ,
  17
Hammett relationships for
  dissociation constants, Table of
  equations, 283
Hammond's postulate, 134
Hammond coefficient (p_x, p_y), 109
Hansch equation, 40
  extended, 96
Hydride ion transfer, quasi-
  symmetrical reaction technique
  and, 194

I, sigma-inductive effect, 75
Identity rate constants, Lewis-
  Kreevoy correlation, 122
Identity reactions, 119
Inductive effects, additivity, 78
Inductive transmission, 75
Interactions, multiple, 79
Intermediates, demonstration of, 163
  demonstration by trapping, 164
Isokinetic relationship, 148
Isopropyl tosylate, solvolysis, 95

Jaffé equation, 80
  computer fitting, 82

k_{ii}, identity rate constants,
  calculation of, 124
Kinetic ambiguity, resolution, 181

Koppel and Palm solvent equation, 96
Kosower's Z scale, 39

Leffler's α and tyrosyl-ᵗRNA synthetase, 186
Leffler's equation for α, derived from Marcus theory, 149
    derivation from intersecting Morse equations and parabolas, 7, 133, 202
Level lines, angles and saddle point shape, 115
    slopes at transition structures, 114
Lewis-Kreevoy correlation, identity rate constants, 122
Lifetime, active complex, 5
    encounter complex, 5
    reaction complex, 5
    transition structure, 5
Log P, examples of calculations, 193
    Table of values, 188
    additive parameter, 41
    definition, 40

m, Grunwald-Winstein solvent parameter, 36
Marcus equation, 8, 121, 131
    derivation of, 133
Mayr equation, 93
Mechanism, by comparison of similarity coefficients, 158
    by effective charge distribution in transition structure, 160
    demonstration of concerted, 169
    diagnosis, 158
    stepwise, 163
Medchem®, 181
Methyl iodide, reaction with nucleophiles, 92
Microscopic medium effects, 140
    criterion for minimal, 144
    proton transfer, 143
    scatter plots, 141
Microscopic reversibility, 29
Minitab®, 202
Molar refractivity, MR, 43
Monophosphate esters ($ROPO_3^{2-}$), hydrolysis, 64

More O'Ferrall-Jencks map, 13, 109, 110
    description, 206
    elimination reaction, 136
    nucleophilic displacement, 119
    reaction of phenoxide ions with phenyl esters, 123
Morse curves, intersecting, 8
Morse equation, 201
MR, molar refractivities, 43
    Table of molar refractivities, 259
Multiple linear regression, 109
Mutarotation of Glucose, 48

v_x (Charton), Table of parameters, 272
n, Swain-Scott nucleophilicity, 32
    Table of parameters, 274
N_+ Ritchie equation, Table of values, 273
    definition, 34
N_{OTs} (Bentley-Schleyer), Table of values, 275
Nitramide decomposition, 150
Nitroalkane anomaly, 139
4-Nitrophenyl acetate, reaction with phenoxide ions, 172
Non-linear free energy relationships, 129
Nucleophilic aliphatic substitution, 32
    front-side attack, 118

Open concerted mechanism, 163
Ortho effect, 146
Ortho substituents, calculation of pK_as for, 152
    treatment of, in Brønsted equations, 147

π, Leo-Hansch equation, Table of parameters, 276
p_x, value from surface shape, 112
p_{xy}, Cordes-Thornton coefficient, 109, 110
    magnitude, 118
    value from surface shape, 112
p_x, p_y, Hammond coefficient, 109

p_y, value from surface shape, 112
P, product state, 56, 206, 217–219
Ps, standard product state, 56, 143, 206, 217–219
Parabolas, intersecting, reaction coordinate and, 133
Parachor, 43, 44
Parallel mechanisms, demonstration by non-linear free energy relationships, 166
from deviant points, 142
Partition coefficients (log P), calculation of, 180
calculation of by fragmentation method, 180
2-Phenyl-2-chloropropane, solvolysis, 85
N-phenylbenzenesulphonamides, dissociation, 80
pK_a values, 99
2-Phenyloxazolin-5-one, formation, 71
Phenols, dissociation with resonance interaction, 84
Phenolysis of phenyl acetates, quasi-symmetrical reaction technique and, 171
Phenyl, 4-hydroxybenzoates, parallel mechanisms of hydrolysis, 168
Phenyl acetates, reaction with imidazole, 59
Phenyl carbamates, alkaline hydrolysis, 87
Phenyl dimethylmethyl chloride, solvolysis, 100
Phenyl diphenylphosphinate esters, reaction with phenoxide ions, 196
Phenyl esters, reaction with phenoxide ions, More O'Ferrall-Jencks map, 123
Phenyl N-benzoylglycinate ester, cyclisation in alkali, 71
Phenyltriethylsilyl ethers, alkaline hydrolysis, 101
Phosphate monoesters, reaction with amines, level lines, 116
N-phosphoisoquinolinium ion, reaction with pyridines, 141

reaction with pyridines, quasi-symmetrical reaction technique, 172
N-Phosphopyridinium ions, reaction with phenoxide ions, 71
Phosphoryl group transfer, 63
pH-Profiles, concave curvature, 168
convex curvature, 166
Physico-chemical constants, calculation of, 174
pK_a (for Edwards equation), Table of values, 274
table of fragmental constants to calculate pK_a of ammonium ions, 279
table of fragmental constants to calculate pK_a of carboxylic acids, 278
pK_a values, examples of calculations, 191, 193
pK_a values, table of, 187
example of additivity, 98
pK_R+ for alcohols, 100
Polar effect, attenuation, 76
influence of distance, 77
Polarity, universal measure, 5
Pseudo-base formation, nitroalkanes, 139
Pyridines, phosphorylation by phosphoramidate ion, 196
Pyrylium ions, hydrolysis, 166, 167

Qbasic®, 202
Quadratic equation, energy surfaces, 112
Quartic equation, energy surfaces, 112
Quasi-symmetrical reaction technique, and hydride ion transfer, 194
breakpoint prediction in, 170
criteria for break, 173
diagnosis of concerted mechanisms, 169
phenolysis of phenyl acetates, 171
use in estimating β_{ii}, 61
use in estimating k_{ii}, 120

%R, equation for, 90
ρ^*, definition of, 20

ρ, calculation of, 77
   dependence on σ, 125
   inversion of sign, 148
r, Swain-Lupton parameter, 89
   Table of values, 289
r, Yukawa-Tsuno parameter
   definition of, 86
R, Reactant state, 56, 206, 217–219.
R (Swain-Lupton), Table of
   parameters, 259
   measurement, 90
R, resonance through polarisation,
   75
Rs, standard reactant state, 56, 143,
   206, 217–219
Rate constants, calculation of, 179
Rate limiting step, 131
   change in as criterion of stepwise
      mechanism, 164
Reaction complex, 4
Reaction coordinate, 4, 8
   as system of intersecting
      parabolas, 133
Reaction rate, diffusion controlled
   limit, 34
Reactivity-Selectivity postulate, 135
Reichardt, E_T scale, 39
Resonance effects, 83
Ritchie equation, 34
Roberts and Moreland equation, σ',
   24

σ', Roberts and Moreland equation,
   24
σ' values, Roberts and Moreland, 89
σ, (*meta* and *para*, Hammett values),
   and Table of constants, 259
   Jaffé's global definition, 27
   primary standard, 27
   secondary standard, 27
   which one to employ?, 26
σ⁻ (Resonance enhanced), Table of
   constants, 259
   definition of, 83
σ* (ortho), Table of constants, 273
   attenuation through carbon chain,
     22
   definition, 20
   example of additivity, 97

Table of constants, 259
σ⁺ (Resonance enhanced), Table of
   constants, 259
   definition of, 86
σ_I, Inductive polar constant, 24
   relationship to σ_R and σ, 83
   Table of constants, 259
σ_I^Q, 24
σ_o values, Table of constants, 280
σ_R, definition of, 83
s, Nucleophile specific parameter
   (Mayr's equation), 93
s, Swain-Scott reaction constant, 32
S, bulk effect, 75
Saddle point, and transition
   structure, 109
   shape and level line angle, 115
   properties, 109
Scatter due to microscopic medium
   effects, 141
Self interaction (p_x or p_y), 107
Semicarbazone formation, 165
Separability, postulate, 9
Software, ClogP®, 181
   Medchem®, 181
Shapes of energy surfaces, 110
   coupling coefficients, 110
Similarity coefficients, mechanism by
   comparison, 158
Similarity concept, 2
   and Leffler's α-parameter, 10
Solvent equations, 35
   multiparameter, 94
Spectroscopic transitions, 41
States, interconversion, 4
Statistical artifact? Free energy
   relationship, 145
Statistical fitting to theoretical
   equations, 202
   to $y = a + bx + cx^2$, 205
   to $y = a + bx_1 + cx_2$, 203
   to $y = ax_1 + bx_2 + cx_1x_2$, 206
   to $y = mx + c$, 203
   to $y_1 = ax_1 + b_1$ and $y_2 = ax_2 + b_2$,
     204
   Brønsted equation, 144
Stepwise mechanism, 163
   change in rate limiting step as
     criterion, 164

diagnosis by convex break, 165
effective charge maps in putative, 173
Steric demand, 23
Structure, definition of molecular, 3
  Kekulé, 3
Substituent effects, cross interaction (p_{xy}), 107
  self interaction (p_x or p_y), 107
N-Sulphopyridinium ion, reaction with phenoxide ions, 72
Sulphyl group transfer, 163
Sultones, effective charge in ring opening, 69
  reactions with phenoxide ions, 68
Sum of bond orders, τ (tightness), 120
Surface tension, 44
Swain equation (solvent), 94
Swain-Lupton equation, 88, 89
Swain-Scott equation, 32
Synchronous diagonal, 120

τ, estimation from Leffler's α, 121
  sum of bond orders (tightness), 120
Table of log P values, 188
Table of pK_a values, 187
Taft's δ, definition of, 21
Taft and Charton relationships for dissociation constants, Table of equations, 284
Taft equation, 19
  solvent equation, 97
Temperature effects, 147
  isokinetic, 148
Theoretical equations, statistical fitting to, 202
Thornton effect, 137
Tightness, τ, 120
Tightness diagonal, 120, 121

Transition Structure, definition of, 1
  and parallel vectors, 137
  and perpendicular vectors, 137
  and saddle point, 109
  bonding in, 12
  polar effect, 12
Trapping of intermediates, 164
Triad, catalytic, model for serine proteinases, 142
Triazines, nucleophilic substitution, 31
  reaction with phenoxide ions, 191
Tyrosyl-^{t}RNA synthetase, Leffler's α, 186

Uncertainties of fit, quasi-symmetrical reaction technique, 171

ω, work terms, 134
Work terms, ω, 134
Worked answers, Chapter, 2, 210
  Chapter, 3, equilibria, 215
  Chapter, 3, rates, 220
  Chapter, 4, 224
  Chapter, 5, 231
  Chapter, 6, 234
  Chapter, 7, 242

Y, Grunwald-Winstein solvent parameter, 36
Y_{G-W} (Grunwald-Winstein), Table of parameters, 275
Y_{OTs} (Bentley-Schleyer), Table of parameters, 275
Young and Jencks equation, 86
Yukawa-Tsuno equation, 86
  Young and Jencks modification, 86
  Table of relationships, 288

Z scale, Kosower, 39